**Serono Symposia, USA**
**Norwell, Massachusetts**

## PROCEEDINGS IN THE SERONO SYMPOSIA, USA SERIES

*IMMUNOBIOLOGY OF REPRODUCTION*
Edited by Joan S. Hunt

*FUNCTION OF SOMATIC CELLS IN THE TESTIS*
Edited by Andrzej Bartke

*GLYCOPROTEIN HORMONES: Structure, Function, and Clinical Implications*
Edited by Joyce W. Lustbader, David Puett, and Raymond W. Ruddon

*GROWTH HORMONE II: Basic and Clinical Aspects*
Edited by Barry B. Bercu and Richard F. Walker

*TROPHOBLAST CELLS: Pathways for Maternal-Embryonic Communication*
Edited by Michael J. Soares, Stuart Handwerger, and Frank Talamantes

*IN VITRO FERTILIZATION AND EMBRYO TRANSFER IN PRIMATES*
Edited by Don P. Wolf, Richard L. Stouffer, and Robert M. Brenner

*OVARIAN CELL INTERACTIONS: Genes to Physiology*
Edited by Aaron J.W. Hsueh and David W. Schomberg

*CELL BIOLOGY AND BIOTECHNOLOGY: Novel Approaches to Increased Cellular Productivity*
Edited by Melvin S. Oka and Randall G. Rupp

*PREIMPLANTATION EMBRYO DEVELOPMENT*
Edited by Barry D. Bavister

*MOLECULAR BASIS OF REPRODUCTIVE ENDOCRINOLOGY*
Edited by Peter C.K. Leung, Aaron J.W. Hsueh, and Henry G. Friesen

*MODES OF ACTION OF GnRH AND GnRH ANALOGS*
Edited by William F. Crowley, Jr., and P. Michael Conn

*FOLLICLE STIMULATING HORMONE: Regulation of Secretion and Molecular Mechanisms of Action*
Edited by Mary Hunzicker-Dunn and Neena B. Schwartz

*SIGNALING MECHANISMS AND GENE EXPRESSION IN THE OVARY*
Edited by Geula Gibori

*GROWTH FACTORS IN REPRODUCTION*
Edited by David W. Schomberg

*UTERINE CONTRACTILITY: Mechanisms of Control*
Edited by Robert E. Garfield

*NEUROENDOCRINE REGULATION OF REPRODUCTION*
Edited by Samuel S.C. Yen and Wylie W. Vale

*FERTILIZATION IN MAMMALS*
Edited by Barry D. Bavister, Jim Cummins, and Eduardo R.S. Roldan

*GAMETE PHYSIOLOGY*
Edited by Ricardo H. Asch, Jose P. Balmaceda, and Ian Johnston

*GLYCOPROTEIN HORMONES: Structure, Synthesis, and Biologic Function*
Edited by William W. Chin and Irving Boime

*THE MENOPAUSE: Biological and Clinical Consequences of Ovarian Failure: Evaluation and Management*
Edited by Stanley G. Korenman

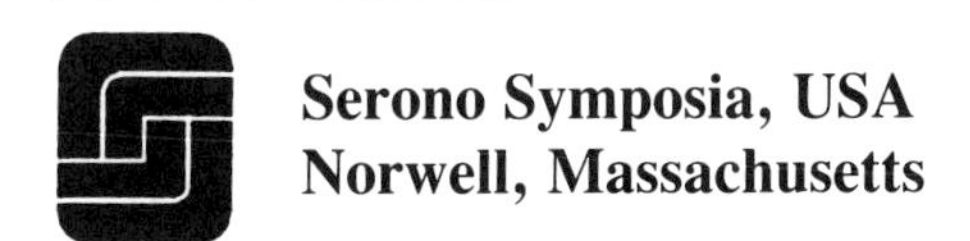


Joan S. Hunt Editor

# Immunobiology of Reproduction

With 80 Figures

Springer-Verlag
New York Berlin Heidelberg London Paris
Tokyo Hong Kong Barcelona Budapest

Joan S. Hunt, Ph.D.
Department of Anatomy and Cell Biology
University of Kansas Medical Center
Kansas City, KS 66160
USA

Proceedings of the Symposium on Immunobiology of Reproduction, sponsored by Serono Symposia, USA, held August 26 to 29, 1993, in Boston, Massachusetts.

For information on previous volumes, please contact Serono Symposia, USA.

Library of Congress Cataloging-in-Publication Data
Immunobiology of reproduction/[edited by] Joan S. Hunt.
p. cm.
"Serono Symposia, USA (Series)"—Series t.p.
Includes bibliographical references and index.
ISBN 0-387-94292-0.—ISBN 3-540-94292-0
1. Reproduction—Immunological aspects—Congresses. 2. Human reproduction—Immunological aspects—Congresses. I. Hunt, Joan S.
[DNLM: 1. Reproduction—immunology—congresses. 2. Growth Substances—immunology—congresses. WQ 205 I31 1994]
QP252.5.I43 1994
612.6—dc20 94-8895

Printed on acid-free paper.



Production coordinated by Marilyn Morrison and managed by Francine McNeill; manufacturing supervised by Gail Simon.
Typeset by Best-set Typesetter Ltd., Hong Kong.
Printed and bound by Braun-Brumfield, Inc., Ann Arbor, MI.

Printed in the United States of America.

9 8 7 6 5 4 3 2 1

ISBN 0-387-94292-0 Springer-Verlag New York Berlin Heidelberg
ISBN 3-540-94292-0 Springer-Verlag Berlin Heidelberg New York

## SYMPOSIUM ON IMMUNOBIOLOGY OF REPRODUCTION

### Scientific Committee



### Organizing Secretary

*This volume is dedicated to the memory of Thomas G. Wegmann, Ph.D., whose fertile imagination and contagious enthusiasm stimulated a generation of researchers to their best efforts, and whose personal warmth and kindness inspired admiration and affection in all who were privileged to know him.*

# Preface

As early as 1953, in a landmark paper by R.E. Billingham, L. Brent, and P.B. Medawar, interactions between the immune system and processes relating to reproduction, fertility, and embryonic development were examined experimentally. Since that time, many biologists have been intrigued with the *immunological paradox* of pregnancy, where maternal and fetal tissues peaceably coexist despite their genetic differences. Others have been caught up in the concept that the products of immune cells might promote pregnancy, in the problems presented by auto- and alloimmunity in men and women, and in the potential associations between the conditions required for successful reproduction and those that pave the way to cancer. As a consequence, special groups operating under the aegis of established societies and new societies composed exclusively of basic and clinical scientists interested in the immunological aspects of reproduction have emerged in the last decade. Highly focused meetings have been held in Canada (T.G. Wegmann and T.J. Gill, organizers) and France (G. Chauout and J. Mowbray, organizers).

With the development of new insights into the interactive components of the immune system and the generation of powerful new molecular and cellular tools, singular progress has been made in understanding the immunological events that culminate in successful pregnancy. In order to bring together the leaders in contemporary aspects of reproductive immunology, the first U.S. symposium on the "Immunobiology of Reproduction" was held from August 26 to 29, 1993, in Boston, Massachusetts. The Scientific Committee (Drs. A.E. Beer, J.R. Head, J.C. Herr, J.A. Hill, and J.S. Hunt, Chairman) and the participants, who were established basic and clinical scientists as well as students and trainees, are grateful to Serono Symposia, USA, which contributed financial and organizational support. We particularly appreciate the efforts of Dr. Bruce Burnett. Financial support for the attendance of students, fellows, and new investigators was received from the National Institutes of Health (1R13-HD-30693), and we are indebted to Drs. Alan Lock (NICHD) and Michele Hogan (NIAID) for their interest and special efforts.

The resounding success of this meeting was due in large part to the twenty outstanding scientists who made up the panel of speakers, served as resourceful session moderators, and contributed the chapters in this volume. Particular praise is due to the poster presenters (Dr. J.R. Scott, Moderator). All of the participants are to be thanked for contributing insightful questions to the lively discussions that characterized this enthusiastic gathering of scientists from universities, government, and industry.

JOAN S. HUNT

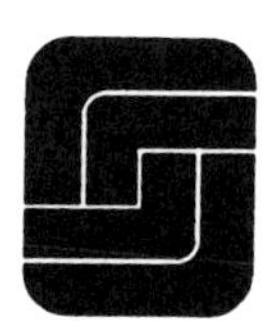

# Contents


Preface ........ ix
Contributors ........ xv

**Part I. Hematopoietic Cells in Reproductive Tissues**

1. Immunobiological Characterization of the Trophoblast-Decidual Interface in Human Pregnancy ........ 3
PETER M. JOHNSON

2. Immunity in the Human Male Reproductive Tract ........ 14
JEFFREY PUDNEY AND DEBORAH J. ANDERSON

3. Endometriosis: Immune Cells and Their Products ........ 23
JOSEPH A. HILL

**Part II. Growth Factors/Cytokines in the Female Reproductive Tract and Placenta**

4. Cytokines and Pregnancy Recognition ........ 37
FULLER W. BAZER, THOMAS E. SPENCER, TROY L. OTT, AND HOWARD M. JOHNSON

5. Role of Locally Produced Growth Factors in Human Placental Growth and Invasion with Special Reference to Transforming Growth Factors ........ 57
PEEYUSH K. LALA AND JEFFREY J. LYSIAK

6. Uterine Epithelial GM-CSF and Its Interlocutory Role During Early Pregnancy in the Mouse ........ 82
SARAH A. ROBERTSON, ANNA C. SEAMARK, AND ROBERT F. SEAMARK

7. Tilt Toward $T_{H2}$ in Successful Pregnancy ................. 99
THOMAS G. WEGMANN, HUI LIN, JANE YUI, MARIA GARCIA-LLORET, TIM MOSMANN, AND LARRY GUILBERT

8. Use of the Osteopetrotic Mouse for Studying Macrophages in the Reproductive Tract ............................. 104
PAULA E. COHEN AND JEFFREY W. POLLARD

**Part III. Growth Factor Networks in Pregnancy Loss and Cancer**

9. Novel Transforming Growth Factor Betas (TGFβ2) in Pregnancy and Cancer ................................ 125
DAVID A. CLARK, KATHLEEN C. FLANDERS, GILL VINCE, PHYLLIS STARKEY, HAL HIRTE, JUSTIN MANUEL, JENNIFER UNDERWOOD, AND JAMES MOWBRAY

10. Tumor Necrosis Factor α: Potential Relationships with Cancers of the Female Reproductive Tract ................ 136
JOAN S. HUNT, HUA-LIN CHEN, YAPING YANG, KATHERINE F. ROBY, AND FERNANDO U. GARCIA

**Part IV. Placental Expression of Major Histocompatibility Complex and Associated Genes**

11. MHC Gene Expression in Placentas of Domestic Animals ........................................ 153
D.F. ANTCZAK, JULI K. MAHER, GABRIELE GRÜNIG, W.L. DONALDSON, JULIA KYDD, AND W.R. ALLEN

12. MHC-Linked Genes and Their Role in Growth and Reproduction ..................................... 170
THOMAS J. GILL III, HONG-NERNG HO, AMAL KANBOUR-SHAKIR, AND HEINZ W. KUNZ

13. Molecular Regulatory Mechanisms That Repress Classical HLA Class I Gene Expression in Human Placenta ......... 184
PHILIPPE LE BOUTEILLER, THIERRY GUILLAUDEUX, MARYSE GIRR, CÉCILE DEMEUR, AND ANNE-MARIE RODRIGUEZ

**Part V. Experimental Models of MHC Gene Expression**

14. Developmental Regulation of MHC Class I Gene Expression in Mice ..................................... 201
Elizabeth K. Bikoff and Elizabeth J. Robertson

15. Overexpression of Class I MHC in Murine Trophoblast and Increased Rates of Spontaneous Abortion ............ 214
Joseph R. Voland, Christopher Becker, and Farideh Hooshmand

**Part VI. Immunological Aspects of Human Infertility**

16. Mechanism of Self-Tolerance and Autoimmune Disease Pathogenesis: Analysis Based on Ovarian Autoimmune Models ................................. 239
Kenneth S.K. Tung, Ya-Huan Lou, and Hedy Smith

17. Immunobiological Effects of Vasectomy and Vasovasostomy in the Rat Model ........................................ 254
John C. Herr, Charles J. Flickinger, and Stuart S. Howards

18. Antiphospholipid Antibodies and Placental Development ... 285
Neal S. Rote, Timothy W. Lyden, Elizabeth Vogt, and Ah Kau Ng

19. Immunotherapy for Recurrent Spontaneous Abortion ...... 303
Carolyn B. Coulam

20. New Horizons in the Evaluation and Treatment of Recurrent Pregnancy Loss ............................. 316
Alan E. Beer, Joanne Y.H. Kwak, Alice Gilman-Sachs, and Kenneth D. Beaman

Author Index ............................................. 335

Subject Index ............................................. 337

# Contributors

W.R. Allen, Thoroughbred Breeders' Association Equine Fertility Unit, Newmarket, Suffolk, UK.

Deborah J. Anderson, Fearing Research Laboratory, Department of Obstetrics, Gynecology, and Reproductive Biology, Brigham and Women's Hospital, Harvard Medical School, Boston, Massachusetts, USA.

D.F. Antczak, James A. Baker Institute for Animal Health, College of Veterinary Medicine, Cornell University, Ithaca, New York, USA.

Fuller W. Bazer, Department of Animal Science, Texas A&M University, College Station, Texas, USA.

Kenneth D. Beaman, Department of Microbiology and Immunology, Clinical Immunology Laboratory, University of Health Sciences/The Chicago Medical School, North Chicago, Illinois, USA.

Christopher Becker, UCSD Cancer Center, University of California at San Diego, La Jolla, California, USA.

Alan E. Beer, Division of Reproductive Medicine, Departments of Microbiology and Immunology and Obstetrics and Gynecology, University of Health Sciences/The Chicago Medical School, North Chicago, Illinois, USA.

Elizabeth K. Bikoff, Departments of Biochemistry, Molecular, Cellular, and Developmental Biology, The Biological Laboratories, Harvard University, Cambridge, Massachusetts, USA.

Hua-Lin Chen, Department of Anatomy and Cell Biology, University of Kansas Medical Center, Kansas City, Kansas, USA.

David A. Clark, Department of Medicine, Obstetrics and Gynecology, and Pathology, McMaster University, Hamilton, Ontario, Canada.

Paula E. Cohen, Department of Developmental and Molecular Biology, Albert Einstein College of Medicine, New York, New York, USA.

Carolyn B. Coulam, Reproductive Immunology Genetics and IVF Institute, Fairfax, Virginia, USA.

Cécile Demeur, Unité INSERM 100, CHU Purpan, Cedex, France.

W.L. Donaldson, James A. Baker Institute for Animal Health, College of Veterinary Medicine, Cornell University, Ithaca, New York, USA.

Kathleen C. Flanders, Laboratory of Chemoprevention, NCI-NIH, Bethesda, Maryland, USA.

Charles J. Flickinger, Department of Anatomy and Cell Biology, The Center for Recombinant Gamete Contraceptive Vaccinogens, Health Sciences Center, University of Virginia, Charlottesville, Virginia, USA.

Fernando U. Garcia, Department of Pathology and Laboratory Medicine, University of Kansas Medical Center, Kansas City, Kansas, USA.

Maria Garcia-Lloret, Department of Immunology, University of Alberta, Edmonton, Alberta, Canada.

Thomas J. Gill III, Department of Pathology, School of Medicine, University of Pittsburgh, Pittsburgh, Pennsylvania, USA.

Alice Gilman-Sachs, Department of Microbiology and Immunology, Clinical Immunology Laboratory, University of Health Sciences/The Chicago Medical School, North Chicago, Illinois, USA.

Maryse Girr, Unité INSERM 100, CHU Purpan, Cedex, France.

Gabriele Grünig, James A. Baker Institute for Animal Health, College of Veterinary Medicine, Cornell University, Ithaca, New York, USA.

Larry Guilbert, Department of Immunology, University of Alberta, and Canadian Red Cross Blood Transfusion Service, Edmonton, Alberta, Canada.

Thierry Guillaudeux, Unité INSERM 100, CHU Purpan, Cedex, France.

JOHN C. HERR, Department of Anatomy and Cell Biology, The Center for Recombinant Gamete Contraceptive Vaccinogens, Health Sciences Center, University of Virginia, Charlottesville, Virginia, USA.

JOSEPH A. HILL, Fearing Research Laboratory, Division of Reproductive Immunology, Department of Obstetrics, Gynecology, and Reproductive Biology, Brigham and Women's Hospital, Harvard Medical School, Boston, Massachusetts, USA.

HAL HIRTE, Ontario Cancer Treatment Research Foundation, Hamilton, Ontario, Canada.

HONG-NERNG HO, Department of Obstetrics and Gynecology, National Taiwan University Hospital, Taipei, Taiwan, ROC.

FARIDEH HOOSHMAND, UCSD Cancer Center, University of California at San Diego, La Jolla, California, USA.

STUART S. HOWARDS, Department of Urology, The Center for Recombinant Gamete Contraceptive Vaccinogens, Health Sciences Center, University of Virginia, Charlottesville, Virginia, USA.

JOAN S. HUNT, Department of Anatomy and Cell Biology and Department of Pathology and Laboratory Medicine, University of Kansas Medical Center, Kansas City, Kansas, USA.

HOWARD M. JOHNSON, Department of Microbiology and Cell Science, University of Florida, Gainesville, Florida, USA.

PETER M. JOHNSON, Department of Immunology, University of Liverpool, Liverpool, UK.

AMAL KANBOUR-SHAKIR, Pathology Department, Magee Women's Hospital, Pittsburgh, Pennsylvania, USA.

HEINZ W. KUNZ, Department of Pathology, School of Medicine, University of Pittsburgh, Pittsburgh, Pennsylvania, USA.

JOANNE Y.H. KWAK, Reproductive Immunology, Division of Reproductive Medicine, Departments of Microbiology and Immunology and Obstetrics and Gynecology, University of Health Sciences/The Chicago Medical School, North Chicago, Illinois, USA.

JULIA KYDD, Thoroughbred Breeders' Association Equine Fertility Unit, Newmarket, Suffolk, UK.

PEEYUSH K. LALA, Department of Anatomy, University of Western Ontario, London, Ontario, Canada.

PHILIPE LE BOUTEILLER, Unité INSERM 100, CHU Purpan, Cedex, France.

HUI LIN, Department of Immunology, University of Alberta, Edmonton, Alberta, Canada.

YA-HUAN LOU, Experimental Pathology, Department of Pathology, Health Sciences Center, University of Virginia, Charlottesville, Virginia, USA.

TIMOTHY W. LYDEN, Department of Microbiology and Immunology, Wright State University School of Medicine, Dayton, Ohio, USA.

JEFFREY J. LYSIAK, Department of Anatomy, University of Western Ontario, London, Ontario, Canada.

JULI K. MAHER, James A. Baker Institute for Animal Health, College of Veterinary Medicine, Cornell University, Ithaca, New York, USA.

JUSTIN MANUEL, McMaster University Medical Centre, Hamilton, Ontario, Canada.

TIM MOSMANN, Department of Immunology, University of Alberta, Edmonton, Alberta, Canada.

JAMES MOWBRAY, St. Mary's Hospital Medical School, London, UK.

AH KAU NG, Department of Applied Medical Sciences, University of Southern Maine/Foundation for Blood Research, Scarborough, Maine, USA.

TROY L. OTT, Department of Animal Science, Texas A&M University, College Station, Texas, USA.

JEFFREY W. POLLARD, Departments of Developmental and Molecular Biology and Obstetrics and Gynecology, Albert Einstein College of Medicine, New York, New York, USA.

JEFFREY PUDNEY, Fearing Research Laboratory, Department of Obstetrics, Gynecology, and Reproductive Biology, Brigham and Women's Hospital, Harvard Medical School, Boston, Massachusetts, USA.

ELIZABETH J. ROBERTSON, Departments of Biochemistry, Molecular, Cellular, and Developmental Biology, The Biological Laboratories, Harvard University, Cambridge, Massachusetts, USA.

SARAH A. ROBERTSON, Department of Obstetrics and Gynaecology, University of Adelaide, South Australia.

KATHERINE F. ROBY, Department of Anatomy and Cell Biology, University of Kansas Medical Center, Kansas City, Kansas, USA.

ANNE-MARIE RODRIGUEZ, Unité INSERM 100, CHU Purpan, Cedex, France.

NEAL S. ROTE, Departments of Microbiology and Immunology and Obstetrics and Gynecology, Wright State University School of Medicine, Dayton, Ohio, USA.

ANNA C. SEAMARK, Department of Obstetrics and Gynaecology, University of Adelaide, South Australia.

ROBERT F. SEAMARK, Department of Obstetrics and Gynaecology, University of Adelaide, South Australia.

HEDY SMITH, Department of Immunology, St. Jude's Hospital and Research Center, Memphis, Tennessee, USA.

THOMAS E. SPENCER, Department of Animal Science, Texas A&M University, College Station, Texas, USA.

PHYLLIS STARKEY, Nuffield Department of Obstetrics and Gynecology, University of Oxford, Maternity Department, John Radcliffe Hospital, Headington, Oxford, UK.

KENNETH S.K. TUNG, Experimental Pathology, Department of Pathology, Health Sciences Center, University of Virginia, Charlottesville, Virginia, USA.

JENNIFER UNDERWOOD, St. Mary's Hospital Medical School, London, UK.

GILL VINCE, Nuffield Department of Obstetrics and Gynecology, University of Oxford, Maternity Department, John Radcliffe Hospital, Headington, Oxford, UK.

Elizabeth Vogt, Department of Microbiology and Immunology, Wright State University School of Medicine, Dayton, Ohio, USA.

Joseph R. Voland, UCSD Cancer Center, University of California at San Diego, La Jolla, California, USA.

Thomas G. Wegmann, Department of Immunology, University of Alberta, Edmonton, Alberta, Canada.

Yaping Yang, Department of Anatomy and Cell Biology, University of Kansas Medical Center, Kansas City, Kansas, USA.

Jane Yui, Department of Immunology, University of Alberta, Edmonton, Alberta, Canada.

# Part I

## Hematopoietic Cells in Reproductive Tissues

# 1

# Immunobiological Characterization of the Trophoblast-Decidual Interface in Human Pregancy

PETER M. JOHNSON

The last decade has witnessed an explosion of new information characterizing the immunobiological features of uteroplacental tissues in pregnancy, most notably in humans, using immunohistochemical, cell culture, cytokine, and other molecular analytical techniques. Many of the cellular and molecular components contributing to successful viviparity in genetically outbred pregnancy have been identified, although not all the functional detail is yet known (1–3). This modern focus has identified specialized immunobiological features of the fetal trophoblastic cells that form the continuous lining of extraembryonic tissue, as well as cells within the maternal decidualized endometrium, which, collectively, may confer transplantation protection for the haplo-nonidentical fetus as an intrauterine allograft throughout pregnancy. This chapter selectively addresses certain of these aspects, notably (i) mechanisms developed by trophoblast involving modulation of the class I *major histocompatibility complex* (MHC) molecule and the complement regulatory protein expression that may critically protect these vital cells from any maternal blood-borne or decidual cytotoxic attack, (ii) the significance of endogenous retroviral expression by syncytiotrophoblast, and (iii) the description of the characteristic $CD3^-$ $CD16^-$ $CD56^{++}$ *decidual granulated leukocyte* (dGL) cell population and their cytokine release profile that may be an important part of cytokine-mediated fetomaternal signaling events in normal pregnancy.

## Classical Class I and II MHC Alloantigens

Classical class I (HLA-A, -B, -C) and class II (HLA-DR, -DQ, -DP) alloantigenic MHC molecules expressed at cell surfaces are fundamental in directing autologous T cell responses and are of central importance in transplantation reactions. Both syncytiotrophoblast and extravillous cytotrophoblast in human uteroplacental tissue are in direct and extensive contact with maternal immune cells in blood and decidua. These fetal cells, however, together with villous cytotrophoblast, do not express any classical polymorphic class I or II MHC alloantigens (2–5). There is also normally no class I or II MHC molecular expression by the human preimplantation embryo (6 and Johnson et al., unpublished observations). In contrast, nontrophoblastic fetal cells within the mesenchymal core of the chorionic villous placenta—for example, Hofbauer cells (macrophages)—express HLA-A and -B alloantigens after the first few weeks of pregnancy, and class II MHC molecules may be expressed in the third trimester (2–5); it is these fetal cells that are presumed to act as immunogenic stimuli for the not-infrequent pregnancy-induced maternal production of anti-HLA alloantibody following tissue damage or hemorrhage, notably at parturition. Villous trophoblast is resistant to up-regulation of expression of MHC molecules by exogenous cytokines, such as *interferon* $\gamma$ (IFN$\gamma$) or *tumor necrosis factor* $\alpha$ (TNF$\alpha$) (5, 7, 8). Trophoblastic expression of classical MHC alloantigens is regulated at the transcriptional level, although the precise mechanism(s) underlying this selective class I MHC gene control remains to be fully elucidated (2, 5, 9).

The fetal trophoblastic cells that arise from the trophectoderm layer of the blastocyst, therefore, constitute a continuous HLA alloantigen-negative cellular cocoon that effectively separates maternal and fetal immunocompetent cells in uteroplacental tissue. This absence of a display of classical class I or II MHC alloantigens by trophoblastic cell populations throughout gestation must be of prime importance in protecting this vital fetal tissue interface from recognition by either maternal $CD8^+$ cytotoxic T lymphocytes or anti-HLA alloantibody, both in the intervillous spaces or in decidual tissue. It also explains why syncytial trophoblast cellular fragments that normally break away from the placenta and may become lodged in the maternal pulmonary vasculature survive for many months without provoking a maternal immunological rejection response (10).

## Endogenous Retroviral Particle Expression by Syncytiotrophoblast

There may be other, more subtle, sequelae to the lack of syncytiotrophoblastic expression of classical polymorphic class I or II MHC molecules. Human syncytiotrophoblast normally expresses assembled intracellular endogenous RNA retroviral particles (11), particularly in early pregnancy; yet it is unable effectively to present endogenously synthesized viral peptides at its cell surface to T cells since these would be required to be bound in association with class I or II MHC molecules.

Much evidence has now accumulated—by ultrastructural studies, reverse transcriptase activity, immunochemical localization of retroviral antigenic epitopes, and detection of retroviral RNA transcripts (4, 11)—that demonstrates normal syncytiotrophoblastic expression of intact non-infective endogenous type-C retroviral particles. It is of current interest that at least some of these endogenous retroviral antigenic epitopes appear to cross-react with those expressed by *human immunodeficiency virus-1* (HIV-1) (12, 13). Expression of individual endogenous retroviral proteins may be relatively common in normal cells, but, unusually, there appears to be significant concordant expression of products of the host genome (proviral DNA) to form assembled intracytoplasmic particles within normal syncytiotrophoblast in extraembryonic tissue. Although still poorly characterized, and in the absence as yet of specific immunological or molecular biology reagents for detection and isolation purposes, various different endogenous retroviral transcripts have been detected to date in placental tissue, including those homologous to the ERV3 (HERV-R) and ERV9 proviruses (11, 14–17; Mwenda et al., unpublished observations). It is most attractive to link this expression in extraembryonic tissue with syncytial cell formation, although other nonexclusive biological advantages could result from the potent immunosuppressive or immunomodulatory potential of certain retroviral-derived peptides (e.g., p15E) (11, 17–19). In addition, whether there is any relationship between transcriptional control of class I MHC genes and endogenous retroviral sequences remains to be determined.

## Nonclassical Class I MHC Molecules

Several studies using biochemical and molecular biological techniques have now shown clearly that human extravillous cytotrophoblast invading maternal decidualized endometrial tissue, as well as certain choriocarcinoma (malignant cytotrophoblast) cell lines, selectively expresses the nonclassical class I MHC molecule HLA-G (4, 20–22). The HLA-G gene is closely homologous with other class I MHC molecules and is located within the MHC region on the short arm of chromosome 6, but lacks the 13-bp upstream core IFN consensus sequence permitting up-regulation of its rate of transcription (23). It encodes a 39-kd MHC heavy chain that is a transmembrane molecule with a truncated cytoplasmic tail and that is associated with β2 microglobulin at the cell surface (23). Some HLA-G may be secreted rather than bound to the cell membrane (24). Although there has been strong evolutionary pressure at the HLA-G gene locus to restrict protein polymorphism, a *Hind*III restriction fragment-length polymorphism has been described, but all known sequence differences in HLA-G cDNA transcripts appear to be silent nucleotide substitutions with invariant inferred amino acid sequences (25).

RFLP analysis using specific probes for both classical and nonclassical (including HLA-G) class I genes, as well as the genes located within the genomic MHC region, has not uncovered any significant differences between cases of unexplained *recurrent spontaneous abortion* (RSA) compared with controls (26). An HLA-C-like sequence has also been consistently isolated from certain choriocarcinoma cell lines with a chromosome 6 trisomy and that—although extremely homologous to HLA-C locus sequences (notably those encoding HLA-Cw4)—is not completely identical with any known single HLA-C allele (27). It still remains to be clarified whether this is an artifact of the BeWo and JEG-3 cell lines or whether this gene is functionally expressed by cytotrophoblast in normal pregnancy.

HLA-G transcripts were reported originally to be found only at high levels of expression by invasive cytotrophoblast in extraembryonic tissue, although at least some HLA-G may be expressed by cells of the fetal liver (28), fetal thymus, and fetal eye (29). It is important to clarify whether or not there is any HLA-G expression in the thymus since this would indicate whether a positive selection of HLA-G-restricted T cells could occur that might recognize a trophoblast-derived peptide bound to HLA-G as a foreign peptide bound to self-class I MHC. HLA-G transcripts without protein expression have also been described in villous cytotrophoblast that might represent preinvasive cells within chorionic villi (30). Nevertheless, the strong HLA-G protein expression by invasive extravillous cytotrophoblast would suggest that this molecule has evolved for a specialized immunogenetic function in human pregnancy.

Speculation on the immunobiological role of HLA-G has centered on the possibility of its being a nonpolymorphic intercellular recognition molecule involved in either cytotrophoblast invasion (31) or the signaling of specific responses (notably cytokine secretion) by the characteristic $CD3^-$ $CD16^-$ $CD56^{++}$ dGL cell population in the maternal decidualized endometrium (see later). It has also been put forward that HLA-G might act as a passive invariant cell surface class I MHC molecule protecting cytotrophoblast from maternal MHC-nonrestricted *natural killer* (NK) cell attack—the *missing self* hypothesis (32)—without presenting a classical polymorphic class I MHC alloantigen that would act as a target for an alloimmune response (2, 4). It is known that cytotrophoblast is resistant to NK cell-mediated cytolysis (31). Initial data in support of this concept have been obtained by study of the response of cloned $CD3^-$ $CD16^-$ $CD56^{++}$ dGL to $HLA\text{-}G^+$ and $HLA\text{-}G^-$ transfectant lymphoblastoid cell lines (33 and Deniz et al., unpublished observations), although such experimental variations as the cell surface HLA-G density may be important in determining the strength of protection to NK cell-mediated cytolysis given by expression of this nonpolymorphic class I MHC molecule.

HLA-G may be able to mediate cell-cell adhesion through binding with CD8 (34), although maternal $CD3^+$ $CD8^+$ T cell numbers are relatively scarce in the decidua (35). However, a fundamental question remains concerning what, if any, antigenic peptides might be functionally presented by HLA-G molecules on the surface of invasive cytotrophoblast. The monomorphism of HLA-G could indicate conservative selection pressures to present an invariant peptide derived from an endogenous trophoblast self-protein. An opposing concept is that trophoblastic HLA-G is a redundant evolutionary relic of the human MHC region. For example, in a separate species, $\beta_2$ microglobulin-deficient mice obtained following homologous recombination in germ-line genes do not express constitutive class I MHC molecules, yet reproduce readily with normal litter size (36); however, these mice lack functional $CD8^+$ cytolytic T cells and are unable to distinguish self from nonself.

## Complement Regulatory Proteins

Increasingly, scientific attention is focusing on the complement system in human reproduction (37, 38). It is now well described that all fetal trophoblast populations throughout gestation, from the preimplantation embryo to the term placenta, express particularly high levels of cell surface complement regulatory proteins—notably, *membrane cofactor protein* ([MCP] CD46), *decay accelerating factor* ([DAF] CD55) and *membrane attack complex* (MAC) inhibitory factor (P18, protectin; CD59) (Table 1.1) (2, 4, 6, 39). Both CD46 and CD55 down-regulate complement cascade activity at the level of complement component C3 conversion. CD46 is a transmembrane glycoprotein that binds C3b (and, to a lesser extent, C4b) and acts as a cofactor for enzymatic cleavage and inactivation by factor I serine protease activity. It appears to be identical to the cell surface antigens formerly designated as *trophoblast-leukocyte common or cross-reactive* (TLX) antigens (40). CD55 has structural

TABLE 1.1. Expression of complement regulatory proteins by human spermatozoa, oocyte, and trophoblast.

| | Spermatozoa | Oocyte | Trophoblast |
|---|---|---|---|
| CD46 (MCP) | ++ (AR) | + | ++ |
| CD55 (DAF) | + (h > t) | + | ++ |
| CD59 | ++ (h, t) | ++ | ++ |

Expression was determined by immunohistochemical staining of acetone-fixed fresh sperm, oocyte, and placental villous tissue sections. (++ = strong staining; + = weak staining; AR = acrosome-reacted spermatozoa only; h = head; t = tail.)

homology with CD46; it prevents formation and accelerates decay of C3 convertases, complementing the function of CD46 (38). CD59 inhibits formation of the terminal complement MAC. Unlike CD46, both CD55 and CD59 are anchored to the cell membrane by a *glycosyl phosphatidylinositol* (GPI) moiety and, hence, could also serve a role in intracellular signal transduction mechanisms.

Nevertheless, the strong trophoblastic expression of these proteins has the net effect of frustrating any potential complement-mediated damage at this fetomaternal tissue interface as a result of either maternal antifetal antibody attack or hemostatic alterations and local tissue remodeling that may promote persistent low-level activation of the complement cascade. This innate resistance to complement-mediated cytolysis, together with selective control of class I MHC molecule expression, represents strategies employed by trophoblast to evade maternal cytotoxic attack.

Although most cells predominantly express one or both of two MCP isoforms of ~65 and ~55 kd (41), substantial polymorphism at the glycoprotein level, as well as in sequences of cDNA clones, has been shown that is a result of alternative splicing of exons encoding a serine-threonine-proline-rich region (42). This molecular variation would appear redundant for a biological role solely as a cell surface complement regulatory protein and could indicate an additional function in reproductive processes. CD46 is also strongly expressed by endometrial glandular epithelium (43) and on the surface of acrosome-reacted, but not non-reacted, spermatozoa (44–46). The spermatozoal CD46 molecule appears to be a unique tissue-specific isoform characterized by delections that result in a shorter structure close to the transmembrane region and an unusual cytoplasmic tail (42). In contrast to CD46, CD55 and CD59 are expressed on both acrosome-reacted and intact spermatozoa (47, 48) (Table 1.1).

As key components of the complement system are absent or present at low levels in seminal plasma, spermatozoal complement regulatory proteins may primarily assist protection from complement-mediated cytolytic damage in the female reproductive tract (49). However, the unusual ultrastructural localization and molecular isoform of spermatozoal CD46 could indicate a separate or additional role in sperm-oocyte interaction other than direct binding and regulation of C3b. For example, a secondary function in restricting localized proteolytic damage at the site of fertilization has been suggested (48). CD46, CD55, and CD59 expression have also been identified on human oocytes (6 and Johnson et al., unpublished observations) (Table 1.1); hence, both sperm-oocyte and blastocyst-endometrial epithelium interactions occur between genetically disparate cells carrying one or more C3b-binding molecules. It has been shown that treatment of spermatozoa with anti-CD46 monoclonal antibody causes a significant decrease in their ability to facilitate hamster egg penetration (44). More recently, we have demonstrated that anti-CD46 monoclonal antibody will significantly inhibit human sperm binding and pronuclear formation in an autologous zona-free oocyte penetration assay (50 and Taylor et al., unpublished observations).

## $CD3^-$ Decidual Granulated Leukocytes

The predominant leukocyte population in the human first-trimester decidualized endometrium consists of phenotypically distinctive $CD3^-$ dGL. Leukocytes account for over 30% of all decidual stromal cells, and dGL account for up to 75% of this stromal leukocyte population, the remaining leukocytes being mostly tissue macrophages with some $CD3^+$ T cells and virtually no B cells (2, 35, 51). CD56 (a marker of NK cells) is expressed very strongly by dGL, as are CD38 and CD69, although these cells lack CD16 and strong naive functional MHC-nonrestricted NK cell activity (35, 51). These cells lack mature T cell markers (CD3, CD4, and CD8), but a significant proportion express CD2; they express the low-affinity p75 chain of the IL-2 receptor, but not the high-affinity p55 chain (CD25) (35, 51).

Phenotypic analysis would suggest that these cells most closely represent an unusual differentiation stage of NK cells (35). Most dGL express the mucosal lymphocyte antigen HML-1, as well as other β1 and β2 integrins and adhesion molecules that are members of the immunoglobulin superfamily (52). The immunobiological function(s) of these $CD3^-$ $CD16^-$ $CD56^{++}$ dGL is unknown, although most speculation focuses on a beneficial influence on fetal trophoblast growth, invasion, or differentiation through mechanisms related to NK cell-type activity, immunosuppression, or generation of maternal cytokine signals (paracrine events) (2, 31, 35).

One experimental approach that we have used is a lymphocyte cloning technique with PHA and IL-2 to obtain short-term homogenous $CD3^-$ dGL clones for initial characterization and functional studies (53). These clones, including $CD16^-$ clones, exhibit MHC-nonrestricted cytolytic activity against susceptible cellular targets that may be protected by constitutive expression of the transfected HLA-G gene (33 and Deniz et al., unpublished observations). Recent work has also shown that these cells produce substantial amounts of TNFα in particuler, but also IFNγ and *granulocyte-macrophage colony stimulating factor* (GM-CSF) (54 and Deniz et al., unpublished observations). Previous work has also shown production of *transforming growth factor beta* (TGFβ) (53). Interestingly, all these cytokines (TNFα, IFNγ, GM-CSF, and TGFβ) could influence the growth and development of extraembryonic tissue by interaction with the specific receptors that are present on human trophoblastic cells (8, 55, 56). Maternally derived cytokines produced by endometrial cells other than $CD3^-$ granulated leukocytes, such as *macrophage colony stimulating factor* ([M-CSF] CSF-I) (57), may act similarly through specific trophoblast cell surface receptors (58).

Indeed, it remains unclear as to the immunobiological advantage to trophoblastic cells in expressing such a range of different cell surface receptors for maternal cytokines produced locally in decidual tissue. One exception would appear to be receptors for c-*kit* ligand (stem cell factor) that are absent from trophoblast (56), whereas the ligand itself is produced by trophoblast (59). It is now clear that in normal pregnancy the $CD3^-$ $CD16^-$ $CD56^{++}$ dGL, together with other uteroplacental cells, take part in a complex cytokine-mediated fetomaternal immunobiological signaling network (60) that is yet to be fully unraveled.

## References

1. Dondero F, Johnson PM, eds. Reproductive immunology. Serono Symposium Publications; vol 97. New York: Raven Press, 1993:1–387.
2. Johnson PM. Reproductive and materno-fetal relations. In: Lachmann PJ, Peters DK, Rosen FS, Walport MJ, eds. Clinical aspects of immunology. 5th ed. Boston: Blackwell Scientific, 1993:755–67.
3. Hunt JS. Immunobiology of pregnancy. Curr Opin Immunol 1992;4:591–6.
4. Johnson PM. Immunology of human extraembryonic fetal membranes. In: Coulam CB, Faulk WP, McIntyre JA, eds. Immunological obstetrics. New York: Norton WW, 1992:177–88.
5. Hunt JS, Orr HT. HLA and maternal-fetal recognition. FASEB J 1992; 6:2344–8.
6. Roberts JM, Taylor CT, Melling GC, Kingsland CR, Johnson PM. Expression of the CD46 antigen, and absence of class I MHC antigen, on the human oocyte and preimplantation blastocyst. Immunology 1992;75:202–5.
7. Hunt JS, Andrews GK, Wood GW. Normal trophoblast resist induction of class I HLA. J Immunol 1987;138:2481–7.
8. Peyman JA, Hammond GL. Localization of IFN-γ receptor in first-trimester placenta to trophoblasts but lack of stimulation of HLA-DRA, -DRB or invariant chain mRNA expression by IFN-γ. J Immunol 1992;149:2675–80.
9. Boucrat J, Hawley S, Robertson K, Bernard D, Loke YW, Le Bouteiller P. Differential nuclear expression of enhancer A DNA-binding proteins in human first trimester trophoblast cells. J Immunol 1993;150:3882–94.
10. Attwood HD, Park WW. Embolism to the lungs by trophoblast. J Obstet Gynaecol Br Commonw 1961;68:611–7.
11. Johnson PM, Lyden TW, Mwenda J. Endogenous retroviral expression in the human placenta. Am J Reprod Immunol 1990;23:115–20.
12. Lyden TW, Johnson PM, Mwenda JM, Rote NS. Immunolocalization of HIV-1 cross-reactive placental proteins in the normal non-infected human decidua and basal plate. Placenta 1993;14:A46.
13. Lyden TW, Johnson PM, Mwenda JM, Rote NS. Anti-HIV monoclonal antibodies cross-react with normal human trophoblast. Trophoblast Res 1993;7.
14. Rabson AB, Hamagishi Y, Steele PE, Tykocinski M, Martin MA. Characterization of human endogenous retroviral envelope RNA transcripts. J Virol 1985;56:176–82.

15. Kato N, Pfeiffer-Ohlsson S, Kato M, et al. Tissue-specific expression of human provirus ERV3 mRNA in human placenta: two of the three ERV3 mRNAs contain human cellular sequences. J Virol 1987;61:2182–92.
16. Mager DI. Polyadenylation function and sequence variability of the long terminal repeats of the human endogenous retrovirus-like family RTVL-H. Virology 1985;173:591–9.
17. Lindeskog M, Medstrand P, Blomberg J. Sequence variation of human endogenous retrovirus ERV9-related elements in an *env* region corresponding to an immunosuppressive peptide: transcription in normal and neoplastic cells. J Virol 1993;67:1122–6.
18. Harris DT, Cianciolo GJ, Synderman R, Argov S, Koren HS. Inhibition of human natural killer cell activity by a synthetic peptide homologous to a conserved region in the retroviral protein, p15E. J Immunol 1987;138:889–94.
19. Banki K, Maceda J, Hurley E, et al. Human T cell lymphotrophic virus (HTLV)-related endogenous sequence, HRES-1, encodes a 28 kDa protein: a possible autoantigen for HTLV-1 *gag*-reactive autoantibodies. Proc Natl Acad Sci USA 1992;89:1939–43.
20. Ellis SA, Palmer MS, McMichael AJ. Human trophoblast and the choriocarcinoma cell line BeWo express a truncated HLA class I molecule. J Immunol 1990;144:731–5.
21. Kovats S, Main EK, Librach C, Stubblebine M, Fisher SJ, DeMars R. A class I antigen, HLA-G, expressed in human trophoblasts. Science 1990;248:220–3.
22. Risk JM, Johnson PM. Northern blot analysis of HLA-G expression by BeWo human choriocarcinoma cells. J Reprod Immunol 1990;18:199–203.
23. Heinrichs H, Orr HT. HLA non-A,B,C class I genes: their structure and expression. Immunol Res 1990;9:265–74.
24. Ishitani A, Geraghty DE. Alternative splicing of HLA-G transcripts yields proteins with primary structures resembling both class I and class II antigens. Proc Natl Acad Sci USA 1992;89:3947–51.
25. Pook MA, Woodcock V, Tassabehji M, et al. Characterization of an expressible nonclassical class I HLA gene. Hum Immunol 1991;32:102–9.
26. Risk JM, Johnson PM. Immunogenetic studies in unexplained recurrent miscarriage of pregnancy. In: Tsuji K, Aizawa M, Sasazuki T, eds. HLA 1991; vol 1. Oxford: Oxford University Press, 1992:1012–7.
27. Ellis SA, Palmer MS, McMichael AJ. Complete nucleotide sequence of a unique HLA class I C-locus product expressed on the human choriocarcinoma cell line BeWo. J Immunol 1989;142:3281–5.
28. Houlihan JM, Biro PA, Fergar-Payne A, Simpson KL, Holmes CH. Evidence for the expression of non-HLA-A, -B, -C class I genes in the human fetal liver. J Immunol 1992;149:668–75.
29. Shukla H, Swaroop A, Srvastava R, Weissman SM. The mRNA of a human class I gene HLA-G/HLA-6.0 exhibits a restricted pattern of expression. Nucleic Acids Res 1990;18:2189.
30. Yelavarthi KK, Fishback JL, Hunt JS. Analysis of HLA-G mRNA in human placental and extraplacental membrane cells by in situ hybridization. J Immunol 1991;146:2847–54.
31. King A, Loke YW. On the nature and function of uterine granular lymphocytes. Immunol Today 1991;12:432–5.
32. Ljunggren G, Karre K. In search of the "missing self"—MHC molecules and NK cell recognition. Immunol Today 1990;11:237–44.

33. Johnson PM, Risk JM, Deniz G, Christmas SE. Immunological studies of HLA-G expression by human trophoblast. In: Tsuji K, Aizawa M, Sasazuki T, eds. HLA 1991; vol 1. Oxford: Oxford University Press, 1992:1039–44.
34. Sanders SK, Giblin PA, Kavathas P. Cell-cell adhesion mediated by CD8 and human histocompatibility leukocyte antigen G, a nonclassical major histocompatibility complex class I molecule on cytotrophoblasts. J Exp Med 1991;174:737–40.
35. Bulmer JN, Morrison L, Longfellow M, Ritson A. Leukocytes in human decidua: investigation of surface markers and function. Colloque INSERM 1991;212:189–96.
36. Zilstra M, Bix M, Simister NE, Loring JM, Raulet DH, Jaenisch R. $\beta_2$-microglobulin-deficient mice lack $CD4^-$ $CD8^+$ cytolytic T cells. Nature 1990;344:742–6.
37. Holmes CH, Simpson KL. Complement in pregnancy: new insights into the immunobiology of the fetomaternal relationship. Baillieres Clin Obstet Gynaecol 1992;6:439–60.
38. Rooney IA, Oglesby TJ, Atkinson JP. Complement in human reproduction: activation and control. Immunol Res 1993.
39. Holmes CH, Simpson KL, Okada H, et al. Complement regulatory proteins at the feto-maternal interface during human placental development: distribution of CD59 by comparison with membrane cofactor protein (CD46) and decay accelerating factor (CD55). Eur J Immunol 1992;22:1579–85.
40. Purcell DFJ, McKenzie IFC, Lublin DM, et al. The human cell surface glycoproteins HuLy-m5, membrane co-factor protein (MCP) of the complement system, and trophoblast-leukocyte common (TLX) antigen are CD46. Immunology 1990;70:155–61.
41. Stern PL, Beresford N, Thompson S, Johnson PM, Webb PD, Hole N. Characterization of the human trophoblast-leukocyte common antigenic molecules defined by a monoclonal antibody. J Immunol 1986;137:1604–9.
42. Russell SM, Sparrow RL, McKenzie IFC, Purcell DFC. Tissue-specific and allelic expression of the complement regulator CD46 is controlled by alternative splicing. Eur J Immunol 1992;22:1513–8.
43. Johnson PM, Bulmer JN. Uterine gland epithelium in human pregnancy lacks maternal MHC antigens but does express fetal trophoblast antigens. J Immunol 1984;132:1608–11.
44. Anderson DJ, Michaelson JS, Johnson PM. Trophoblast/leukocyte common antigen is expressed by human testicular germ cells and appears on the surface of acrosome-related sperm. Biol Reprod 1989;41:285–93.
45. Cervani F, Oglesby TJ, Adams EM, et al. Identification and characterization of membrane cofactor protein of human spermatozoa. J Immunol 1992;148: 1431–7.
46. Okabe M, Ying X, Nagira M, et al. Homology of an acrosome-reacted sperm-specific antigen to CD46. J Pharmacobiodyn 1992;15:455–9.
47. Rooney IA, Davies A, Morgan BP. Membrane attack complex (MAC)-mediated damage to spermatozoa: protection of cells by the presence on their membranes of MAC inhibitory proteins. Immunology 1992;75:499–506.
48. D'Cruz OJ, Haas GG. The expression of the complement regulators CD46, CD55 and CD59 by human sperm does not protect them from anti-sperm antibody- and complement-mediated immune injury. Fertil Steril 1993;59: 876–84.

49. Bozas SE, Kirszbaum L, Sparrow RL, Walker ID. Several vascular complement inhibitors are present on human sperm. Biol Reprod 1993;48:503–11.
50. Taylor CT, Melling GC, Biljan MM, Kingsland CR, Johnson PM. A possible role for CD46 in human fertilisation? Fertil Steril 1993;8(suppl 1):63.
51. Bulmer JN. Decidual cellular responses. Curr Opin Immunol 1989;1:1141–7.
52. Burrows TD, King A, Loke YW. Expression of adhesion molecules by human decidualized large granular lymphocytes. Cell Immunol 1993;147: 81–94.
53. Christmas SE, Bulmer JN, Meager A, Johnson PM. Phenotypic and functional analysis of human $CD3^-$ decidual leukocyte clones. Immunology 1990;71:182–9.
54. Johnson PM, Deniz G, McLaughlin PJ, Hampson J, Christmas SE. Functional properties of cloned $CD3^-$ decidual granular leukocytes and low-affinity cytokine receptor expression on human trophoblast. In: Dondero F, Johnson PM, eds. Reproductive Immunology. Serono Symposium Publications; vol 97. New York: Raven Press, 1993:141–4.
55. Mitchell EJ, Fitz-Gibbon L, O'Connor-McCourt MD. Subtypes of betaglycan and of type I and type II transforming growth factor-β (TGF-β) receptors with different affinities for TGF-β1 and TGF-β2 are exhibited by human placental trophoblast cells. J Cell Physiol 1992;150:334–43.
56. Hampson J, McLaughlin PJ, Johnson PM. Low-affinity receptors for tumour necrosis factor-α, interferon-γ and granulocyte-macrophage colony stimulating factor are expressed on human placental syncytiotrophoblast. Immunology 1993;79:485–90.
57. Daiter E, Pampfer S, Yeung YG, Barad D, Stanley ER, Pollard JW. Expression of colony-stimulating factor-1 in the human uterus and placenta. J Clin Endocrinol Metab 1992;74:850–8.
58. Jokhi PP, Chumbley G, King A, Gardner L, Loke YW. Expression of the colony stimulating factor-1 receptor (c-*fms* product) by cells at the human uteroplacental interface. Lab Invest, 1993:308–20.
59. Sharkey A, Jones DSC, Brown KD, Smith SK. Expression of messenger RNA for *kit*-ligand in human placenta: localization by in situ hybridization and identification of alternatively spliced variants. Mol Endocrinol 1991; 6:1235–41.
60. Mitchell MD, Trautman MS, Dudley DJ. Cytokine networking in the placenta. Placenta 1993;14:249–75.

# 2

# Immunity in the Human Male Reproductive Tract

Jeffrey Pudney and Deborah J. Anderson

In contrast to the female, little information is available concerning the immunobiology of the male reproductive tract (1, 2). There are several reasons for this discrepancy. First, due to differences in the anatomy of the reproductive tract between males and females, women tend to be more susceptible to infection by sexually transmitted pathogens than men. Although a certain degree of mechanical protection is afforded by the stratified squamous epithelium of the vagina, this can be abraded and/or damaged during sexual intercourse, allowing portals of entry for organisms responsible for venereal diseases. The cervix is at risk for infection from semen that may contain venereal pathogens. A certain degree of protection for this region, however, is provided by mucous secretions released by the cervical glands. Also, the lower reproductive tract of healthy women sustains a normal population of flora that can play a nonspecific role in resisting invasion by sexually transmitted pathogens. If the balance of this flora is disturbed for any reason, this can allow colonization by other, different organisms that may result in a pathological condition.

Interest in the immunobiology of the female reproductive tract, therefore, arises from a need to understand the immune response to sexually transmitted diseases to facilitate clinical management. An added concern for females is the effect that infection with a venereal pathogen may have on fertility and the outcome of pregnancy. Also, tissue analysis, either morphological or physiological, is facilitated by the relative ease with which biopsies can be obtained from the female reproductive tract.

For the healthy male the principal area of exposure and, therefore, potential site of invasion by many sexually transmitted pathogens (particularly those established by mucosal transmission) is the tip of the penile urethra (meatus). With less tissue area exposed, men tend to be less vulnerable to infection than women, resulting in fewer investigations into the immunobiology of the male reproductive tract. The male reproductive tract is difficult to study because invasive methods must be used to acquire tissue samples. It is possible, however, to obtain an indirect measure of the immune status of reproductive organs in men through analysis of sexual secretions. Semen invariably contains leukocytes. Recent studies have shown that men with large numbers of seminal white blood cells, a condition called *leukocytospermia*, have decreased semen quality (3, 4). It has been suggested that leukocytospermia reflects a subclinical infection or inflammation of the reproductive tract.

Preejaculatory fluid is secreted by the bulbourethral glands of Cowper and periurethral glands of Littre and is thought to act as a lubricant during sexual intercourse. Little is known about preejaculatory fluid, but recent studies have shown that this secretion also contains abundant white blood cells (5). This secretion could be involved in the sexual transmission of venereal pathogens since preejaculatory fluid from HIV-1 seropositive men has been shown to contain host cells positive for the presence of HIV (5, 6).

The presence of leukocytes in sexual secretions from the male reproductive tract provides an avenue for assessing the immunobiology of male reproductive organs. The presence and relative abundance of white blood cells and macrophages in semen and preejaculatory fluid may be used to diagnose infection or inflammatory conditions, and, following collection of semen into large volumes of media to dilute out the toxic effect of seminal plasma, lymphocytes from these secretions can be isolated and studied functionally.

## Immunobiology of the Testis

The testis consists of a germinal compartment, the seminiferous tubules, responsible for the production of spermatozoa, and the interstitial tissue that contains Leydig cells, which are responsible for the secretion of testosterone. The immunodynamics of the testis in normal, healthy men is unusual in that few lymphocytes reside in this organ. The most abundant hematopoietic cells are macrophages, which are located in the interstitial tissue (7, 8). This immunological profile can change during testicular disease. For instance, although lymphocytes are virtually absent from the healthy testis, they can appear in large numbers in the murine testis during immune orchitis (9) and in the testes of HIV-1 seropositive men (10). The profile of infiltrating lymphocytes differs according to the underlying pathology of the testis. In men with oligospermia and obstructive azoospermia, interstitial $CD8^+$ suppressor/cytotoxic lymphocytes predominate, whereas in unilateral testicular obstruction and postvasectomy patients, there is a predominance of $CD4^+$ helper/inducer lymphocytes (8).

The testis is said to comprise an immunologically privileged site because developing germ cells are protected from autoimmune attack. This may be due to the presence of testicular immunosuppressive factors (11) and a blood-testis barrier that anatomically (12) and physiologically (13) sequesters germ cells from the immune system. Furthermore, the tight junctional barriers developed by adjacent Sertoli cells, which form the blood-testis barrier, prevent the entry of macromolecules, such as immunoglobulins, into the seminiferous epithelium. Studies on animal models have indicated that potentially autoantigenic germ cells can exist outside the blood-testis barrier (14). This suggests that other immunoprotective mechanisms operate in the testis to prevent immune response development against the nonsequestered autoantigens (15).

## Immunobiology of the Excurrent Duct System

The reproductive excurrent ducts consist of the rete testis that conveys sperm from the testis and the efferent ducts that transport the germ cells to the epididymis. Spermatozoa undergo final maturational changes during passage through the epididymis and are eventually stored in the tail or cauda epididymis. During ejaculation the stored sperm are forced into the vas deferens for eventual release.

The excurrent ducts are lined with a typical mucosa consisting of an epithelium resting on a basal lamina with an underlying lamina propria. Along the length of the excurrent ducts, a normal complement of leukocytes can be detected. As with other mucosa, intraepithelial lymphocytes have been described in all regions of the excurrent duct system. Intraepithelial lymphocytes are $CD8^+$, whereas $CD4^+$ lymphocytes are located exclusively in the ductal connective tissue (16, 17).

The immunological privileged status of the excurrent duct system is probably not as secure as that of the testis since the epithelial junctions are less tight than those creating the blood-testis barrier (18). The excurrent ducts, however, must protect the spermatozoa that still bear antigens capable of inducing autoimmune reactions, but how this is accomplished is not known. Immunosuppressive activity has been detected in epididymal fluid (19, 20). Also, it has been suggested that the $CD8^+$ intraepithelial lymphocytes present along the length of the excurrent ducts represent a population of suppressor cells that down-regulate immunological reactions directed against spermatozoa (16).

The excurrent duct system contains a population of immune cells, T lymphocytes, and antigen-presenting macrophages that could mount a cell-mediated immune response against invading infectious pathogens. Few data are available, however, on humoral mucosal immunity in the excurrent ducts; secretory IgA plasma cells have not been identified in the excurrent duct system. Studies in our laboratory have shown that the epithelium of all regions of the human excurrent ducts are capable of expressing secretory component (21). *Secretory component* is the receptor for secretory IgA and is responsible for transporting this immunoglobulin across the epithelial lining for secretion into the lumen. It has been shown that secretory component is up-regulated in excurrent ducts from HIV-1 seropositive men (21). This suggests, therefore, that excurrent ducts have the ability to mount a mucosal immune response using secretory IgA either derived as a transudate from blood or locally produced by plasma cells. Macrophages appear to be the most abundant hematopoietic cells in the lamina propria along the length of the excurrent duct system. These cells are occasionally detected in the epithelium and lumen of the epididymis, where they appear to phagocytose dead and dying sperm. The number of intraepithelial and lumenal macrophages is up-regulated in the epididymides from HIV-1 seropositive men.

The immune response of the excurrent ducts to infection in the human male is essentially unknown. Initial experiments in our laboratory using a primate model, the bonnet monkey (*Macaca radiata*), demonstrated that experimental epididymal infection with *E. coli* significantly increased the abundance of macrophages, $CD8^+$ lymphocytes, and expression of secretory component compared to control monkeys. This suggests that normal regulatory and immunoprotective functions exist and operate at the level of the excurrent duct system that would facilitate resistance to invasion and infection by venereal pathogens.

## Immunobiology of the Accessory Glands

The accessory glands consist of the seminal vesicles and prostate gland. These glands supply secretions involved in maintaining spermatozoa both in the ejaculate and during their sojourn in the female reproductive tract. It appears that approximately 70% of the seminal fluid is obtained from the seminal vesicles and approximately 30% from the prostate gland (22).

Little is known concerning the immunobiology of the seminal vesicles. Studies in our laboratory show that the mucosa of the seminal vesicle contains macrophages as well as extra- and intraepithelial lymphocytes. The epithelium of the seminal vesicles is capable of expressing secretory component, and expression is up-regulated in this tissue from HIV-1 seropositive men (21). The normal human prostate contains abundant lymphocytes and macrophages both in the stroma and the epithelium. The epithelium of the prostate of healthy men does not express secretory component (23), but our experiments have demonstrated that the protein is present in prostates of HIV-1 seropositive men (21).

Concentrations of IgG and IgA are greater in the first portion of the ejaculate, which is derived mainly from the prostate, than in the terminal portions, which are composed primarily of secretions from the seminal vesicles (24, 25). Pure, sterile prostatic fluid obtained by digital massage of the prostate from men with no history of urogenital tract infection was found to contain high levels of IgG, moderate levels of IgA, and almost no IgM (26). Most of the IgA was found to occur in the form of secretory IgA. Studies on total class-specific Ig and class-specific/antigen-specific Ig in men with reproductive tract infections (gonococcal urethritis and bacterial prostatitis) demonstrated that the most profound local mucosal immune response takes place in the prostate (26, 27). An interesting finding is that asymptomatic bacterial colonization of the prostate can stimulate a mucosal immune response. Furthermore, a local secretory immune response of the prostate has been reported for men with bacteriuria (26).

It has been suggested that venereal infection of the reproductive tract can be largely prevented or contained by a secretory immune response in the prostate (28). This would account for the resistance of the prostate to sexually transmitted pathogens and the relative infrequency of invasion of the deep reproductive organs in men suffering from infection of the urethra or the urinary bladder (28).

## Immunobiology of the Urethra

In the normal, healthy male, the urethra is the first site of exposure to many sexually transmitted diseases. It is extremely important, therefore, to understand the immunocompetence of the penile urethra and the immune responses it can develop to resist invasion by venereal pathogens. Studies on the human penile urethra in our laboratory have shown this tissue is extremely immunodynamic. Numerous lymphocytes of both $CD4^+$ and $CD8^+$ phenotypes, as well as antigen-presenting cells in the form of macrophages and dendritic cells, are present in the lamina propria and epithelium of the urethral mucosa. The urethra may also be capable of mounting a local mucosal immune response since abundant plasma cells positive for both IgA and J-chain are detectable in the lamina propria. The epithelium is also positive for the presence of IgA and J-chain, as well as secretory component.

In several cases an inflammatory condition has been identified in the urethra by the presence of extremely large numbers of granulocytes in the mucosa, most of which are positive for neutrophil elastase. Many of these granulocytes are located in the epithelium lining the urethra. For all these cases there was a dramatic up-regulation in the number of lymphocytes and macrophages present in the urethral mucosa, plus an enhancement of secretory component expression by the epithelium. These findings clearly indicate that the urethral mucosa can mount a vigorous immune defense against infectious pathogens involving both cell-mediated and mucosal secretory responses.

The presence of numerous immunocompetent cells in the epithelium of the urethral mucosa would suggest this is a major site of release for these cells into the preejaculatory fluid. The bulbourethral glands of Cowpers are also involved in the secretion of preejaculatory fluid. Studies on these glands, however, have shown that they contain few leukocytes and would therefore be unlikely to contribute these cells to the preejaculatory fluid (29).

The urethral mucosa is not only an important region for the generation of resistance to venereal pathogens, but could also play a role in the sexual transmission of these organisms. As previously mentioned, studies have detected HIV-positive host cells (macrophages and/or $CD4^+$ lymphocytes) in preejaculatory fluid from HIV-1 seropositive men. It is quite possible that these virally infected cells originated from the epithelium of the urethral mucosa. This underscores the importance of understanding the immunobiology of the penile urethra in health and disease so that better strategies in terms of drugs and/or vaccines can be designed to limit or prevent sexual transmission of venereal pathogens.

## Conclusion

Although a fair amount of information is available concerning the immunobiology of the male reproductive tract, it is fragmentary and incomplete. A comprehensive analysis of the immunocompetence of reproductive organs and their potential for mounting immune responses in the face of invading venereal pathogens is urgently needed. Descriptive information—that is, the presence, location, and abundance of specific types of leukocytes—must be acquired before the role(s) these cells play in the normal functions of the reproductive tract can be understood. Then, these studies can be extended and supplemented to include analyses of their activities. Is their presence essential for normal reproductive function, and can they, through either hypo- or hyperactivity, cause pathological conditions that result in infertility?

There are several reports to suggest that secretions of immunocompetent cells (i.e., cytokines) may be necessary for normal functional activity of the reproductive tract. For instance, human *interleukin-1α* (IL-1α) has been shown to stimulate spermatogonial proliferation in the rat testis in vivo (30). It has also been reported that various human interferons can inhibit steroidogenesis in porcine Leydig cells (31). These examples illustrate the fact that the function of immunocompetent cells present in the male reproductive tract goes beyond their known role in immune defense. Lymphocytes and macrophages are abundant in reproductive tissue and may secrete products essential for normal spermatogensis and sperm maturation. If these cells occur in either abnormally high or low concentrations or if they are inappropriately activated or deactivated, they could affect male fertility. For example, if the normal functions of testicular macrophages are disturbed by an inflammatory or autoimmune reaction, the production of beneficial products may cease; instead, secretion of toxic factors, such as free oxygen radicals, hydrogen peroxide, or monokines, may occur. This situation could adversely affect steroidogenesis by Leydig cells that in turn would impair spermatogenesis.

Another example would be the epididymis, where there is evidence that immunocompetent cells can have adverse effects on reproductive potential. When activated by antigens, immune complexes, or infectious organisms, lymphocytes and macrophages release cytokines, such as *interferon* $\gamma$ (IFN$\gamma$) (lymphocytes) and *tumor necrosis factor* $\alpha$ (TNF$\alpha$) (lymphocytes and macrophages), that have been shown to reduce sperm motility (32, 33) and the ability of sperm to penetrate hamster eggs in vivo (34). Recent studies have also shown that epithelial cells in many regions of the male reproductive tract are capable of expressing HLA-DR (2). *HLA-DR* is a molecule involved in antigen presentation and activation and is normally expressed by macrophages, B cells, and activated T cells. Inappropriate expression of HLA-DR by reproductive tissues, therefore, could play a role in the development of autoimmune responses in reproductive organs by activation and amplification of T cell responses at these sites.

In summary, many studies indicate that the male reproductive tract is immunologically competent and capable of mounting protective cellular and antibody-mediated immunological responses. It is now of the utmost importance to develop a better understanding of the interaction that can occur between immunocompetent cells and reproductive tract cells and the consequences of this interaction to male fertility.

*Acknowledgments.* This research was supported by Grant RO1-AI25305 from the National Institutes of Health and by a contract from the United States Agency for International Development (USAID) Contraceptive Research and Development (CONRAD) Program, which receives funds for AIDS research from an interagency agreement with the National Institute of Child Health and Human Devleopment (NICHD). The views expressed by the authors do not necessarily reflect the views of the NIH, USAID, or CONRAD.

## References

1. Barratt CLR, Bolton EA, Cooke ID. Functional significance of white blood cells in the male and female reproductive tract. Hum Reprod 1990;5:639–48.
2. Pudney JA, Anderson DJ. Organization of immunocompetent cells and their function in the male reproductive tract. In: Griffin PD, Johnson PM, eds.

Local immunity in reproductive tract tissues. Oxford: Oxford University Press, 1993:131–45.
3. Wolff H, Politch JA, Martinez A, Haimovici F, Hill JA, Anderson DJ. Leukocytospermia is associated with poor semen quality. Fertil Steril 1991; 53:528–36.
4. Politch JA, Wolff H, Hill JA, Anderson DJ. Comparison of methods to enumerate white blood cells in semen. Fertil Steril 1993;60:372–5.
5. Pudney J, Oneta M, Mayer K, Seage G, Anderson D. Pre-ejaculatory fluid as potential vector for the sexual transmission of HIV-1. Lancet 1992;340:1470.
6. Ilaria G, Jacobs JL, Polsky B, et al. Detection of HIV-1 DNA sequences in pre-ejaculatory fluid. Lancet 1992;340:1469.
7. Pollanen P, Niemi M. Immunohistochemical identification of macrophages, lymphoid cells, and HLA antigens in the human testis. Int J Androl 1987; 10:37–47.
8. El-Demiry MIM, Hargreave TB, Busuttil A, Elton RE, James K, Chisholm GD. Immunocompetent cells in human testis in health and disease. Fertil Steril 1987;48:470–9.
9. Tung KSK, Yule TD, Mahi-Brown CA, Listrom MB. Distribution of histopathology and Ia positive cells in actively induced and passively transferred experimental autoimmune orchitis. J Immunol 1987;138:752–9.
10. Pudney J, Anderson D. Orchitis and human immunodeficiency virus type 1 infected cells in reproductive tissues from men with the Acquired Immunodeficiency Syndrome (AIDS). Am J Pathol 1991;139:149–60.
11. Pollanen P, von Euler M, Soder O. Testicular immunoregulatory factors. J Reprod Immunol 1990;18:51–76.
12. Dym M, Fawcett DW. The blood-testis barrier in the rat and the physiological compartment of the seminiferous epithelium. Biol Reprod 1970;3:308–26.
13. Johnson MH, Setchell BP. Protein and immunoglobulin content of rete testis fluid of rams. J Reprod Fertil 1968;17:403–6.
14. Yule TD, Montoya GD, Russell LD, Williams TM, Tung KSK. Autoantigenic germ cells exist outside the blood-testis barrier. J Immunol 1988;141:1161–7.
15. Yule TD, Mahi-Brown CA, Tung KSK. Role of testicular autoantigens and influence of lymphokines in testicular autoimmune disease. J Reprod Immunol 1990;18:89–103.
16. Ritchie AWS, Hargreave TB, James K, Chisholm GD. Intra-epithelial lymphocytes in the normal epididymis: a mechanism for tolerance to sperm auto-antigens. Br J Urol 1984;56:79–83.
17. El-Demiry MIM, James K. Lymphocyte subsets and macrophages in the male genital tract in health and disease. Eur J Urol 1988;14:226–35.
18. Suzuki F, Nagamo T. Regional differences of cell junctions in the excurrent duct epithelium of the rat as revealed by freeze-fracture. Anat Rec 1978; 191:503–20.
19. Anderson DJ, Tartar TH. Immunosuppressive effects of mouse seminal plasma components in vivo and in vitro. J Immunol 1982;128:535–9.
20. Harkins H, Anderson DJ. Seminal plasma factor(s) released by sperm inhibit cytotoxic activity of "natural killer" (NK) lymphoid cells. Biol Reprod 1984;30(suppl):139.
21. Anderson DJ, Pudney J. Mucosal immune defense against HIV-1 in the male urogenital tract. Vac Res 1992;1:143–50.

22. Lundquist F. Aspects of the biochemistry of human semen. Acta Physiol Scand 1949;19(suppl 66):1–105.
23. Brandtzaeg P, Christiansen E, Muller E, Purvis K. Humoral immune response patterns of human mucosa, including the reproductive tracts. In: Griffin PD, Johnson PM, eds. Local immunity in reproductive tract tissues. Delhi: Oxford University Press, 1993:97–130.
24. Rumke P. The origin of immunoglobulins in semen. Clin Exp Immunol 1974;17:287–97.
25. Tauber PF, Zaneveld LJD, Propping D, Schumacher GFB. Components of human split ejaculates: spermatozoa fructose, immunoglobulins, albumin, lactoferrin, transferrin, and other plasma proteins. J Reprod Fertil 1975; 43:249–67.
26. Fowler JE Jr, Keiser DL, Mariano M. Immunologic response of the prostate to bacteriuria and bacterial prostatitis; Part 1, immunoglobulin concentrations in prostatic fluids. J Urol 1982;128:158–64.
27. McMillan A, McNeillage G, Young H. Antibodies to *Neisseria gonorrhoeae*: a study of the urethral exudates of 232 men. J Infect Dis 1979;140:89–95.
28. Fowler JE Jr. Immunity to infection in the male reproductive tract. In: Griffin PD, Johnson PM, eds. Local immunity in reproductive tract tissues. Delhi: Oxford University Press, 1993:341–55.
29. Migliari R, Riva A, Lantini MS, Melis M, Usai E. Diffuse lymphoid tissue associated with the human bulbourethral gland: an immunohistologic characterization. J Androl 1992;13:337–41.
30. Pollanen P, Soder O, Parvinen M. Interleukin-1α stimulation of spermatogonial proliferation in vivo. Reprod Fertil Dev 1989;1:85–7.
31. Orava M. Comparison of the inhibitory effects of alpha and gamma interferons on testosterone production in porcine Leydig cells. 5th Eur workshop Mol Cell Endocrinol of the Testis. Brighton, UK, 1988:1317.
32. Hill JA, Haimovici F, Politch JA, Anderson DJ. Effects of soluble products of activated lymphocytes and macrophages (lymphokines and monokines) on human sperm parameters. Fertil Steril 1987;47:460–5.
33. Eisermann J, Register KB, Strickler RC, Collins JL. The effects of tumor necrosis factor on human sperm motility in vitro. J Androl 1989;10:270–6.
34. Hill JA, Cohen J, Anderson DJ. The effects of lymphokines and monokines on human sperm fertilizing ability in the zona-free hamster egg penetration test. Am J Obstet Gynecol 1989;160:1154–9.

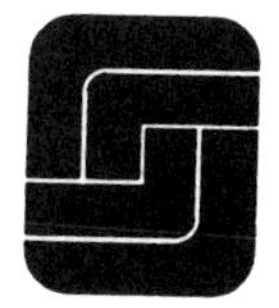

# 3

# Endometriosis: Immune Cells and Their Products

JOSEPH A. HILL

*Endometriosis*, defined as the occurrence of functioning endometrial tissue in ectopic (outside of the uterus) locations, is among the most enigmatic and problematic diseases affecting the reproductive health of women. Endometriosis is associated with pelvic pain and is believed to be the cause of infertility in up to 20% of subfertile women. Approximately 4 per 1000 women aged 15 to 64 are hospitalized with endometriosis each year (1). Endometriosis-related costs in terms of health care expenditures and time away from work of afflicted women are enormous, while the costs in terms of pain and suffering cannot be estimated.

Despite nearly 50 years of research, the pathophysiologic mechanisms responsible for the establishment and progression of endometriosis and endometriosis-associated reproductive failure have not been fully defined. There is a general consensus that endometriosis most probably develops as a result of retrograde menstruation (2), although extrapelvic cases of endometriosis (reviewed in 3) certainly cannot be explained by this mechanism. Retrograde menstruation appears to be a common occurrence. It has been observed in approximately 90% of women having a diagnostic laparoscopy performed while menstruating (4) and in up to 75% of women having laparoscopic sterilization during their menses (5). Endometrial tissue has been histologically observed in women with (19%) and without (11%) endometriosis who were experiencing retrograde menstruation at the time of laparoscopy (6). It can never be determined whether these observations were confounded by the surgical procedure necessitating anesthesia and paralytic medication, which could potentially facilitate retrograde menstruation. However, retrograde menstruation may be an important cause of pelvic endometriosis in susceptible individuals. What defines this susceptibility is still largely unknown, although accumulating evidence suggests that inflammatory/immunologic factors may play an important role in the development of endometriosis and endometriosis-associated reproductive failure.

## Inflammatory Aspects of Endometriosis

Inflammation is a localized protective response to tissue injury that can destroy both the injurious agent and the injured tissue. The inflammatory and immune responses are intimately connected through cytokine and antibody-mediating mechanisms. Peritoneal fluid lysozyme levels indicate that localized intraperitoneal inflammation occurs in women with active endometriosis (7). Macrophage numbers and concentrations are abundant in the peritoneal fluid of women with active endometriosis (8–11).

In addition to being increased, peritoneal macrophages in women with active endometriosis also have altered maturational characteristics (12) and increased capping ability for *major histocompatibility complex* (MHC) class II antigens, unlike peritoneal macrophages from women without endometriosis (13). Increased macrophage activation in response to endometriosis causes secretion of growth factors and cytokines that can influence ectopic endometrial growth and further recruitment of the immune/inflammatory cells initiating the cytokine cascade. Fibroblast- and macrophage-secreted growth factors, such as *epidermal growth factor* (EGF) and *platelet-derived growth factor* (PDGF), have been shown to stimulate endometrial stromal cells in vitro (14) and may influence the growth of endometriosis through paracrine mechanisms via EGF receptors on endometriotic implants (15). The peritoneal environment is also immunostimulatory (16), suggesting that in the presence of antigenic stimulation caused by active endometriosis, proliferation and activation of recruited leukocytes could be enhanced.

The concept that ectopic endometrium is antigenic in some but not all women was proposed over 20 years ago following the observation that complement's biologically active fragment, C3, was present around ectopic endometrial glands (17). Other investigators have not demonstrated a difference in the specific deposition of complement in ectopic as compared to eutopic endometrial glands (18). Earlier, other investigators (19), using immunofluorescence techniques, reported no difference in the presence of either C3 or C4 in eutopic endometrium, regardless of menstrual cycle phase or surgical diagnosis (endometriosis, pelvic inflammatory disease, and normal pelvis).

Tissue implants of endometriosis have been demonstrated to be able to produce and secrete C3 in vitro (20). Proliferative endometrial biopsies from women with minimal (stage I) endometriosis (21) have also been shown to produce significantly greater amounts of C3 in vitro than biopsies from women with either no endometriosis or women with severe (stage IV) endometriosis (22). The propensity to secrete biologically active C3 into the peritoneal environment has important inflammatory and immunologic implications (23). Complement is known to (i) be chemotactic for leukocytes; (ii) activate immune and inflammatory cells; (iii) mediate the release of cytokines, growth factors, and prostaglandin; (iv) cause fibroblast proliferation, potentially facilitating adhesion formation; and (v) inhibit fertilization (24).

## Humoral and Cellular Autoimmunity in Endometriosis

Several investigators have ascribed increased levels of circulating autoantibodies to undefined endometrial and ovarian antigens and to antiphospholipid in women with endometriosis (25–27). The association of antiendometrial antibodies with endometriosis is controversial, as not all investigators have been able to substantiate their existence (28). Very few, if any, B lymphocytes and no plasma cells have been identified in peritoneal fluid of women with endometriosis (8). Therefore, local (within the peritoneal environment) humoral (B lymphocyte-mediated) immunity appears unlikely in women with endometriosis. However, increased specific *immunoglobulin G* (IgG) autoantibody levels have been reported in peritoneal fluid from women with endometriosis as compared to fertile controls (29). Whether peritoneal fluid IgG autoantibodies in this study represented transudation from the peripheral circulation was not determined. The cause and effect relationships of any autoimmune phenomena and endometriosis remain to be established; however, antiendometrial antibody concentrations have been reported to be directly proportional to the amount of active disease (30).

If antiendometrial antibodies are present in the peritoneal environment, they could facilitate the development of endometriosis by coating endometrial cells, thus blocking potential antigenic sites required for cytotoxic T cell removal. Clearly, more work is needed to understand the potential role of humoral immunity in the pathophysiology of endometriosis.

Cellular immunity to autologous endometrium has been reported to be suppressed in rhesus monkeys with spontaneous endometriosis (31). This observation led to speculation that deficient cellular immunity may be involved in the establishment of endometriosis in women. A cellular immune basis for the development of endometriosis was further supported by the observation of diminished T cell-mediated cytotoxicity to autologous but uncharacterized and nonstandardized endometrial cells from infertile women with endometriosis as compared to infertile women without the disease (32).

Decreased *natural killer* (NK) cell activity has also been reported to be involved in the pathophysiology of endometriosis. NK cell-mediated cytotoxicity from peripheral blood against nonstandardized autologous endometrium has been reported to be decreased in women with endometriosis as compared to peripheral blood NK reactivities of women without endometriosis (33, 34). NK cell activity from cells in peritoneal fluid against K562 target cells has also been reported to be decreased in women with endometriosis as compared to women without the disease (35). These data suggest that impaired clearance of ectopic endometrium by NK cells may contribute to the development of endometriosis. Studies in baboons have also shown decreased antiendometrial cytotoxicity in animals with endometriosis; however, this deficiency could not be explained by alterations in NK cell activity (36).

The precise role of NK cells in the development of endometriosis remains to be defined. Human studies to date, although informative, are still phenomenological, as cause-versus-effect associations remain unresolved (37). The involvement of the immune system in the development of endometriosis is supported by other studies in rhesus monkeys where agents with immunosuppressive capabilities, such as proton irradiation, polychlorinated biphenyls, and dioxin, have been shown to increase the incidence of endometriosis in exposed animals (38–40). Immunosuppression can also increase the progression of spontaneous endometriosis in baboons (41).

## Consequences of Immune Reactivity

Activated immune and inflammatory cells in the peritoneal environment may either directly or indirectly affect reproduction since coelomic ovulation in humans ensures that the oocyte and sperm are exposed to peritoneal fluid within patent fallopian tubes. Direct inflammatory/immunologic effects include phagocytosis and cytotoxicity, while indirect effects include complement and cytokine secretion (20, 42, 43). Cytokines derived from macrophages—*tumor necrosis factor* (TNF) and *interleukin-1* (IL-1)—and lymphocytes—IL-2 and TNF—have been reported to be increased in the peritoneal fluid of many women with endometriosis compared to those without the disease (16, 44–47) and may be involved in endometriosis-associated reproductive failure (43, 48).

The immunologic cytokines, IL-1, TNF, and *interferon* $\gamma$ (IFN$\gamma$), could mediate reproductive failure through their effects on ovarian steroidogenesis (49, 50), sperm motion parameters (51, 52), fertilization (53), embryo development (54), implantation events (55), and trophoblast growth (56). Interestingly, peritoneal fluids from many women with endometriosis have also been shown to adversely affect in vitro sperm motion parameters (57, 58), fertilization (59, 60), and embryo development (47, 61), although not by all investigators (62–65). Potential explanations for these conflicting reports include differences in methodology among studies, small sample sizes, and inadequate controls. Comparisons between women with and without visible evidence of endometriosis alone are insufficient to arrive at meaningful conclusions regarding fecundity. Therefore, inclusion of a fertile control population should be a fundamental component of any study design addressing fertility issues in endometriosis since similar peritoneal factors may be operable in subfertile women with and without endometriosis. Another variable potentially confounding endometriosis studies is that endometriosis may represent a heterogenous disease. Similarly, causes for endometriosis establishment, progression, and adverse effects may also be heterogenous.

Studies regarding fertility in women with endometriosis are controversial. Reduced fecundity in women with moderate (stage III) and severe (stage IV) disease (21) is readily appreciated due to alterations in pelvic anatomy brought about by endometriomas and adhesion formation. Fertility issues in women with stage I and stage II disease (21) are less clear (66–70), although the monthly fecundity rate in these women is less than that of the general population (71). As endometriosis may represent a heterogenous disease, the mechanisms of reduced fecundity may similarly be heterogenous.

Another problematic variable is that many of the diagnostic modalities used in investigating infertility may have fertility-enhancing effects within the peritoneal environment (72). For instance, fecundity has been reported to be transiently enhanced in subfertile women after *hysterosalpingography* (HSG) (73–79) and is also commonly believed to be enhanced following tubal lavage. In vitro macrophage phagocytosis has been shown to be inhibited by HSG contrast media (72, 80–86). We have recently reported that both oil and water-soluble HSG contrast media and the commonly used chromotubation dyes (methylene blue and indigo carmine) can inhibit not only peritoneal macrophage phagocytosis, but also peritoneal lymphocyte proliferation in a dose-dependent manner in vitro (72). We have proposed that these agents could cause a transient decrease in peritoneal leukocyte numbers and function, resulting in lowered concentrations of potentially reproductive toxic cytokines, thus creating a more favorable peritoneal milieu for conception and pregnancy maintenance (72).

Conventional therapeutic agents for endometriosis may also have diverse immunologic effects that potentially extend their mechanisms for fertility enhancement. Danazol, the isoxazole derivative of the synthetic steroid 17α-ethinyl-testosterone, is not only efficacious in treating endometriosis due to its effects on the endocrine system (reviewed in 82), but may also be beneficial because of its ability to decrease macrophage phagocytosis and lymphocyte proliferation (83), immunologic cytokine secretion (47, 84), and autoantibody production (85, 86). *Gonadotropin releasing hormone* (GnRH) agonists, such as buserelin, have also been reported to decrease cytokine secretion (47), but have had no effect on autoantibody production (85).

## Summary

There is substantial evidence that immunologic factors play a role in the pathogenesis of endometriosis and endometriosis-associated reproductive failure. Evidence that these immunologic factors predispose, are merely coincidental, or are a direct result of endometriosis requires further investigation. Retrograde menstruation appears to be etiologically important in susceptible women. Alterations in both inflammatory and immune responsiveness may define this susceptibility. Macrophages appear to be the pivotal players, although T lymphocytes, NK cells, and their secreted products may also be important not only in the establishment and growth of endometriosis, but also in the sequelae of endometriosis, including inflammation, adhesion formation, and reproductive failure.

## *References*

1. Cramer DW. Epidemiology of endometriosis. In: Wilson EA, ed. Endometriosis. New York: Alan R. Liss, 1987:5.
2. Sampson JA. Peritoneal endometriosis due to the menstrual dissemination of endometrial tissue into the peritoneal cavity. Am J Obstet Gynecol 1927; 14:422.
3. Rock JA, Markham SM. Extra pelvic endometriosis. In: Wilson EA, ed. Endometriosis. New York: Alan R. Liss, 1987:183–206.
4. Halme J, Hammond MG, Hulka JF, Raj SG, Talbert LM. Retrograde menstruation in healthy women and in patients with endometriosis. Obstet Gynecol 1984;64:151.
5. Liu DTY, Hitchcock A. Endometriosis: its association with retrograde menstruation, dysmenorrhea and tubal pathology. Br J Obstet Gynecol 1986; 93:859.
6. Bartosik D, Jacobs SG, Kelly LJ. Endometrial tissue in peritoneal fluid. Fertil Steril 1986;46:796.
7. Olive DL, Haney AF, Weinberg AF. The nature of the intraperitoneal exudate associated with infertility: peritoneal fluid and serum lysozyme activity. Fertil Steril 1987;48:802.
8. Hill JA, Farris HMP, Schiff I, Anderson DJ. Characterization of leukocyte subpopulations in the peritoneal fluid of women with endometriosis. Fertil Steril 1988;50:216.
9. Haney AF, Muscato JJ, Weinberg JB. Peritoneal fluid cell populations in infertility patients. Fertil Steril 1983;35:696.
10. Halme J, Becker S, Hammond MC, Talbert L. Pelvic macrophages in normal and infertile women: the role of patent tubes. Am J Obstet Gynecol 1982; 142:890.
11. Badaway SZA, Cuenca V, Marshall L, Munchback R, Rinas AC, Coble DA. Cellular components in peritoneal fluid in infertile patients with and without endometriosis. Fertil Steril 1984;42:704.

12. Halme J, Becker S, Hammond MG, Raj S. Increased activation of pelvic macrophage: a possible cause of infertility in endometriosis. Am J Obstet Gynecol 1983;145:333.
13. Haskill S, Becker S, Juntell M, Sporn S, Halme J. Normal human peritoneal macrophages are unable to cap and internalize class II antigens. Cell Immunol 1988;115:100.
14. Surrey ES, Halme J. Effect of peritoneal fluid from endometriosis patients on endometrial stromal cell proliferation in vitro. Obstet Gynecol 1990;76:792.
15. Melega C, Galducci M, Bulletti C, Galassi A, Jasonni VM. Tissue factors influencing growth and maintenance of endometriosis. Ann NY Acad Sci 1991;622:256.
16. Hill JA, Anderson DJ. Lymphocyte activity in the presence of peritoneal fluid from fertile women and infertile women with and without endometriosis. Am J Obstet Gynecol 1989;160:154.
17. Weed JC, Arquenbourg PC. Endometriosis: can it produce an autoimmune response resulting in infertility? Clin Obstet Gynecol 1980;23:885.
18. D'Cruz OJ, Wild RA. Evaluation of endometrial tissue specific complement activation in women with endometriosis. Fertil Steril 1992;57:787.
19. Bartosik D, Damjanov I, Viscasello RR, Riley JA. Immunoproteins in the endometrium: clinical correlates of the presence of complement fractions C3 and C4. Am J Obstet Gynecol 1987;156:11.
20. Isaacson KB, Coutifaris C, Garcia CR, Lyttle CR. Production and secretion of complement component 3 by endometriotic tissue. J Clin Endocrinol Metab 1989;69:1003.
21. The American Fertility Society: revised American Fertility Society classification of endometriosis: 1985. Fertil Steril 1985;43:351.
22. Isaacson KB, Galman M, Coutifaris C, Lyttle CR. Endometrial synthesis and secretion of complement component-3 by patients with and without endometriosis. Fertil Steril 1990;53:836.
23. Whaley K. The complement system. In: Whaley K, ed. Complement in health and disease. Lancaster, UK: MTP Press, 1987:1.
24. Anderson DJ, Abbott AF, Wang HA, Jack RA. The role of the complement component C3b and its receptors in sperm-oocyte interaction. PNAS (in press).
25. Mathur S, Perress MR, Williamson HO, Youmans CD, Maney SA. Autoimmunity to endometrium and ovary in patients with endometriosis. Clin Exp Immunol 1982;50:259.
26. Wild RA, Shivers CA. Antiendometrial antibodies in patients with endometriosis. Am J Reprod Immunol Microbiol 1985;8:84.
27. Gleicher N, El-Roey A, Confino E. Is endometriosis an autoimmune disease? Obstet Gynecol 1987;70:115.
28. Switchenko AC, Kauffman RS, Becker M. Are there endometrial antibodies in sera of women with endometriosis? Fertil Steril 1991;56:235.
29. Confino E, Harlow L, Gleiche N. Peritoneal fluid and serum autoantibody levels in patients with endometriosis. Fertil Steril 1990;53:242.
30. Homm RJ, Garza DE, Mathur S, Austin M, Bagget B, Williamson HO. Immunological aspects of surgically induced experimental endometriosis: variation in response to therapy. Fertil Steril 1989;52:132.

31. Damouski WP, Stede RW, Baker GF. Deficient cellular immunity in endometriosis. Am J Obstet Gynecol 1981;141:377.
32. Steele RW, Dmowski WP, Marmer DJ. Immunologic aspects of human endometriosis. Am J Reprod Immunol 1984;6:33.
33. Oosterlynck DJ, Cornillie FJ, Waer M, Vanderputte M, Koninckx PR. Women with endometriosis show a defect in natural killer activity resulting in a decreased cytotoxicity to autologous endometrium. Fertil Steril 1991;56:45.
34. Vigano P, Vercellini P, DiBlasio AM, Colombo A, Candianin GB, Vignali M. Deficient antiendometrium lymphocyte-mediated cytotoxicity in patients with endometriosis. Fertil Steril 1991;56:894.
35. Oosterlynck DJ, Meuleman C, Waer M, Vandeputte M, Koninckx PR. The natural killer activity of peritoneal fluid lymphocytes is decreased in women with endometriosis. Fertil Steril 1992;58:292.
36. D'Hooghe TM, Scheerlinck JP, Koninckx PR, Bambra CS. Deficient antiendometrium lymphocyte-mediated cytotoxicity but normal natural killer activity in baboons with endometriosis. [Abstract]. Symposium on Immunobiology of Reproduction. Serono Symposia, USA, Boston, August 26–29, 1993.
37. Hill JA. Immunology and endometriosis. Fertil Steril 1992;58:262.
38. Yochmowitz MG, Wood DH, Salmon YL. Seventeen-year mortality experience of proton radiation in *Macaca mulatta*. Radiat Res 1985;102:14–34.
39. Fanton JW, Golden JG. Radiation-induced endometriosis in *Macaca mulatta*. Radiat Res 1991;126:141–6.
40. Allen JR, Barbotti DA, Lambrecht LK, Van Miller JP. Reproductive effects of halogenated aromatic hydrocarbons on nonhuman primates. Ann NY Acad Sci 1979;320:419–25.
41. D'Hooghe TM, Bambra CS, Raeymaekers BM, Hill JA, Koninckx PR. Immunosuppression can increase progression of spontaneous endometriosis in baboons with endometriosis [Abstract]. Symposium on Immunobiology of Reproduction. Serono Symposia, USA, Boston, August 26–29, 1993.
42. Haney AF, Misukonis MA, Weinberg JB. Macrophages and infertility: oviductal macrophages are potential mediators of infertility. Fertil Steril 1983;39:310.
43. Hill JA. Immunologic factors in endometriosis and endometriosis-associated reproductive failure. Inf Reprod Med Clin N Am 1992;3:583.
44. Fakih H, Beggett B, Holtz, et al. Interleukin-1: a possible role in the infertility associated with endometriosis. Fertil Steril 1987;47:213.
45. Eiserman J, Grant MJ, Pineda J, Collins J. Tumor necrosis factor in peritoneal fluid of women undergoing laparoscopic surgery. Fertil Steril 1988;50:573.
46. Halme J. Release of tumor necrosis factor by human peritoneal macrophages in vivo and in vitro. Am J Obstet Gynecol 1989;161:1718.
47. Teketani Y, Kuo TM, Mizuno M. Comparison of cytokine levels and embryo toxicity in peritoneal fluid in infertile women with untreated or treated endometriosis. Am J Obstet Gynecol 1992;167:265.
48. Fukuoka M, Yasuda K, Emi N, et al. Cytokine modulation of progesterone and estradiol secretion in cultures of luteinized human granulosa cells. J Clin Endocrinol Metab 1992;75:254.

49. Hill JA. Cytokines considered critical in pregnancy. Am J Reprod Immunol 1992;28:123.
50. Wang HZ, Lu SH, Han XJ, et al. Inhibitory effect of interferon and tumor necrosis factor on human luteal function in vitro. Fertil Steril 1992;58:941–5.
51. Hill JA, Haimovici F, Politch JA, Anderson DJ. Effects of soluble products of activated lymphocytes and macrophages (lymphokines and monokines) on human sperm motion parameters. Fertil Steril 1987;47:460.
52. Eiserman J, Register K, Strickler R, Collins J. The effects of tumor necrosis factor on human sperm motility in vitro. J Androl 1989;10:270.
53. Hill JA, Cohen J, Anderson DJ. The effects of lymphokines and monokines on sperm fertilizing ability in the zona-free hamster egg penetration test. Am J Obstet Gynecol 1989;160:1154.
54. Hill J, Haimovici F, Anderson DJ. Products of activated lymphocytes and macrophages inhibit mouse embryo development in vitro. J Immunol 1987; 139:2250.
55. Haimovici F, Hill JA, Anderson DJ. The effects of immunologic cytokines on mouse blastocyst implantation in vivo. Biol Reprod 1991;44:69.
56. Berkowitz RS, Hill JA, Kurtz CB, Anderson DJ. Effects of products of activated leukocytes (lymphokines and monokines) on the growth of malignant trophoblast cells in vitro. Am J Obstet Gynecol 1988;158:199–204.
57. Burke RK. Effects of peritoneal washings from women with endometriosis on sperm velocity. J Reprod Med 1987;32:743.
58. Oak MK, Chantler EN, Vaughn Williams, et al. Sperm survival studies in peritoneal fluid from women with endometriosis and unexplained infertility. Clin Reprod Fertil 1985;3:297.
59. Chacho KJ, Chacho MS, Anderson PJ, Scommegna A. Peritoneal fluid in patients with and without endometriosis: prostanoids and macrophages and their effect on the spermatozoa penetration assay. Am J Obstet Gynecol 1986;154:1290.
60. Sueldo CE, Lambert H, Steinlitner A, Rathnick G, Swanson J. The effect of peritoneal fluid from patients with endometriosis on murine sperm-oocyte interaction. Fertil Steril 1987;48:697.
61. Marcos RN, Gibbons WE, Findley WE. Effect of peritoneal fluid on in vitro cleavage of 2-cell mouse embryos: possible role in infertility associated with endometriosis. Fertil Steril 1985;44:678.
62. Muse KN, Estes S, Vernon M, Zavros P, Wilson EA. Effect of endometriosis on sperm motility in peritoneal fluid in vitro [Abstract #286]. Am Fertil Soc, Sept 27–Oct 2, 1986.
63. Halme J, Hall JL. Effect of pelvic fluid from endometriosis patients on human sperm penetration of zona-free hamster ova. Fertil Steril 1982;37:573.
64. Prough SG, Askel S, Gilmore SM, et al. Peritoneal fluid fractions from patients with endometriosis do not promote two-cell mouse embryo growth. Fertil Steril 1990;54:927.
65. Awadalla SG, Friedman CI, Haq AU, Roh SI, Chin N, Kim MH. Local peritoneal factors: their role in infertility associated with endometriosis. Am J Obstet Gynecol 1987;157:1207.
66. Buttram VC. Conservative surgery for endometriosis in the infertile female: a study of 206 patients with implications for both medical and surgical therapy. Fertil Steril 1979;31:117.

67. Muse KN, Wilson EA. How does mild endometriosis cause infertility? Fertil Steril 1982;38:145.
68. Schenken RS, Malinak LR. Conservative surgery versus expectant management for the infertile patient with mild endometriosis. Fertil Steril 1982; 37:183.
69. Seibel MM, Berger MJ, Weinstein FJ, Taymore ML. The effectiveness of danazol on subsequent fertility in minimal endometriosis. Fertil Steril 1982; 38:534.
70. Seibel MM. Minimal pelvic endometriosis and infertility. Semin Reprod Endocrinol 1985;3:307.
71. Olive DL, Haney AF. Endometriosis-associated infertility: a critical review of therapeutic approaches. Obstet Gynecol Surv 1986;41:538.
72. Goodman SB, Rein MS, Hill JA. Hysterosalpingography contrast media and chromotubation dye inhibit peritoneal lymphocyte and macrophage function in vitro: a potential mechanism for fertility enhancement. Fertil Steril 1993; 59:1022.
73. King EL, Herring JS. Sterility studies in private practice. Am J Obstet Gynecol 1949;58:258.
74. Weir WC, Weir DR. Therapeutic value of salpingograms in infertility. Fertil Steril 1951;2:514.
75. Mackey RA, Glass RH, Olson LE, Vaidya R. Pregnancy following hysterosalpingography with oil and water soluble dye. Fertil Steril 1971;22:504.
76. DeCherney AH, Kort H, Barrey JB, DeVae GR. Increased pregnancy rate with oil-soluble hysterosalpingography dye. Fertil Steril 1980;33:407.
77. Shuaba MG, Shapiro SS, Haning RV Jr. Hysterosalpingography with oil contrast media enhances fertility in patients with infertility of unknown etiology. Fertil Steril 1985;40:604.
78. Alper MM, Garner PR, Spence JEH, Quarrington AM. Pregnancy rates after hysterosalpingography with oil and water soluble contrast media. Obstet Gynecol 1986;68:6.
79. DeBoer AD, Vermer HM, Williamson WN, Sanders FB. Oil or aqueous contrast media for HSG: a perspective, randomized, clinical study. Eur J Obstet Gynecol Reprod Biol 1988;28:65.
80. Boyer P, Territo MC, de Ziegler D, Meldsum DR. Ethiodol inhibits phagocytosis by pelvic peritoneal macrophages. Fertil Steril 1986;46:715.
81. Johnson JV, Montoya IA, Olive DL. Ethiodal oil contrast medium inhibits macrophage phagocytose and adherence by altering membrane electronegativity and microviscosity. Fertil Steril 1992;58:511.
82. Barbieri RL, Ryan KJ. Danazol: endocrine pharmacology and therapeutic applications. Am J Obstet Gynecol 1981;141:453.
83. Hill JA, Barbieri RL, Anderson DJ. Immunosuppressive effects of danazol in vitro. Fertil Steril 1984;48:414.
84. Mori H, Nakagawa M, Itoh W, Wada K, Tamaya T. Danazol suppresses the production of interleukin 1-beta and tumor necrosis factor by human monocytes. Am J Reprod Immunol 1990;24:45.
85. El Roeiy A, Dmowski WP, Gleicher N. Danazol but not gonadotropin releasing hormone against suppresses autoantibodies in endometriosis. Fertil Steril 1988;50:864.

86. Ota H, Maki M, Shidara Y, et al. Effects of danazol at the immunologic level in patients with adenomyosis, with special reference to autoantibodies: a multi-center cooperative study. Am J Obstet Gynecol 1992;167:481.

# Part II

## Growth Factors/Cytokines in the Female Reproductive Tract and Placenta

# 4

# Cytokines and Pregnancy Recognition

Fuller W. Bazer, Thomas E. Spencer, Troy L. Ott, and Howard M. Johnson

Maternal recognition of pregnancy results from signaling between the trophoblast of the conceptus (embryo and associated membranes) and the maternal system (1). These signals ensure maintenance of the structural and functional integrity of the *corpus luteum* (CL) that would otherwise regress at the end of the estrous or menstrual cycle. The CL produces progesterone, the hormone of pregnancy that is required to stimulate and maintain endometrial functions that are permissive to early embryonic development, implantation, placentation, and successful fetoplacental development. The terms *luteotrophic*, *luteal protective*, *antiluteolytic*, and *luteolytic* will be defined before mechanisms for maternal recognition of pregnancy are discussed. A *luteotrophic* signal, *chorionic gonadotropin* (CG), is produced by primate conceptuses and is believed to act directly on the CL, via receptors for *luteinizing hormone* (LH), to ensure maintenance of its structural and functional integrity (2). The ovarian cycle of primates is uterine independent; that is, luteolytic events responsible for regression of the CL and cessation of progesterone secretion at the end of a menstrual cycle result from the intraovarian effects of prostaglandins (3), oxytocin (4), or other, as yet undefined, luteolytic agents. A *luteolytic* agent causes the structural and functional demise of the CL, or luteolysis. *Prostaglandin* $F_{2\alpha}$ (PGF) is the luteolytic signal common to most, if not all, mammals. There may also be *luteal protective* signals—for example, $PGE_2$ (PGE)—that antagonize the potential luteolytic effects of PGF.

The ovarian cycle of subprimate mammals is uterine dependent, and hysterectomy extends CL maintenance for a period characteristic of the species' gestational period. The uterine endometrium, primarily surface epithelium and perhaps superficial glandular epithelium, releases PGF in pulses that are responsible for luteolysis. Signals from conceptuses of subprimate mammals are termed *antiluteolytic* because they either inhibit endometrial release of luteolytic amounts of PGF or alter the pulsatile pattern of endometrial secretion of PGF to abrogate its luteolytic effects. In subprimate mammals, conceptus signals responsible for maternal recognition of pregnancy appear to act in a paracrine manner to interrupt endometrial production of luteolytic PGF, but do not act directly on the CL. Antiluteolytic signals include estrogen and lactogenic hormones in the pig and *interferon* $\tau$ (IFN$\tau$) from trophoblast of ruminant conceptuses.

## Endocrine Requirements for Luteolysis

### *Primates*

The menstrual cycle in humans is from the onset of menses in one cycle to the onset of menses in the subsequent cycle and averages 28 days. Ovulation occurs in response to a surge of LH released on about day 14, and CL regression, a prelude to the onset of menses, results from the intra-ovarian effects of PGF (3), oxytocin (4), and/or other unidentified hormones acting independently or in concert. The uterine-independent luteolytic mechanism in primates remains poorly understood, but available information has been reviewed recently (2).

Maternal recognition of pregnancy in primates appears to involve independent interactions between the conceptus and uterus, as well as between the conceptus and ovary. Interactions between the conceptus and uterus result in the maintenance of a progesterone-responsive uterus and a reduction in the numbers of endometrial *estrogen receptors* (ER) (5–8). Progesterone, interacting with *progesterone receptors* (PR) in stromal cells, maintains an endometrium that is permissive to conceptus development, implantation, decidualization, placentation, and feto-placental development to term. The significance of the absence of ER in the endometrial epithelium and stroma of primates (baboons) during early pregnancy is not known. Finally, CG, produced by the trophoblast/chorion, appears to be responsible for maternal recognition of pregnancy in primates. Production of CG begins during the periimplantation period, and it exerts its luteotrophic effect directly on the CL. However, growth factors, prostaglandins other than PGF, or other hormones (e.g., relaxin or lactogenic hormones) may influence luteal function directly or indirectly during pregnancy (2). Results are not available to indicate a requirement for cytokines in pregnancy recognition in primates.

### *Ruminants*

Sheep, cattle, and goats have estrous cycles of about 17, 21, and 20 days, respectively. An ovulatory surge of LH coincident with the onset of estrus (day 0) initiates events that culminate in ovulation about 30 h later. With maturation of the CL, concentrations of progesterone in peripheral blood are maximum in mid-diestrus (days 12–14), and in cyclic females luteolysis is induced by pulsatile release of PGF from endometrial epithelium during late diestrus (day 15). The antiluteolytic signal for pregnancy recognition in ruminants is IFNτ produced by trophectoderm (9). IFNτ exerts a paracrine, antiluteolytic effect on the endometrium to inhibit endometrial production of luteolytic pulses of PGF. Other conceptus and/or uterine products secreted during pregnancy—for example, PGE and *platelet activating factor* (PAF)—may exert luteal protective effects (9).

Endometrial production of luteolytic pulses of PGF in ruminants is dependent upon the effects of progesterone, estrogen, and oxytocin on uterine epithelium. Endocrine regulation of luteolysis is best understood in sheep, but mechanisms appear common to cows and goats. Endometrium of ewes is stimulated by progesterone to increase phospholipid stores (arachidonic acid source) and cyclooxygenase enzymatic activity necessary for conversion of arachidonic acid to PGF (10). Oxytocin secreted by the CL and posterior pituitary stimulates release of luteolytic pulses of PGF from endometrial epithelium (9). Oxytocin acts through receptors on endometrial epithelium to stimulate the *inositol phospholipid* (IP) second-messenger system that in turn releases luteolytic pulses of PGF (11). In sheep (11) and cows (12–15), endometrial receptors for oxytocin are present between estrus and about day 4 of the cycle, are low or undetectable between days 5 and 13, and then increase rapidly between days 14 and 16 (sheep) or days 17 and 20 (cows and goats). *Oxytocin receptors* (OTR) are localized to the endometrial surface epithelium during the luteolytic period of sheep (16).

Progesterone, binding to PR, initiates unknown events that inhibit synthesis of OTR by endometrial epithelium for 10–12 days, the *progesterone block* period (17). Progesterone down-regulates its own receptors after days 12–14 of the cycle to end the progesterone block, and there is a rapid increase in endometrial OTR that is enhanced by estrogen (12). During the estrous cycle of cattle, endometrial ER and PR are highest during the first 10–12 days after onset of estrus; then both decline to their lowest levels on about day 13 (12). The ER then increase between days 14 and 21, with OTR increasing between days 17 and 21. Similarly, in sheep and goats endometrial OTR increase rapidly 48–72h prior to estrus. During the luteolytic period the endometrium is characterized by low PR, increasing ER, and increasing OTR. In cyclic ewes endometrial PR and PR mRNA decrease from days 10 to 14 and then increase on day 16; that is, immediately after luteolysis (18). This coincides with increasing ER, ER mRNA, OTR, and OTR mRNA between day 14 and estrus (18–21). During the luteolytic period about 98% of the PGF pulses are associated with coinciding pulses of oxytocin. However, only about 50% of the oxytocin pulses result in a pulse of PGF (22), which suggests that oxytocin is responsible for coordinating luteolytic events. The PGF pulse frequency may be less than that for oxytocin because of the time required for replenishment of pools of phospholipids from which arachidonic acid can be mobilized for synthesis of PGF (23). Treatment of ewes (24) or goats (25) with an oxytocin antagonist, passive or active immunization of ewes against oxytocin (26, 27), or continuous infusion of oxytocin to down-regulate OTR (28) prevents or significantly delays luteolysis. These results indicate a central role for oxytocin in the luteolytic mechanism that is dependent upon pulsatile release of PGF (29).

Uterine release of about 5 pulses of PGF per 25 h is required to initiate luteolysis (17). Low-amplitude pulses of PGF from the uterus act on large luteal cells to cause trafficking of secretory granules containing oxytocin to the cell surface and exocytosis of oxytocin, which then induces a pulse of uterine PGF (11). This mechanism is repeated at 4- to 5-h intervals between days 14 and 17 or until the CL is depleted of its finite stores of oxytocin. Other mechanisms to explain the episodic release of oxytocin and PGF have been proposed because the release of oxytocin from CL—even when there is one CL on each ovary—and from the posterior pituitary are synchronous in sheep (22). Control of oxytocin release by hormones other than PGF or PGE is possible, but such a factor has not been defined.

## Pregnancy Recognition Signals

### *Sheep*

The presence of the conceptus in the uterus prevents luteolysis because an antiluteolytic signal(s) produced by the conceptus prevents uterine production of luteolytic pulses of PGF. IFNτ is the antiluteolytic signal produced by conceptuses of ruminants. Potential mechanisms of action include (i) stabilization or up-regulation of endometrial PR to extend the progesterone block and prevent endometrial synthesis of OTR and/or ER, (ii) direct inhibition of ER to attenuate the episodic release of PGF required for luteolysis, (iii) direct inhibition of synthesis of endometrial OTR, and (iv) inhibition of postreceptor mechanisms that prevent oxytocin induction of pulsatile release of PGF (9). Available results indicate that endometrial epithelium of pregnant ewes, cows, and goats have few or no receptors for either oxytocin or estrogen (21).

Pregnant ewes fail to experience luteolysis in response to doses of exogenous oxytocin and estradiol that cause luteolysis in cyclic ewes and cows (9). Release of oxytocin and oxytocin-neurophysin may be reduced, increased, or not different in pregnant compared to cyclic ewes between days 13 and 16 after estrus. However, uterine receptors for oxytocin are very low or absent in pregnant ewes (11, 17). Basal secretion of PGF by sheep endometrium is higher for pregnant than cyclic ewes; however, the pulsatile release of PGF required for luteolysis is abolished during pregnancy (9).

Homogenates of sheep conceptuses extend the interestrous interval in ewes when infused into the uterine lumen, but not the uteroovarian venous drainage (9). Sheep conceptus homogenates do not contain either CG-like or prolactin-like activity (30). Through in vitro culture of sheep conceptuses and analysis of radiolabeled proteins released into the culture medium, the first major protein secreted by mononuclear cells of ovine trophectoderm was identified as trophoblastin, then ovine trophoblast protein 1, then type I trophoblast interferon, and now, IFNτ (9, 31).

## Structure of Interferon Genes

*Tau* (τ) IFNs are a unique subclass of the 172-amino acid α2 (omega [ω]) interferons. Identification of this reproductive hormone as an interferon was surprising primarily because of its massive production by the trophectoderm and apparent minimal virus inducibility (31–33). *Ovine IFNτ* (oIFNτ) and *bovine IFNτ* (bIFNτ) were initially identified as type I IFNs following cDNA cloning and amino acid sequencing (34–37) and later given the Greek letter designation τ (31). IFNτs were confirmed as functional interferons on the basis of their potent antiviral and antiproliferative activities (38–40).

Support for IFNτ as a distinct IFN gene subtype came from the high amino acid sequence homology of IFNτ across ruminant species (41) and its apparently unique antiluteolytic biological activity (9). The IFNτs of cattle and sheep arise from multiple mRNAs of approximately 1 kb in length that probably arose from multiple genes (41, 42). The functional significance of multiple IFNτ isoforms is equivocal since individual oIFNτ isoforms, produced using recombinant DNA technology, are sufficient to extend CL function when injected into the uterine lumen of cyclic ewes (43–45).

Like other type I IFN genes, IFNτ genes are intronless. A 595-bp open reading frame codes for a 195-amino acid preprotein containing a 23-amino acid signal sequence that is cleaved to yield a mature protein of 172 amino acids. Secreted forms of oIFNτ are not glycosylated, although one of the sequenced genes contains a potential N-glycosylation site (Asn-Thr-Thr) (46). Bovine IFNτ transcripts contain a potential site for N-glycosylation at Asn78, and multiple glycosylation variants are present in the secreted proteins (47). Periimplantation goat conceptuses also express multiple isoforms of a *caprine IFNτ* (cIFNτ) that cross-react with antisera to oIFNτ and bIFNτ (48). Both N-glycosylated and unglycosylated forms of cIFNτ are secreted by goat conceptuses (48), and the genes for cIFNτ and oIFNτ share greater than 95% identity (41).

The structural relatedness of the IFNτs of domestic ruminants is supported by their cross-species functional relatedness. Early experiments demonstrated that ovine trophoblastic vesicles could extend the interestrous interval when placed in the uterine lumen of cattle (49), suggesting that the signals for maternal recognition of pregnancy were similar. Recently, it was demonstrated that purified *recombinant oIFNτ* (roIFNτ) produced from a synthetic gene in yeast (45) suppressed oxytocin-induced uterine PGF production and extended the interestrous interval in cattle (50). Recombinant oIFNτ is apparently fully functional in goats as well since twice-daily intrauterine injections of 100-μg roIFNτ from days 14 to 18 postestrus extended CL function approximately 8 days (51).

Southern blotting of ovine and bovine genomic DNA detected 4–5 IFNτ genes using probes designed to distinguish between IFNω and IFNτ (42, 52, 53). IFNτ genes were also identified in related ruminants, including musk ox, gazelle, and giraffe (41, 52). Using those same probes, Leaman and Roberts (41) were not able to detect IFNτ genes in horse, pig, llama, dolphin, mouse, rabbit, and human, suggesting that IFNτ genes diverged from IFNω genes recently (30–50 million years ago) (31). The oIFNτ and bIFNτ cDNAs share a greater degree of homology than bIFNτ does with bIFNω (52). This cross-species homology and presence within a limited subset of mammals (*Artiodactyla*) strongly implicates evolutionary divergence of the IFNτ from IFNω. This is manifest as a unique mechanism for maternal recognition of pregnancy that relies on massive production of IFNτ during a defined period of early pregnancy.

Within the coding region, oIFNτ and bIFNτ transcripts exhibit ~90% identity (54), and their inferred amino acid sequences share ~80% identity. The predicted amino acid identity between bIFNτ and bIFNα1 and IFNω is ~50% and ~72%, respectively (54). *Human IFNα* (hIFNα) displaces oIFNτ from ovine endometrial receptors (55–57). The IFNτ composite surface profile (40) and predicted relative hydrophilicity profiles (58, 59) both support a conserved three-dimensional structure for IFNαs and IFNτs. Pontzer et al. (40) demonstrated that a synthetic peptide corresponding to the carboxyterminus (AA 139–172) of oIFNτ blocked the antiviral activity of oIFNτ as well as that of natural oIFNα, recombinant bIFNα, and recombinant hIFNα, but not recombinant bIFNγ. A synthetic peptide corresponding to the aminoterminus (AA 1–37) of oIFNτ only blocked antiviral activity without affecting the antiviral activity of the IFNαs (40). These results suggest that the C-terminus may contain a receptor binding epitope common to type I IFNs, whereas the aminoterminus may mediate the unique and characteristic biologic activities of the IFNτ. Interestingly, the aminoterminus peptide (AA 1–37) of oIFNτ has antiluteolytic agonist activity (60).

The IFNτs are only ~30% identical to IFNβ; however, IFNα, -β, and -τ are all believed to act through the type I interferon receptor. Competition for the same receptor dictates that portions of the epitope involved in receptor binding of IFNβ are shared by IFNα, -β, and -τ. If the locations of hydrophobic residues between IFNτ and IFNβ are compared, they are

75% identical, and it is the hydrophobic residues that appear to critically influence protein folding (61).

Secondary structure predictions indicate substantial regions of helical structure and interchain disulfide bonds between conserved cysteine residues at positions 1, 29, 99, and 139 in ovine and bovine IFNτ (41, 58, 61). Using CD spectra obtained from roIFNτ (45), it was predicted that oIFNτ was approximately 70% α-helix, with the remainder being random or a combination of β-sheet and turn (61). Five regions of helical structure (designated A through E) are connected by loop regions in an antiparallel arrangement thought to provide IFNτ with a three-dimensional structure remarkably similar to that for IFNβ, IL-1, IL-4, growth hormone, and *granulocyte-macrophage colony stimulating factor* (GM-CSF) (31, 41, 61). Although the precise packing of the helical regions is not known, predictions based on the crystal structure of murine IFNβ (62) suggest that the α-helices form a 4-helical bundle motif that is thought to mediate receptor binding (61). This model is especially appealing because it brings previously identified functional domains (40, 63) into close proximity in the three-dimensional structure of IFNτ (61). Experiments are ongoing to determine the crystal structure of roIFNτ.

The 5′ (~76%) and 3′ (~92%) noncoding regions of ovine and bovine IFNτ mRNAs share a high degree of cross-species conservation (54). Ovine and bovine IFNτ genes also share considerable identity in 5′ and 3′ flanking regions (41, 42, 46, 53). The degree of conservation cross-species, which is greater than 90% for bovine, ovine, caprine, and musk ox in the 100 bases upstream of the transcription start site, and the apparent divergence from related IFNα1 and IFNω genes support the unique quantitative and qualitative pattern of IFNτ gene expression (41, 46). Structures of the IFNτ gene promoters have been reviewed extensively (31, 41, 42, 46, 53). The IFNτ genes share considerable within- and cross-species homology up to position −400 (41, 53). A consensus IFNτ promoter contains GAAANN sequences and putative IRF-1 binding hexamers, although the arrangement of these motifs differs from the viral response elements present in other type I IFNs (41, 42, 53). Functionally, this is supported by the poor virus inducibility of the IFNτ genes (33, 64). Perhaps more interesting is the cross-species sequence identity of IFNτ promoters to approximately −400 upstream of the transcription start site; promoters of other type I IFNs typically diverge beyond −150 (41). Because IFNτ promoter constructs are expressed in uninduced cells of trophoblast origin (JAR and BeWo cells), but not in nontrophoblast cells, it is postulated that trophoblast cell-specific factors activate transcription of the IFNτ genes via distal enhancer elements (31, 33).

Nephew et al. (46) provided evidence for just such an element in a recent report that presented sequence data for four additional oIFNτ genes. Only one of those genes (clone 010) was expressed during the period of maternal recognition of pregnancy at levels comparable to those for oIFNτ. The promoter region for this gene (to −175) is greater than 95% identical to the oTP-p7 gene of Leaman and Roberts (41). Clone 010 had an AP-1-like regulatory response element (ATGGGTCAGA) starting at −929, suggesting that factors (i.e., cytokines) that affect AP-1 enhancer activity, such as GM-CSF (65), may regulate oIFNτ gene expression (46). A subsequent report by that group provided evidence that not only was GM-CSF expressed by the ovine endometrium during early pregnancy, but that the addition of GM-CSF to cultured ovine conceptuses resulted in an approximate doubling of antiviral activity secreted into the culture medium (indicator of oIFNτ production) (66). In addition, Xavier et al. (67) demonstrated coexpression of IFNτ and c-*fos* in sheep conceptuses with maternal expression occurring on days 14–15. These results support the idea that the growing family of cytokines resident in the uterus during early pregnancy may interact to regulate recognition of pregnancy and subsequent fetoplacental development.

## Antiluteolytic Activity of oIFNτ

Ovine IFNτ is secreted between days 10 and 21 of pregnancy, has a molecular weight of 19,000, and binds to type I interferon receptors (9). Ovine IFNτ has potent antiviral, antiproliferative, and immunosuppressive biological activities in addition to its antiluteolytic activity (9). Infusion of highly purified native IFNτ (68) or recombinant IFNτ (roIFNτ) (45, 69) into the uterine lumen from days 12 to 14 extends the interestrous interval and CL lifespan; therefore, IFNτ alone is assumed to be the antiluteolytic factor produced by sheep conceptuses. Similarly, intrauterine injections of roIFNτ into the uterine lumen of cows (50) and goats (51) extends the interestrous interval significantly. Ovine IFNτ probably acts as a paracrine antiluteolytic hormone on endometrium since there is no evidence for its transport from the uterus to affect the CL directly (9).

Using endometrium taken on day 15 of the estrous cycle (OTR present), it was determined that oIFNτ does not compete with oxytocin for its receptor, inhibit oxytocin stimulation of endometrial IP metabolism, or inhibit oxytocin stimulation of endometrial secretion of PGF (9). The antiluteolytic effect of oIFNτ must, therefore, prevent development of the luteolytic mechanism. Secretion of oIFNτ (ng/uterine flushing) begins on about day 10 and increases as conceptuses change morphologically from spherical (312 ng), to tubular (1380 ng) and filamentous (4455 ng) forms on days 12–13. Successful transfer of embryos to cyclic ewes can be accomplished only as late as day 12; that is, 48–72 h prior to the luteolytic period. This suggests that oIFNτ is secreted prior to the luteolytic period to inhibit directly or indirectly the endometrial synthesis of OTR and the uterine release of oxytocin-dependent luteolytic pulses of PGF (9).

Functional endometrial OTR are present in low numbers in pregnant ewes (9). Similarly, intrauterine injections of oIFNτ or roIFNτ between days 11 and 15 of the estrous cycle reduce the numbers of OTR, the affinity of the OTR for oxytocin, and endometrial ER protein; however, the effects of oIFNτ on endometrial PR were not detected (18, 19). Both IP metabolism and PGF secretion in response to oxytocin are reduced significantly when endometrium of cyclic ewes is exposed to oIFNτ on days 12–14 (9). These results indicate the absence of functional endometrial OTR in ewes treated with oIFNτ. Oxytocin receptor affinity decreases in the absence of estrogenic stimulation of rat myometrium (70), and OTR affinities tend to be lower for endometrium from pregnant cows ($1.5 \pm 0.5$ vs. $0.9 \pm 0.1$ nM) (14). Ovine IFNτ may inhibit synthesis of OTR and reduce their affinity for oxytocin directly or perhaps indirectly by down-regulating endometrial ER and/or stabilizing endometrial PR.

Rapid replenishment of endometrial ER in ewes follows withdrawal of progesterone, which suggests that failure of endometrial ER to increase during pregnancy is associated with either stabilization of PR or direct inhibition of ER synthesis by oIFNτ (9). During pregnancy endometrial ER protein and ER mRNA are significantly lower for pregnant than cyclic ewes on day 16 (18, 19, 71) and for cyclic ewes receiving intrauterine infusions of oIFNτ on days 11–15 and hysterectomized on day 16 (19). In addition, immunocytochemical studies indicate the absence of ER in endometrial surface and superficial glandular epithelium of day 15 pregnant ewes (21). Interferons can inhibit synthesis, turnover, or movement of receptors within membranes, and treatment of patients having steroid-dependent adenocarcinoma with IFNβ increased PR and decreased ER in tumor cells (9).

## Type I Interferon Receptors

Ovine IFNτ shares the highest amino acid homology with interferon α2 (IFNα) (31) and can be cross-linked to both 100-kd and 70-kd proteins in ovine endometrial membrane preparations. This can be competed by recombinant bovine and human α1 interferons (31). The human type I IFN receptor, present on essentially all human cell types (72, 73), is a 95- to 100-kd transmembrane glycoprotein (74) that is generated from a 2.7-kb mRNA encoding a 65-kd protein with 15 potential asparagine-linked glycosylation sites (75). Thus, ovine endometrial tissues may express both unglycosylated (70 kd) and glycosylated (100 kd) forms of the type I IFN receptor to which IFNτ may bind to elicit its cellular actions. The actions of IFNτ on ovine endometrium do not involve increases in cAMP, cGMP, or IP turnover (9); however, effects of IFNτ may be mediated through a signal transduction system similar, if not identical, to that of IFNα and type I IFN receptor.

The type I IFNs act by increasing rates of transcription through *cis*-acting DNA elements called *IFN-stimulated response elements* (ISREs) present in IFN-responsive genes (76, 77). A complex of three proteins, termed *interferon-stimulated gene factor 3α* (ISGF3α), is responsible for the transcriptional effects of IFNα (78, 79). Within minutes of binding to type I IFN receptors on the cell surface, type I IFNs activate an intracellular tyrosine kinase termed *tyk 2* that rapidly phosphorylates the three cytoplasmic ISGF3α proteins (80). These ISGF3α proteins (84, 91, and 113 kd) are present in latent, unphosphorylated forms in the cytoplasm of unstimulated cells. Upon activation by phosphorylation the ISGF3α proteins aggregate with a fourth protein termed *ISGF3γ*, a 48-kd DNA binding protein. The complex of ISGF3γ and ISGF3α proteins then translocates to the nucleus and interacts directly with ISREs to affect the transcription of specific genes by RNA polymerase II. The ISRE consensus motif (GGAAANNGAAACT) comprises a common motif, GGAAA, found in a number of viral enhancers and a second, unique motif, GAAACT, found in the promoter DNA of IFN-stimulated genes (76).

Among the proteins for which transcription is affected by type I interferons are PR and ER. In endometrial adenocarcinoma cells, levels of ER and PR protein are increased by IFNα2b (81). IFNα enhances levels of PR, but not ER, in AE-7 endometrial cancer cells (82). In human breast cancer tissue (83, 84) and human and rabbit endometrium (83), IFNα increases ER expression. The amounts of mRNA for PR and ER were not measured in the above studies, but the effects of IFNα were probably due to increased transcription of the PR and/or ER genes. The organization of the rabbit PR (85) and human PR (86) and ER (87) genes has been described, but functional ISREs have not been characterized in the 5′ flanking regions of these genes. Computer-assisted analyses of these regions for homology to the consensus ISRE sequence indicated putative ISREs in rabbit PR (−2976 and −2605) and human PR (−1324) and ER (−1838), with the transcription initiation site as (+1). It has been demonstrated that intrauterine injections of oIFNτ in ovariectomized ewes prevents development of endometrial sensitivity to oxytocin only when progesterone replacement therapy is provided, which strongly suggests that the antiluteolytic effect of oIFNτ is dependent on the presence of progesterone and endometrial PR (88).

The ovine PR and ER genes have not been cloned, but nucleotide sequences of their partial cDNA clones display high homology (>80%) with those of other mammals (89). If the signal transduction system of IFNτ is similar to that of IFNα and functional ISREs are present in the genomic DNA of the ovine PR and ER, the antiluteolytic effects of IFNτ could be due to the direct effects on transcription of ovine endometrial PR and/or ER genes. The available results (9, 18, 19) indicate that IFNτ inhibits increases in ER mRNA and protein, whereas PR protein remains stable, and PR mRNA tends to decrease between days 12 and 16 of pregnancy.

Temporal changes in endometrial receptors for progesterone during the estrous cycle and early pregnancy of sheep (18) indicated that (i) endometrial PR is lower on days 12 and 14 than on days 10 and 16 of the estrous cycle; (ii) endometrial PR did not change between days 10 and 16 of pregnancy, indicative of stabilization of PR, despite a gradual decrease in PR mRNA; (iii) changes in PR mRNA differed between cyclic and pregnant ewes, tending to increase between days 12 and 16 of the estrous cycle and decrease during the same period for pregnant ewes; and (iv) the ratios for PR:ER and PR mRNA:ER mRNA were higher for pregnant ewes. In vivo the antiluteolytic effects of oIFNτ are dependent on the presence of progesterone, and in the absence of progesterone, oIFNτ actually stimulated PGFM release in response to exogenous oxytocin while inhibiting this response in the presence of progesterone (88). Altering relative concentrations of endometrial PR and ER or the PR:ER ratios may influence conceptus-mediated antiluteolytic mechanisms. At present, consistent inhibitory effects of oIFNτ on ER and ER mRNA have been demonstrated, but the effects of oIFNτ on PR and PR mRNA are equivocal.

High-affinity, low-capacity binding sites, presumably type I IFN receptors, for oIFNτ are present in endometrial membranes, and hIFNα will displace oTP-1 from those receptors (9). Unoccupied type I receptors are similar for cyclic and pregnant ewes on days 8 and 12, but decrease thereafter for pregnant ewes (56). Interactions between type I endometrial receptors and oIFNτ appear to differ from those between *recombinant bovine IFNα1* (rbIFNα) or *recombinant human IFNα* (rhuIFNα) since antiluteolytic activity is at least 7-fold greater than that of rbIFNα and rhuIFNα (9). For example, daily intrauterine infusion of rbIFNα extended interestrous intervals of ewes to greater than 19 days when 2000 μg, but not 200 g, was infused over each 24-h period from days 9 through 19 (89).

Ovine IFNτ, the antiluteolytic pregnancy recognition signal in sheep, prevents uterine secretion of luteolytic pulses of PGF. Our current working hypothesis is that it stabilizes endometrial PR and prevents up-regulation of endometrial ER and OTR. Failure of pregnant ewes to respond to the potential luteolytic effects of estradiol and oxytocin can be explained by the absence of uterine ER and OTR. This working hypothesis also applies to cows and goats since intrauterine injections of roIFNτ extend the interestrous interval in both cows (50) and goats (51) and inhibit oxytocin-induced secretion of PGF by the uterus in cows (50).

## *Cattle*

Bovine IFNτ cross-reacts immunologically with oIFNτ, has high amino acid sequence homology with oIFNτ, and possesses potent antiviral activity. Secretion of bIFNτ is maximal around days 16–19 of pregnancy; however, mRNA for bIFNτ can be detected as early as day 12. Secretion of bIFNτ by trophectoderm/chorion may continue until at least day 38 of pregnancy (9).

During maternal recognition of pregnancy, ovarian follicular populations are altered, and follicular waves on the ovary bearing the CL, but not the contralateral ovary, are suppressed in cattle (9). These effects may be supportive of the antiluteolytic mechanism whereby local suppression of follicular development reduces secretion of estradiol that could otherwise enhance uterine secretion of luteolytic pulses of PGF. Endometrial receptors for oxytocin are significantly reduced in pregnant compared to cyclic cattle during the luteolytic period (14, 15); for example, 563 ± 117 versus 18 ± 5 fmol/mg protein for day 18 cyclic and pregnant cows, respectively (15). As with sheep, bIFNτ inhibits synthesis of endometrial ER and OTR to abrogate uterine production of luteolytic pulses of PGF.

### *Goat*

Pregnancy recognition in goats occurs around day 17, and goat conceptuses secrete cIFNτ between days 16 and 21 that can be immunoprecipitated with antiserum to oIFNτ. Therefore, cIFNτ is assumed to be antiluteolytic protein in goats (90). Pulsatile release of oxytocin and PGF is suppressed in pregnant compared to cyclic goats between days 10–12 and estrus or day 20 of pregnancy (91), suggesting that antiluteolytic mechanisms in the goat are similar to those for sheep and cows.

## Pigs and Horses

### *Luteolytic Events*

Endocrine requirements for luteolysis in pigs and horses have not been clearly delineated. However, it is known that luteolysis occurs during late diestrus; that is, following stimulation of the uterine endometrium by progesterone for 10–12 days. Luteolysis occurs when pulsatile release of uterine PGF into the uterine venous drainage begins on about day 15 or 16 of the estrous cycle (9). The CL of pigs contains very low levels of oxytocin and vasopressin and undetectable levels of oxytocin mRNA, but the potential role of these neuropeptides of ovarian or posterior pituitary origin in luteolysis in pigs has not been established (9). The endometrium of pigs must contain receptors for oxytocin because it responds in vitro to oxytocin with increased secretion of PGF and IP turnover (9).

The uterine endometrium of mares releases PGF, which results in luteolysis; however, neither the pattern of release of PGF required for luteolysis nor endocrine regulation of uterine production of luteolytic PGF is established. It is known that cervical stimulation of oxytocin release via the Ferguson reflex stimulates uterine secretion of PGF and that administration of exogenous oxytocin stimulates uterine release of PGF in mares (9).

The CL of pigs is refractory to luteolytic effects of PGF until about day 13 of the estrous cycle because luteal receptors for PGF are insufficient to allow PGF to exert a luteolytic effect until days 12–14 of the estrous cycle (92). The CL of mares, however, is sensitive to the luteolytic effects of PGF after about day 5 postovulation, as is the case for sheep, cattle, and goats (9).

## *Pregnancy Recognition in Pigs*

The theory of maternal recognition of pregnancy in pigs has been reviewed extensively (9). The major assumptions are that uterine endometrium secretes the luteolysin PGF and that the conceptuses secrete estrogens that are antiluteolytic. The present theory is that PGF is secreted in an endocrine direction, toward the uterine vasculature, in cyclic gilts and transported to the CL to exert its luteolytic effect. However, in pregnant pigs the direction of secretion of PGF is exocrine, into the uterine lumen, where it is sequestered to exert its biological effects in utero and/or be metabolized to prevent luteolysis.

*Pig conceptus secretory proteins* (pCSP) recovered from culture medium of day 15 conceptuses (118) have antiviral activity due to the presence of both IFN$\alpha$ (25%) and IFN$\gamma$ (75%) between days 15 and 21 of gestation (9). Intrauterine infusion of pCSP on days 12–15 of the estrous cycle has no effect on interestrous interval or temporal changes in concentrations of progesterone in plasma, but PGE in peripheral plasma, presumably from the uterus, increases. At present, the roles of IFNs produced by pig trophoblast are unknown.

# Summary

Assisted reproductive technologies result in a rate of pregnancy failure of about 84% compared to an estimated failure rate of 39% in women following natural conception. The major cause of reproductive wastage in humans is the failure of conceptuses to develop normally and implant between days 17 to 28 of pregnancy. In animal agriculture high rates of embryonic loss are also a major cause of reproductive failure. These losses range from 15% to 30% or higher, with most losses occurring between days 12 to 25 after mating. Considerable efforts are being made to understand the factors responsible for the high periimplantation embryonic death losses and to develop means to prevent or ameliorate this loss. The failure of conceptuses to provide adequate pregnancy recognition signaling may account for much of this pregnancy wastage.

Pregnancy recognition requires the presence of a functional CL. Signals from the trophoblast that ensure maintenance of functional CL are luteotrophic in primates and antiluteolytic in subprimate mammals. The primate conceptus secretes CG to affect CL development and function directly. However, trophectoderm of ruminant conceptuses secrete IFNτ, a unique antiluteolytic cytokine that abrogates the luteolytic mechanism. In pigs the secretion of estrogen and its effect on uterine receptors for prolactin appear essential for CL maintenance (93). However, pig conceptuses also secrete a unique IFNα and IFNγ whose functions are not known. In subprimate mammals a common thread appears to exist due to the fact that receptors for prolactin and type I IFNs are members of the IgG superfamily (94). Certainly, evidence is strong for a common cytokine-mediated antiluteolytic mechanism for pregnancy recognition in ruminants. The redundancy in the production of cytokines and lymphokines by cells of the immune system and those of the reproductive system is also of interest, but is not understood. Therefore, an explanation of mechanisms whereby cytokines can influence pregnancy recognition will provide information critical to improving reproductive efficiency and will undoubtedly increase our understanding of the complex interactions of the soluble cytokine network.

## *References*


1. Short RV. Implantation and the maternal recognition of pregnancy. In: Heap RB, ed. Foetal autonomy, Ciba Foundation Symposium. London: Churchill, 1969:377–86.
2. Hearn JP, Webley GE, Gidley-Baird AA. Chorionic gonadotrophin and embryo-maternal recognition during the peri-implantation period in primates. J Reprod Fertil 1992;92:497–509.
3. Zelinski-Wooten MB, Stouffer RL. Intraluteal infusions of prostaglandins of the E, D, I, and A series prevent PGF2α-induced, but not spontaneous luteal regression in rhesus monkeys. Biol Reprod 1990;43:507–16.
4. Khan-Dawood FS, Marut EL, Dawood MY. Oxytocin in the corpus luteum of the cynomolgus monkey (*Macaca fascicularis*). Endocrinology 1984;115:570–4.
5. Lessey BA, Killam AP, Metzger DA, Haney AF, Greene GL, McCarty KS. Immunohistochemical analysis of human uterine estrogen and progesterone receptors throughout the menstrual cycle. J Clin Endocrinol 1988;67:334–40.
6. Clarke CL. Cell-specific regulation of progesterone receptor in the female reproductive system. Mol Cell Endocrinol 1990;70:C29–33.
7. Okulicz WC, Savasta AM, Hoberg LM, Longcope C. Biochemical and immunohistochemical analyses of estrogen and progesterone receptors in the rhesus monkey uterus during the proliferative and secretory phases of artificial menstrual cycles. Fertil Steril 1990;53:913–20.
8. Hild-Petito S, Verhage HG, Fazleabas AT. Estrogen and progestin receptor localization during implantation and early pregnancy in the baboon (*Papio anubis*) uterus. Biol Reprod 1991;44(suppl 1):185.

9. Bazer FW. Mediators of maternal recognition of pregnancy in mammals. Proc Soc Exp Biol Med 1992;199:373–84.
10. Eggleston DL, Wilken C, Van Kirk EA, Slaughter RG, Ji TH, Murdoch WJ. Progesterone induces expression of endometrial messenger RNA encoding for cyclooxygenase (sheep). Prostaglandins 1990;39:675–83.
11. Flint APF, Sheldrick EL. Ovarian oxytocin and maternal recognition of pregnancy. J Reprod Fertil 1986;76:831–9.
12. Meyer HHD, Mittermeier T, Schams D. Dynamics of oxytocin, estrogen and progestin receptors in the bovine endometrium during the estrous cycle. Acta Endocrinol (Copenh) 1986;118:96–104.
13. Soloff MS, Fields MJ. Changes in oxytocin receptor concentrations throughout the estrous cycle of the cow. Biol Reprod 1989;40:283–7.
14. Fuchs AR, Behrens O, Helmer H, Liu CH, Barros CM, Fields MJ. Oxytocin and vasopressin receptors in bovine endometrium and myometrium during the estrous cycle and early pregnancy. Endocrinology 1990;127:629–36.
15. Jenner LJ, Parkinson TJ, Lamming GE. Uterine oxytocin receptors in cyclic and pregnant cows. J Reprod Fertil 1991;91:49–58.
16. Wallace JM, Helliwell R, Morgan PJ. Autoradiographical localization of oxytocin binding sites on ovine oviduct and uterus throughout the oestrous cycle. Reprod Fertil Dev 1991;3:127–35.
17. McCracken JA, Schramm W, Okulicz WC. Hormone receptor control of pulsatile secretion of $PGF_{2a}$ from ovine uterus during luteolysis and its abrogation in early pregnancy. Anim Reprod Sci 1984;7:31–56.
18. Ott TL, Zhou Y, Mirando MA, et al. Changes in progesterone and oestrogen receptor messenger ribonucleic acid and protein during maternal recognition of pregnancy and luteolysis in ewes. J Mol Endocrinol 1993;10:171–83.
19. Mirando MA, Harney JP, Zhou Y, Ogle TF, Ott TL, Bazer FW. Changes in progesterone and oestrogen receptor messenger ribonucleic acid and protein and oxytocin receptors in endometrium of ewes after intrauterine injection of ovine trophoblast interferon. J Mol Endocrinol 1993;10:185–92.
20. Stewart HJ, Stevenson KR, Flint APF. Isolation and structure of a partial sheep oxytocin receptor cDNA and its use as a probe for northern analysis of endometrial RNA. J Mol Endocrinol 1993.
21. Cherny RA, Salamonsen LA, Findlay JK. Immunocytochemical localization of oestrogen receptors in the endometrium of the ewe. Reprod Fertil Dev 1991;3:321–31.
22. Hooper SB, Watkins WB, Thorburn GD. Oxytocin, oxytocin associated neurophysin, and prostaglandin $F_2\alpha$ concentrations in the utero-ovarian vein of pregnant and nonpregnant sheep. Endocrinology 1986;119:2590–7.
23. Poyser N. A possible explanation for the refractoriness of uterine prostaglandin production. J Reprod Fertil 1991;91:374–84.
24. Jenkin G. The interaction between oxytocin and prostaglandin $F2\alpha$ during luteal regression and early pregnancy in sheep. Reprod Fertil Dev 1992;4:321–8.
25. Homeida AM, Khalafalla AE. Effects of oxytocin-antagonist injections on luteal regression in the goat. Br J Pharmacol 1987;90:281–4.
26. Schams D, Prokopp S, Barth D. The effect of active and passive immunization against oxytocin on ovarian cyclicity in ewes. Acta Endocrinol (Copenh) 1983;103:337–44.

27. Wathes DC, Ayad VJ, McGoff SA, Morgan KL. Effect of active immunization against oxytocin on gonadotrophin secretion and the establishment of pregnancy in the ewe. J Reprod Fertil 1989;86:653–64.
28. Flint APF, Sheldrick EL. Continuous infusion of oxytocin prevents induction of oxytocin receptors and blocks luteal regression in cyclic ewes. J Reprod Fertil 1985;75:623–31.
29. Schramm WL, Bovaird ME, Glew ME, Schramm G, McCracken JA. Corpus luteum regression induced by ultra-low pulses of prostaglandin F2α. Prostaglandins 1983;26:347–64.
30. Ellinwood WE, Nett TM, Niswender GD. Maintenance of the corpus luteum of early pregnancy in the ewe, I. Luteotropic properties of embryonic homogenates. Biol Reprod 1979;21:281–8.
31. Roberts RM, Cross JC, Leaman DW. Interferons as hormones of pregnancy. Endocr Rev 1992;13:432–52.
32. Godkin JD, Bazer FW, Moffatt J, Sessions F, Roberts RM. Purification and properties of a major, low molecular weight protein released by the trophoblast of sheep blastocysts at day 13–21. J Reprod Fertil 1982;65:141–50.
33. Cross JC, Roberts RM. Constitutive and trophoblast-specific expression of a class of bovine interferon genes. Proc Natl Acad Sci USA 1991;88:3817–21.
34. Imakawa K, Anthony RV, Kazemi M, Marotti KR, Polites HG, Roberts RM. Interferon-like sequence of ovine trophoblast protein secreted by embryonic trophectoderm. Nature 1987;330:377–9.
35. Stewart HJ, McCann SHE, Northrop AJ, Lamming GE, Flint APF. Sheep antiluteolytic interferon: cDNA sequence and analysis of mRNA levels. J Mol Endocrinol 1989;2:65–71.
36. Charpigny G, Reinaud P, Heut J-C, et al. High homology between a trophoblast protein (trophoblastin) isolated from ovine embryo and α-interferons. FEBS Lett 1988;228:12–6.
37. Charlier M, Hue D, Martal J, Gaye P. Cloning and expression of cDNA encoding ovine trophoblastin: its identity with a class-II alpha interferon. Gene 1989;77:341–8.
38. Pontzer CH, Torres BA, Vallet JL, Bazer FW, Johnson HM. Antiviral activity of the pregnancy recognition hormone ovine trophoblast protein-1. Biochem Biophys Res Commun 1988;152:801–7.
39. Roberts RM, Imakawa K, Niwano Y, et al. Interferon production by the preimplantation sheep embryo. J Interferon Res 1989;9:175–87.
40. Pontzer CH, Bazer FW, Johnson HM. Antiproliferative activity of a pregnancy recognition hormone, ovine trophoblast protein-1. Cancer Res 1991; 51:5304–7.
41. Leaman DW, Roberts RM. Genes for the trophoblast interferons in sheep, goat, and musk ox and distribution of related genes among mammals. J Interferon Res 1992;12:1–11.
42. Charlier M, Hue D, Boisnard M, Martal J, Gaye P. Cloning and structural analysis of two distinct families of ovine interferon-α genes encoding functional class II and trophoblast (oTP) α-interferons. Mol Cell Endocrinol 1991;76:161–71.
43. Martal J, Degryse E, Charpigny G, et al. Evidence for extended maintenance of the corpus luteum by uterine infusion of a recombinant α-interferon (trophoblastin) in sheep. J Endocrinol 1990;127:R5.

44. Ott TL, Van Heeke G, Hostetler CE, et al. Intrauterine injection of recombinant ovine interferon-tau extends the interestrous interval in sheep. Theriogenology 1993.
45. Ott TL, Van Heeke G, Johnson HM, Bazer FW. Cloning and expression in *Saccharomyces cerevisiae* of a synthetic gene for type I trophoblast interferon ovine trophoblast protein-1: purification and antiviral activity. J Interferon Res 1991;11:357–64.
46. Nephew KP, Whaley AE, Christenson RK, Imakawa K. Differential expression of distinct mRNAs for ovine trophoblast protein-1 and related sheep type I interferons. Biol Reprod 1993;48:768–78.
47. Helmer SD, Hansen PJ, Anthony RV, Thatcher WW, Bazer FW, Roberts RM. Identification of bovine trophoblast protein-1, a secretory protein immunologically related to ovine trophoblast protein-1. J Reprod Fertil 1987;79: 83–91.
48. Baumbach GA, Duby RT, Godkin JD. N-glycosylated and unglycosylated forms of caprine trophoblast protein-1 are secreted by preimplantation goat conceptuses. Biochem Biophys Res Commun 1990;172:16–21.
49. Heyman Y, Camous S, Fevre J, Meziou W, Martal J. Maintenance of the corpus luteum after uterine transfer of trophoblastic vesicles to cyclic cows and ewes. J Reprod Fertil 1984;70:533–40.
50. Meyer MD, Drost M, Ott TL, et al. Recombinant ovine trophoblast protein-1 extends corpus luteum lifespan and reduced uterine secretion of prostaglandin $F_{2\alpha}$ in cattle. J Anim Sci 1992;70(suppl 1):270.
51. Ott TL, Newton GR. Intrauterine injection of recombinant ovine interferon τ (roIFNτ) blocks luteolysis and extends CL lifespan in goats. Biol Reprod 1993;48(suppl 1):173.
52. Leaman DW, Cross JC, Roberts RM. Genes for the trophoblast interferons and their distribution among mammals. Reprod Fertil Dev 1992;4:349–53.
53. Hansen TR, Leaman DW, Cross JC, Mathailagan N, Bixby JA, Roberts RM. The genes for the trophoblast interferons and the related interferon-αII possess distinct 5′-promoter and 3′-flanking sequences. J Biol Chem 1991;266: 3060–7.
54. Imakawa K, Hansen TR, Malathy P-V, et al. Molecular cloning and characterization of complementary deoxyribonucleic acids corresponding to bovine trophoblast protein-1: a comparison with ovine trophoblast protein-1 and bovine interferon-αII. Mol Endocrinol 1989;3:127–39.
55. Stewart HJ, McCann SHE, Barker PJ, Lee KE, Lamming GE, Flint APF. Interferon sequence homology and receptor binding activity of ovine trophoblast antiluteolytic protein. J Endocrinol 1987;115:R13–5.
56. Knickerbocker JJ, Niswender GD. Characterization of endometrial receptors for ovine trophoblast protein-1 during the estrous cycle and early pregnancy in sheep. Biol Reprod 1989;40:361–70.
57. Hansen TR, Kazemi M, Keisler DH, Futhan-Veedu M, Imakawa K, Roberts RM. Complex binding of the embryonic interferon, ovine trophoblast protein-1, to endometrial receptors. J Interferon Res 1989;9:215–25.
58. Roberts RM, Farin CE, Cross JC. Trophoblast proteins and maternal recognition of pregnancy. In: Milligan SR, ed. Oxford reviews in reproductive biology. Oxford: Oxford University Press, 1990;12:147–80.

59. Whaley AE, Carroll RS, Nephew KP, Imakawa K. Molecular cloning of unique interferons from human placenta. Biol Reprod 1991;44(suppl 1):186.
60. Schalue TK. Effects of intrauterine infusion of synthetic peptide fragments corresponding to ovine trophoblast interferon (otIFN) on oxytocin-induced endometrial inositol phosphate turnover in ewes. Biol Reprod 1992;46(suppl 1):70.
61. Jarpe MA, Johnson HM, Bazer FW, Ott TL, Pontzer CH. Predicted structural motif of IFNτ. Protein Eng (in review).
62. Senda T, Shimazu T, Matsuda S, et al. Three-dimensional crystal structure of recombinant murine interferon-beta. EMBO J 1992;11:3193–201.
63. Pontzer CH, Ott TL, Bazer FW, Johnson HM. Localization of an antiviral site on the pregnancy recognition hormone, ovine trophoblast protein 1. Proc Natl Acad Sci USA 1990;87:5945–9.
64. Farin CE, Cross JC, Tindle NA, Murphy CN, Farin PW, Roberts RM. Induction of trophoblastic interferon expression in ovine blastocysts after treatment with double-stranded RNA. J Interferon Res 1991;11:151–7.
65. Adunyah SE, Unlap TM, Wagner F, Kraft AS. Regulation of c-jun expression and AP-1 enhancer activity of granulocyte-macrophage colony stimulating factor. J Biol Chem 1991;266:5670–5.
66. Imakawa K, Helmer SD, Nephew KP, Meka CSR, Christenson RK. A novel role for GM-CSF: enhancement of pregnancy specific interferon production, ovine trophoblast protein-1. Endocrinology 1993;132:1869–71.
67. Xavier F, Guillomot M, Charlier M, Martal J, Gaye P. Co-expression of the protooncogene FOS (cfos) and an embryonic interferon (ovine trophoblastin) by sheep conceptuses during implantation. Biol Cell 1991;73:27–33.
68. Vallet JL, Bazer FW, Fliss MFV, Thatcher WW. The effect of ovine conceptus secretory proteins and purified ovine trophoblast protein-one on interestrous interval and plasma concentrations of prostaglandins $F_{2\alpha}$, E and 13,14-dihydro-15-keto-prostaglandin $F_{2\alpha}$ in cyclic ewes. J Reprod Fertil 1988; 84:493–504.
69. Martal J, Degryse E, Charpigny G, et al. Evidence for extended maintenance of the corpus luteum by uterine infusion of recombinant trophoblast α-interferon (trophoblastin) in sheep. J Endocrinol 1990;127:R5–8.
70. Soloff MS. Uterine receptor for oxytocin: effects of estrogen. Biochem Biophys Res Commun 1975;65:205–12.
71. Findlay JK, Clarke IJ, Swaney J, Colvin N, Doughton B. Oestrogen receptors and protein synthesis in caruncular and intercaruncular endometrium of sheep before implantation. J Reprod Fertil 1982;64:329–39.
72. Langer JA, Pestka S. Interferon receptors. Immunol Today 1989;9:393–400.
73. Mogensen KE, Uze' G, Eid P. The cellular receptor of alpha-beta interferons. Experientia 1989;45:500–8.
74. Schabe M, Princler GL, Faltynek CR. Characterization of the human type I interferon receptor by ligand blotting. Eur J Immunol 1988;18:2009–14.
75. Uze' G, Lutfalla G, Gresser I. Genetic transfer of a functional human interferon α-receptor into mouse cells: cloning and expression of its cDNA. Cell 1990;60:225–34.
76. Williams BRG. Signal transduction and transcriptional regulation of interferon-α-stimulated genes. J Interferon Res 1991;11:207–13.

77. Stark GR, Kerr IM. Interferon-dependent signaling pathways: DNA elements, transcription factors, mutations, and effects of viral proteins. J Interferon Res 1992;12:147–51.
78. Fu X-Y, Schindler C, Improta T, Aebersold R, Darnell JE Jr. The proteins of ISGF-3, the interferon α-induced transcriptional activator, define a gene family involved in signal transduction. Proc Natl Acad Sci 1992;89:7840–3.
79. Schindler C, Shuai K, Prezioso VR, Darnell JE Jr. Interferon-dependent tyrosine phosphorylation of a latent cytoplasmic transcription factor. Science 1992;257:809–13.
80. Velazquez L, Fellous M, Stark G, Pellegrini S. A protein tyrosine kinase in the interferon α/β signaling pathway. Cell 1992;70:313–22.
81. Scambia G, Panici PB, Battaglia F, et al. Effect of recombinant human interferon $alpha_{2b}$ on receptors for steroid hormones and epidermal growth factor in patients with endometrial cancer. Eur J Cancer 1991;27:51–3.
82. Angioli R, Untch M, Bernd-Uwe S, et al. Enhancement of progesterone receptor levels by interferons in AE-7 endometrial cancer cells. Cancer 1993;71:2776–81.
83. Dimitrov NV, Meyer CJ, Strander H, Einhorn S, Cantell K. Interferons as a modifier of estrogen receptors. Ann Clin Lab Sci 1984;14:32–9.
84. van den Berg HW, Leahey WJ, Lynch M, Clarke R, Nelson J. Recombinant human interferon alpha increases oestrogen receptor expression in human breast cancer cells (ZR-75-1) and sensitizes them to the anti-proliferative effects of tamoxifen. Br J Cancer 1987;55:255–7.
85. Milgrom E, Dessen P, Zerah V, et al. Organization of the entire rabbit progesterone receptor mRNA and of the promoter and 5′ flanking region of the gene. Nucleic Acids Res 1988;16:5459–72.
86. Kastner P, Krust A, Turcotte B, et al. Two distinct estrogen-regulated promoters generate transcripts encoding the two functionally different human progesterone receptor forms A and B. EMBO J 1990;9:1603–14.
87. Ponglikitmongkol M, Green S, Chambon P. Genomic organization of the human oestrogen receptor gene. EMBO J 1988;7:3385–8.
88. Ott TL, Mirando MA, Davis MA, Bazer FW. Effects of ovine conceptus secretory proteins and progesterone on oxytocin-stimulated endometrial production of prostaglandin and turnover of inositol phosphate in ovariectomized ewes. J Reprod Fertil 1992;95:19–29.
89. Flint APF, Parkinson RJ, Stewart HJ, Vallet JL, Lamming GE. Molecular biology of trophoblast interferons and studies of their effects in vivo. J Reprod Fertil 1991;43(suppl 1):13–25.
90. Spencer TE, Ing NH, Bazer FW. Partial cloning of the ovine estrogen receptor (oER) mRNA and level of uterine ER mRNA during the estrous cycle and early pregnancy in ewes. Biol Reprod 1993;48(suppl 1):189.
91. Gnatek GG, Smith LD, Duby RT, Godkin JD. Maternal recognition of pregnancy in the goat: effects of conceptus removal on interestrous intervals and characterization of conceptus protein production during early pregnancy. Biol Reprod 1989;41:655–64.
92. Homeida AM. Role of oxytocin during the oestrous cycle of ruminants with particular reference to the goat. Anim Breed Abstr 1986;54:263–8.

93. Gadsby JE, Balapure AK, Britt JH, Fitz FA. Prostaglandin F2α receptors on enzyme-dissociated pig luteal cells throughout the estrous cycle. Endocrinology 1990;126:787–95.
94. Thoreau E, Petridou B, Kelly PA, Djiane J, Mornon JP. Structural symmetry of the extracellular domain of the cytokine/growth hormone/prolactin receptor family and interferon receptors revealed by hydrophobic cluster analysis. FEBS Lett 1991;282:26–31.

5

# Role of Locally Produced Growth Factors in Human Placental Growth and Invasion with Special Reference to Transforming Growth Factors

PEEYUSH K. LALA AND JEFFREY J. LYSIAK

Anatomically, the human fetomaternal interface consists of the placenta, a fetally derived organ, and the decidua, a maternally derived tissue. Physical as well as molecular interactions at this interface hold the secrets to two important biological riddles: (i) What protects the placenta, a fetally derived organ and thus genetically disparate from the mother, from destruction by the mother's immune system? and (ii) What protects the uterus from overinvasion by the placenta, which is a highly invasive tumorlike structure? The present chapter focuses largely on our studies related to the second riddle.

## Human Fetomaternal Interface

### *Placenta*

The placenta is an organ essential for the proper development of the embryo or the fetus. The architecture of the placenta brings maternal and fetal blood into close proximity with each other to allow for an efficient exchange of molecules. Nutrients and oxygen are transferred from the maternal to the fetal blood, while fetal waste products are passed from the fetus to the mother. The placenta is also the largest endocrine organ, secreting numerous steroid and protein hormones as well as a large number of locally active growth factors essential for the maintenance of pregnancy.

The structure of the human placenta is obtained by the ability of trophoblast cells rapidly to proliferate, migrate, and invade the endometrium, inclusive of capillaries and glands, and, finally, to tap into the spiral (uteroplacental) arteries of the uterus. In humans placental development begins approximately 6 days after fertilization as the blastocyst attaches to the apical surface of the uterine epithelium, usually on the posterior wall of the uterus (1). In subsequent steps the blastocyst breaches the epithelial barrier, and trophoblast cells begin to invade the

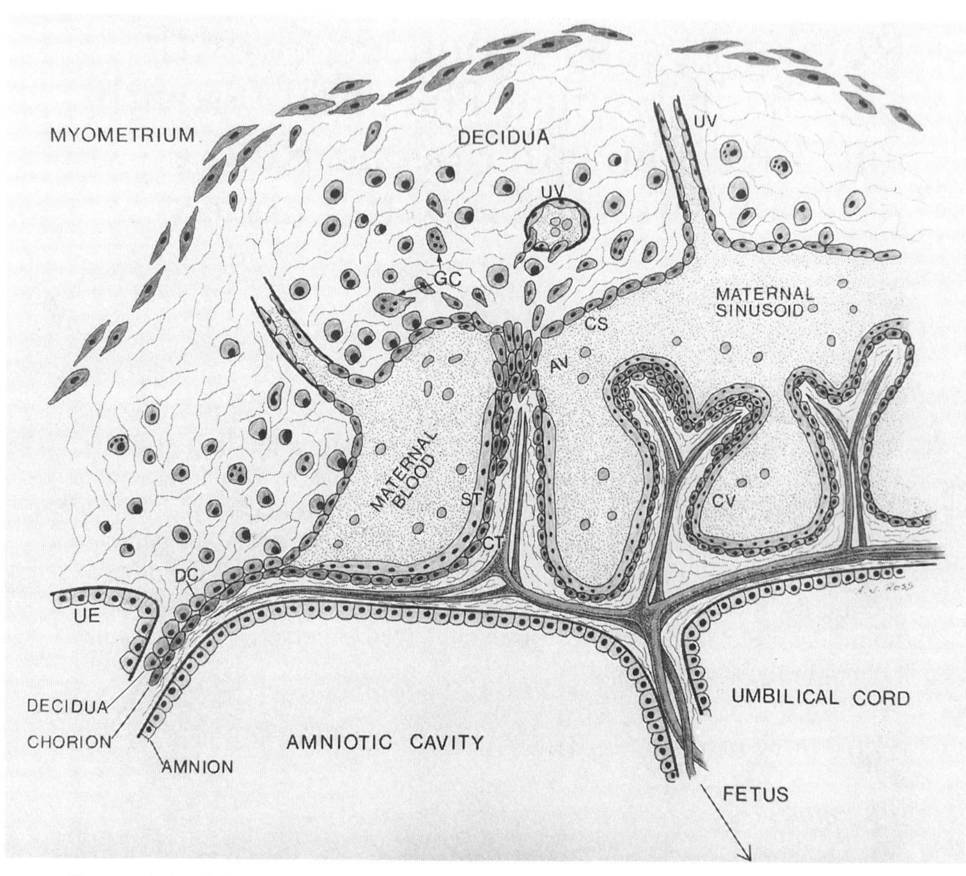


FIGURE 5.1. Schematic diagram illustrating the relationships of the placenta and the uterus at approximately the end of the 1st trimester. A population of chorionic villi, called anchoring villi (AV), serve to maintain the attachment of the placenta to the uterine wall. Cells migrating out from anchoring villi, extravillous trophoblast (ET), give rise to the cytotrophoblastic shell (CS) or remain embedded in the decidual tissue as isolated cells. These isolated trophoblast cells are highly invasive; some reach as far as the muscular layer of the uterus (myometrium), and some invade maternal blood vessels. When these cells differentiate, they fuse and form placental bed giant cells (GC) that are presumably noninvasive cells. (CV = chorionic villus; CT = cytotrophoblast; ST = syncytiotrophoblast; DC = decidual cell; UE = uterine epithelium; UV = uterine blood vessel.) Reprinted with permission from Graham and Lala (7), © National Research Council of Canada 1992.

stroma in the endometrial lining of the uterus. The integrity of the uterine epithelium is soon reestablished, thus positioning the blastocyst completely in the endometrial stroma. As implantation proceeds, trophoblast cells continue to invade endometrial capillaries, glands, and the decidua. The invasion of capillaries creates sinusoidal spaces surrounding villous projections of the trophoblast (primary villi) consisting of an inner layer of cytotrophoblast cells and an outer syncytiotrophoblast layer. Secondary villi form after extraembryonic mesoderm penetrates the villus core. Shortly thereafter, fetal vessels develop in the mesodermal core of the villi, transforming them into tertiary villi that grow and branch, forming treelike structures bathed in the maternal blood known as *floating chorionic villi* (Fig. 5.1).

In some areas cytotrophoblast cells sprout from the tips of the chorionic villi (anchoring villi) through the syncytial layer to extend as columns invading the maternal decidual tissue. This trophoblast cell population is commonly described as the *intermediate* or *extravillous trophoblast*. Within the decidual tissue some of these cells form the cytotrophoblastic shell that anchors the placenta to the uterus (2). Others remain isolated or fuse to form placental bed giant cells. Yet other intermediate trophoblast cells seek out spiral arteries of the uterus, erode their tunica media, replace their endothelium, and prevent them from responding to vasoactive substances in the maternal blood. These cells are often described as the *endovascular trophoblast*. This process is essential for maintaining an adequate perfusion of maternal blood to the placenta (3). A schematic diagram of the human placenta inclusive of various trophoblast cell populations is presented in Figure 5.1.

Many similarities are shared between the placenta and invasive tumors (4–8). In both situations rapidly growing cells are capable of transgressing normal tissue barriers and invading blood vessels. The formation of new blood vessels, *angiogenesis*, is also common to both the placenta and malignant tumors. However, unlike the uncontrolled growth and invasion by cancer cells, trophoblast cell growth and invasion are highly regulated events restricted both spatially and temporally. This chapter summarizes our studies of mechanisms regulating these events.

## Decidua

The decidua is a maternally derived tissue comprising decidual cells, lymphocytes, monocytes, macrophages, granulocytes, stromal fibroblasts, and blood vessels (9). Decidual cells represent a distinct cell class that arises in the endometrium during pregnancy through the proliferation and differentiation of endometrial stromal cells (10–12). The formation of the decidua, *decidualization*, during pregnancy succeeds the implantation of the blastocyst. However, in nonpregnant women a small amount of decidualization may occur in the late secretory phase of the menstrual cycle (13). In the human the decidua can be subdivided into three distinct zones: (i) the *decidua basalis*, in the endometrium directly apposed to the villous trophoblast of the placenta; (ii) the *decidua capsularis*, in the endometrium adjacent to the amnio-chorion (nonvillous trophoblast); and (iii) the *decidua parietalis*, in the remaining endometrium of the uterus facing the lumen.

The life history of decidual cells has been studied extensively in the mouse and rat models. It has been shown that decidual cells arise from proliferation and differentiation of their immediate precursors, the uterine stromal fibroblastlike cells (11, 12). Because of the local immunoregulatory role of decidual cells in the uterus, Kearns and Lala (14) hypothesized that predecidual stem cells may originate from the bone marrow. They addressed this question in the mouse by making radiation bone marrow chimeras and then screening single-cell suspensions of *deciduoma*—that is, decidualization produced in pseudopregnant animals—for the presence of donor-derived cells of disparate H-2 phenotype. They found that at least a subpopulation of decidual cells was derived from ultimate progenitors present in the bone marrow.

Johnson et al. (15) confirmed these findings for the decidua of normal mouse pregnancy by transplanting H-2 disparate donor hemopoietic cells into the yolk sac of developing embryos to produce fertile bone marrow chimeras. By taking advantage of this technique devised by Johnson et al. (15) to make fertile chimeras, but employing donor bone marrow cells bearing a transgenic marker (1000 copies of β-globin gene), Lysiak and Lala (16) recently identified and characterized in situ the decidual cells of normal mouse pregnancy as having a hemopoietic descent. These experiments in mice clearly revealed that predecidual stem cells are hemopoietic in origin and migrate to the uterus at some point in ontogeny. Whether this is also the case in the human remains to be determined.

The decidua has been shown to play an important role in the maintenance of pregnancy. Decidual cells in the human produce prolactin (17, 18) and short-range biological mediators; for example, growth factors (19, 20) and *prostaglandins of the E series* (PGE) (21). In both the human and the mouse, the decidua has been reported to exert a local immunoprotective function for the conceptus through the release of $PGE_2$ (4, 21–27). Decidua-derived $PGE_2$ has been found to inactivate T cells (21, 25) as well as *natural killer* (NK) cells (24, 25) in situ in the human and the murine decidua. This inactivation results, at least in part, from a down-regulation of *interleukin-2* (IL-2) receptors on effector cells and an inhibition of IL-2 production (22, 25), mechanisms that would prevent accidental activation of decidual lymphocytes into *lymphokine-activated killer* (LAK) cells that express trophoblast killer ability (25). *Transfoming growth factor β2* (TGFβ2)-like molecules produced by the murine decidua have also been reported to have an immunoprotective role (28).

We have recently discovered that the decidua also plays a major role in the regulation of trophoblast growth (19, 20) and invasion (6–8) by secreting a number of growth factors. This chapter summarizes some of these studies.

## Growth Factors

A large number of growth factors have been reported to be produced by the human placenta and/or the decidua that may be functionally important for the conceptus. We have recently examined some of these growth factors relative to their role in trophoblast proliferation, differentiation, and invasion. They include *epidermal growth factor* (EGF), *transforming growth factor* $\alpha$ (TGF$\alpha$), and *amphiregulin* (AR)—all binding to the EGF receptor—as well as TGF$\beta$ and *insulin-like growth factor II* (IGF-II).

### *EGF Receptor Ligands*

Proteins belonging to the EGF family all have a common consensus sequence of $CX_7CX_{4-5}CX_{10-13}CXCX_8C$ contained within 36–40 amino acid residues. Members of this family have been shown to be powerful mitogens for a variety of epithelial cells. They exert their mitogenic actions through binding to the EGF receptor, which possesses tyrosine kinase activity (29).

#### EGF

EGF was initially isolated from mouse maxillary glands and termed *tooth-lid factor* because of its ability to cause precocious eyelid opening and incisor eruption in mice (30). Mature EGF is a 53-amino acid polypeptide derived by cleavage of a 1200-amino acid transmembrane precursor molecule (29, 31). EGF has a broad distribution in the body and has been implicated in many normal physiological events, including its important role in epithelial cell renewal (29).

#### TGF$\alpha$

TGF$\alpha$ is a 50-amino acid polypeptide derived from a larger membrane-bound precursor molecule, pro-TGF$\alpha$, by proteolytic cleavage (32–35). Using mutated noncleavable forms of pro-TGF$\alpha$, it has been demonstrated that a proteolytic cleavage is not required for mediation of biological activity (34, 36). This cleavage may represent a regulatory step for transition between membrane-bound pro-TGF$\alpha$ and diffusible TGF$\alpha$ (37).

#### Amphiregulin

Amphiregulin is an 84-amino acid glycoprotein originally purified from the conditioned media of a human breast cancer cell line treated with 12-*O*-tetradecanoylphorbol-13-acetate (38). Like TGF$\alpha$, the mature AR peptide is believed to be the proteolytic cleavage product of a 252-amino acid transmembrane precursor (39). The carboxyterminal of AR shares 38% homology with EGF and 32% homology with TGFα (40). The biological activity of AR is believed to be mediated through the EGF receptor since AR is able to compete with $^{125}$I-EGF for binding. However, two putative nuclear targeting sequences have also been described in its $NH_2$-terminal region (40), and immunoreactive AR has been found in the nucleus of both normal and malignant ovarian and colonic epithelial cells (41, 42).

## *TGFβ*

TGFβ is a 25-kd disulphide-linked homodimer with subunits of 112 amino acids that is derived from a larger precursor molecule (43) and is secreted by most cells in an inactive or latent form (44). The structure of this latent form has been found to consist of a complex of three components: (i) a TGFβ binding protein, (ii) a 40-kd protein that is derived from the TGFβ precursor, and (iii) the dimeric TGFβ protein (45–47). Activation of this complex may occur by a number of mechanisms: (i) cleavage by proteases, such as plasmin and cathepsin D (48); (ii) cleavage by endoglycosidase F or sialidase in the presence of sialic acid or mannose-6-phosphate (45); and (iii) dissociation at pH values below 3.5 or above 12.5 (46). Control of the activation of latent TGFβ may be an important step in regulating its biological activity (49). TGFβ may also be associated with α2 macroglobulin (50) or with matrix proteoglycans, such as decorin, that may limit its biological activity (51, 52). TGFβ may bind to type I, type II, or type III (betaglycan) receptors (37). Of these receptors, type I and type II have been implicated in signal transduction (53, 54), whereas the precise function of type III receptors remains unknown.

## *IGF-II*

IGF-II is a 67-amino acid single-chain polypeptide that shares homology with IGF-I and human insulin. Unlike IGF-I, IGF-II is minimally growth hormone dependent (55). The biological effects of both IGF-I and IGF-II are predominantly mediated through the type I IGF receptor (55). IGF-II is also capable of binding to the IGF-II/mannose-6-phosphate (type II) receptor; however, its role in the biological actions of IGF-II is still uncertain (56, 57). In the blood and extracellular fluid, the IGFs are associated with a family of *binding proteins* (BPs) termed *IGFBPs* (55). To date, 6 BPs have been characterized and labeled as IGFBP-1 to -6 (58). The BPs appear to regulate the biological actions of the IGFs by either enhancing or inhibiting their binding to the IGF receptor (59–61). It is interesting to note that IGFBP-1 and IGFBP-2 have an RGD sequence (58) that can bind to RGD binding sites on the cell surface and

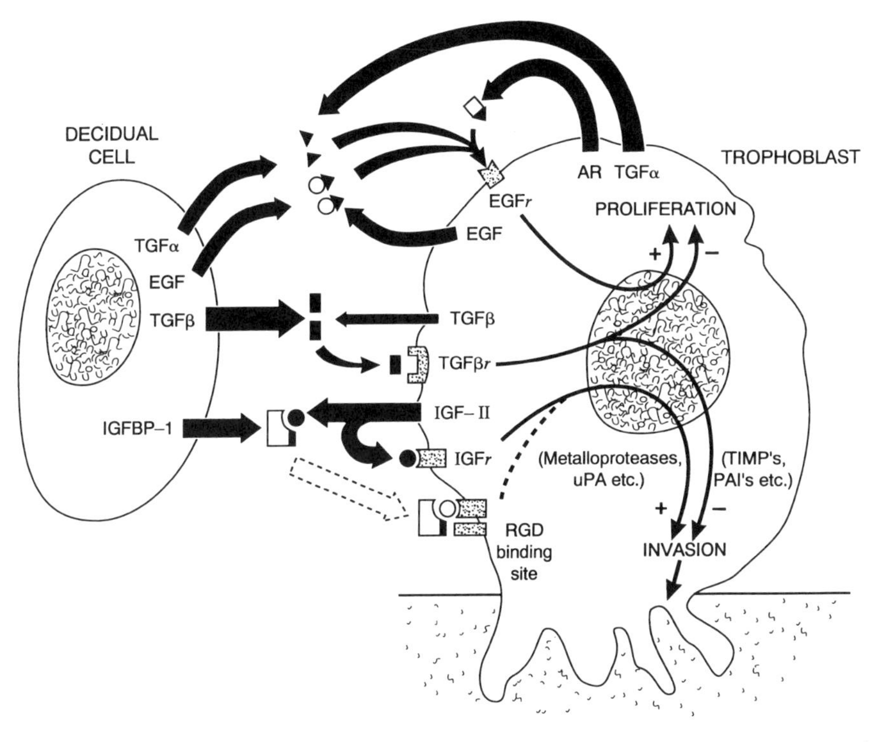


FIGURE 5.2. Schematic diagram of autocrine/paracrine regulation of trophoblast proliferation and invasion by locally produced growth factors. All EGF receptor ligands (EGF and TGFα: autocrine and paracrine; AR: autocrine) stimulate proliferation without affecting invasion. TGFβ (paracrine and to a small extent autocrine) down-regulates proliferation and invasion. IGF-II (autocrine) stimulates invasion with no effect on proliferation. IGFBP-1 augments IGF-II action, possibly by binding to RGD binding sites on the cell surface. Stimulation or inhibition of invasion results from a shift in the balance between degradative enzymes and their natural inhibitors.

thus potentiate binding of certain IGFs to their receptors (as illustrated in Fig. 5.2, discussed below).

## Localization of Growth Factors in the Human Placenta and Decidua

### *EGF and TGFα*

TGFα was first isolated from acid-ethanol extracts of term human placentas (62). Haining et al. (63) employed *reverse transcriptase-polymerase chain reaction* (RT-PCR) to demonstrate the presence of EGF and TGFα mRNA in the human endometrium and term decidua, although the exact cellular source of these peptides remains undetermined. They suggested that these peptides may be required for epithelial regeneration after menstruation and may also be mitogenic for trophoblast growth since trophoblast cells express EGF receptors (64–66). Immunohistochemical analysis of EGF in the human placenta and the decidua throughout gestation has localized EGF to uterine epithelial cells and decidual cells, as well as to both cytotrophoblast and syncytiotrophoblast (67).

Recently, mRNA, immunoreactive protein, and biologically active levels of TGFα and EGF were measured in pooled samples of human placenta from early, mid-, and late gestations (68). These experiments revealed that placentas contained high levels of TGFα and low levels of EGF mRNA—as well as proteins—throughout gestation. The levels of biologically active EGF receptor ligands (proteins able to compete with radiolabeled EGF for EGF receptors) were, however, similar in early and late gestational placentas. These investigators suggested that EGF receptor ligands produced by the placenta, most significantly TGFα and to a lesser extent EGF, may be required for placental development and function (68).

We have examined the cellular location of TGFα by an immunocytochemical method using an anti-TGFα monoclonal antibody (MF9, a gift from Dr. Kudlow, University of Alabama, Birmingham) in a large sample ($n = 29$) of normal human placentas and decidua obtained from different stages of gestation (20). Immunoreactive TGFα was identified in typical decidual cells of the decidua basalis, parietalis, and capsularis throughout gestation, whereas decidual leukocytes, inclusive of macrophages, were negative. Villous trophoblast cells (syncytiotrophoblast and cytotrophoblast) and extravillous cytotrophoblast cells (intermediate trophoblast and cytotrophoblastic shell) also exhibited TGFα immunoreactivity at all gestational ages examined. Filla et al. (69) have also immunolocalized TGFα in the human trophoblast and the decidua. Taken together, these results indicate that TGFα peptide is abundantly present at the human fetomaternal interface throughout gestation and thus may be functionally important. In the rat TGFα mRNA and protein have been localized in situ in the decidua (70, 71), but not in the trophoblast (70), indicating that the decidua is the primary source for this peptide in that species. The precise contribution of various cell types in the human placenta and the decidua to TGFα production remains to be determined.

## *Amphiregulin*

We have recently examined the location of immunoreactive AR in the human placenta and the decidua of various gestational ages (72) using a protein A/G purified rabbit polyclonal antibody, AR-Ab1 (41, 42), and the ABC-peroxidase technique on fixed, paraffin-embedded sections. Specific AR immunoreactivity was localized to the nucleus as well as to the cytoplasm of the syncytiotrophoblast cell layer of chorionic villi in early gestation until approximately week 18, after which time no staining was detectable. Other trophoblast cells—that is, villous cytotrophoblast and extravillous trophoblast cells—and decidual cells from all gestational ages examined showed no immunoreactivity. This unique spatial and temporal distribution of AR in the human placenta may be of biological significance early in human pregnancy (72).

## *TGFβ*

TGFβ was originally purified from normal-term human placentas (73). TGFβ mRNA has been detected in samples of total RNA isolated from human placentas at various stages of gestation, with peak expression at 17 and 34 weeks of gestation (74). Using a rabbit polyclonal anti-TGFβ antibody, these investigators also found immunoreactive TGFβ peptide confined to the syncytiotrophoblast layer of chorionic villi at various gestational ages (74). In situ hybridization studies indicate that syncytiotrophoblast cells also express TGFβ mRNA (Hunt, personal communication).

We have recently examined the ontogeny of the location of TGFβ in the human placenta and the decidua throughout gestation using a mouse monoclonal antibody, 1D11.16.8, capable of recognizing TGFβ1 and TGFβ2 (19, 75). Immunoreactive TGFβ was detected in the cytoplasm of the villous syncytiotrophoblast cell layer throughout gestation. Villous cytotrophoblast cells were negative at all gestational ages, whereas extravillous cytotrophoblast cells of the cytotrophoblastic shell (identified in semiserial sections by cytokeratin staining) showed strong cytoplasmic staining. The mesenchymal core of the chorionic villi also displayed moderate labeling throughout gestation. First-trimester decidual tissue showed intense TGFβ immunoreactivity in the *extracellular matrix* (ECM), with few decidual cells staining. Decidual leukocytes and glandular epithelium were negative. At later gestational ages the frequency of decidual cells with cytoplasmic staining increased, and the staining seen in the ECM declined. At term the majority of decidual cells showed cytoplasmic staining, and the ECM was negative (19, 75). Thus, TGFβ had a selective localization within the placenta. In the decidua TGFβ was mostly confined to the ECM during the first trimester, whereas at term it was mostly intracellular in location. This can be explained by a higher rate of synthesis and/or a slower rate of release of TGFβ by term decidual cells, as well as by a relative decrease in the amount of decidual ECM at term (19). We have further shown that decidua-derived TGFβ is released in its inactive form, whereas TGFβ released by trophoblast cells grown in culture is in its bioactive form (6). Type I and type II, as well as type III (betaglycan), receptors for TGFβ have been demonstrated in freshly isolated trophoblast cells and in first-trimester trophoblast cells in culture (76), indicating that locally derived TGFβ may be important for trophoblast function.

Recently, our laboratory has demonstrated the presence of decorin in the human placenta and the decidua throughout gestation. *Decorin* is a heparan-sulfate proteoglycan that has been shown to be a natural inhibitor of TGFβ (51, 52). A strong colocalization of decorin with TGFβ in the ECM of first-trimester decidua may indicate that decorin may be limiting the biological activity of TGFβ (77).

## *IGF-II*

IGF-II has been located in the human placenta and the decidua throughout gestation. Syncytiotrophoblast, cytotrophoblast, intermediate trophoblast, chorion, amnion, and decidual cells all had weak immunoreactivity (78). IGF-II mRNA is expressed in situ by the villous cytotrophoblast (79) and intermediate trophoblast cells (Han, personal communication) of the human placenta. Only the latter cells retain this expression throughout gestation, indicating that IGF-II production may be an important function of the invasive trophoblast (Han, personal communication). As mentioned earlier, IGFs may bind to a family of BPs. IGFBP-1 peptide is broadly distributed in the human placenta and the decidua (78, 80–82). However, IGFBP-1 mRNA is expressed only by the decidual cells (81 and Han, personal communication). Thus, the biological activity of IGF-II produced

TABLE 5.1. Summary of the location of growth factors in the human placenta and decidua.

| Growth factor | Placenta | Decidua |
|---|---|---|
| EGF | Villous cytotrophoblast and syncytiotrophoblast (protein) | Uterine epithelial and decidual cells (protein and mRNA) |
| TGFα | Villous cytotrophoblast, syncytiotrophoblast, and intermediate trophoblast (protein) | Uterine epithelial and decidual cells (protein and mRNA) |
| AR | Syncytiotrophoblast until week 18 (protein) | Not detected |
| TGFβ | Syncytiotrophoblast (protein and mRNA) and intermediate trophoblast (protein) | First trimester: primarily ECM, few decidual cells; later gestation: decidual cells (protein) |
| IGF-II | All trophoblast subsets (protein), first-trimester villous cytotrophoblast, and intermediate trophoblast of all ages (mRNA) | Decidual cells (protein) (IGFBP-1 mRNA in decidual cells adjacent to intermediate trophoblast) |

by the invasive intermediate trophoblast cells may be influenced in situ by interaction with IGFBP-1 produced by decidual cells in their vicinity (Fig. 5.2). Table 5.1 provides a summary of the location of the above-listed growth factors and their mRNA in the human placenta and the decidua.

## Propagation and Characterization of Pure First-Trimester Invasive Human Trophoblast Cells

An understanding of the roles of specific growth factors on placental growth and invasion had largely been hindered by methodological limitations in propagating pure human trophoblast cells in culture. These limitations have recently been overcome in our laboratory by the development of an explant culture method (83) that has been further refined (19). This method involves the culturing of mechanically derived fragments of first-trimester chorionic villi, which allows proliferative and invasive trophoblast cells to migrate out of the these explants. The migrant adherent trophoblast cells are subsquently passaged. Approximately 30% of these villous explants result in migrant cells containing 100% pure trophoblast cells (as identified by cytokeratin immunostaining). The remaining explants give rise to migrant cells representing a mixture of fibroblasts and trophoblasts (as identifed by cytokeratin and vimentin immunostaining, respectively) (84) with either cell type predominating. Pure normal trophoblast cell lines have been derived from a passage of cells from the 100% cytokeratin-positive cultures.

Early passage trophoblast cells used for functional studies were also extensively phenotyped for a number of other markers. One hundred percent of these cells expressed placental alkaline phosphatase, *urokinase-type plasminogen activator* (uPA) (19), high-affinity uPA receptors (85), and HLA antigen framework region (as recognized by W6/32 antibody) (86). These cells were negative for 63D3 antigen (a macrophage marker) and factor VIII (an endothelial cell marker) (19). Approximately half of the cells expressed NDOG-5 antigen (86), a marker reported to be specific for intermediate trophoblast cells (87). During early passages hCG production declines rapidly (19); however, a minority of the cells show hPL immunoreactivity after several passages (86). Finally, these cells are immunoreactive for IGF-II peptide and express IGF-II mRNA as demonstrated by Northern analysis and in situ hybridization (86). Human intermediate trophoblast cells in situ also express high levels of IGF-II mRNA throughout gestation (Han, personal communication). Taken together, the phenotype of the trophoblast cells grown in culture indicates that they belong to the invasive intermediate trophoblast subpopulation.

The trophoblast cells propagated in vitro senesce after 12–14 passages. However, our laboratory has recently succeeded in extending their life span after transfection with the SV40 large T-antigen (88).

## Effects of Local Growth Factors on Trophoblast Proliferation

The effects of endogenous or exogenous growth factors on the proliferation of early passage first-trimester trophoblast cells were measured from *$^{3}$H-thymidine* ($^{3}$H-TdR) incorporation by cells exposed to $^{3}$H-TdR during the last 6h of 24-h culture in the presence of specific growth factors or their neutralizing antibodies. All EGF receptor ligands, EGF, TGFα, and AR, stimulated $^{3}$H-TdR uptake in trophoblast cells in a dose-dependent manner (20, 72, 89). The growth-promoting functions of EGF (Lysiak, Lala, unpublished observations) and TGFα (20) were also confirmed from a significant increase in the incidence of nuclei immunoreactive for the *proliferating cell nuclear antigen* (PCNA). These studies have recently been corroborated by Filla et al. (69).

To investigate if trophoblast cells in vitro were responding to endogenous TGFα production, $^{3}$H-TdR incorporation was also measured in the presence of a neutralizing anti-TGFα antibody (20). While this antibody neutralized the growth-promoting effects of exogenous TGFα, the antibody alone had no effect on trophoblast proliferation. This finding may be explained in two ways: (i) Trophoblast cells maintained in culture by the present methodology did not produce significantly high levels of TGFα to affect their proliferation; or (ii) TGFα produced by these cells may be in a form not neutralized by the currently used antibody (20). A dramatic decline in trophoblast proliferation, however, was observed when cells were cultured in the presence of an EGF receptor-blocking antibody alone. This effect was not reversed by the addition of EGF or TGFα to the cultures. Thus, proliferation of first-trimester trophoblast cells maintained in vitro appears to be dependent on an autocrine production of one or more of the EGF receptor ligands; for example, EGF, TGFα, AR, cripto, or heparin-binding EGF (89).

The effects of exogenous and endogenous TGFβ were also studied on the proliferation of first-trimester human trophoblast cells (19). The addition of increasing doses of TGFβ resulted in a significant dose-dependent decline in $^{3}$H-TdR incorporation, which was abrogated with a neutralizing anti-TGFβ antibody. The antibody alone led to a small increase in trophoblast proliferation, suggesting the presence of endogenous TGFβ (19). These results indicate that TGFβ provides an antiproliferative signal to first-trimester trophoblast cells and that trophoblast themselves produce small amounts of TGFβ in vitro (19). Increasing doses of IGF-II added to the trophoblast cultures had no significant effect on $^{3}$H-TdR uptake (90).

## Effects of TGFα and TGFβ on Trophoblast Multinucleate Cell Formation

In vivo both villous and extravillous trophoblast cells may differentiate into multinucleate cells, villous syncytiotrophoblast, and placental bed giant cells, respectively. The multinucleate cells formed in first-trimester invasive trophoblast cell cultures are possibly the in vivo equivalents of placental bed giant cells. Since TGFα and TGFβ provide positive and negative signals, respectively, for the proliferation of first-trimester trophoblast cells in culture, we tested whether these growth factors influenced the incidence of multinucleate cell formation in vitro.

Cultures were treated with TGFα or TGFβ and their respective neutralizing antibodies for 72 h and subsequently scored for the incidence of multinucleate cells (19, 20). Exogenous TGFα decreased and TGFβ increased the incidence of multinucleate cells in these cultures. While the addition of anti-TGFα neutralizing antibody had no effect on multinucleate cell formation (20), anti-TGFβ neutralizing antibody significantly reduced the number of multinucleated cells, indicating the presence of endogenous TGFβ in the trophoblast cultures (19). Thus, both TGFs were able to influence first-trimester trophoblast differentiation. In the case of TGFα, this may be either a direct effect by inhibiting trophoblast cell fusion or an indirect effect by stimulating the proliferation of mononuclear cells (20). On the other hand, since TGFβ induced the formation of multinucleate cells in both first-trimester and term trophoblast—and the latter cells sparsely divided—this enhancement of multinucleate cell formation is not necessarily linked with the antiproliferative function of TGFβ (19).

## Role of Local Growth Factors on Trophoblast Invasion

Studies in our laboratory were designed to answer the following questions: (i) Is invasiveness an inherent property of human trophoblast cells independent of the uterine microenvironment? (ii) If so, do they lose their invasive ability later in gestation in a preprogrammed manner, or is their invasiveness controlled by the uterine microenvironment? (iii) Is trophoblast invasion or its control dependent on growth factors produced in an autocrine or paracrine manner? and (iv) What are the molecular mechanisms responsible for trophoblast invasiveness and its control?

Invasion is a property shared by certain normal cells—in particular, embryonic cells—and malignant tumor cells. Tumor biologists have employed a variety of in vitro invasion assays using natural tissues—for example, chick chorioallantoic membrane (91, 92), chick embryonic heart fragments (93), bovine lens capsule (94), and epithelium-free human amnion (95–97)—or reconstituted gels of basement membrane components, such as matrigel (98), to measure the invasiveness of tumor cells. Our laboratory initially employed an amnion invasion assay (6, 97) and later a matrigel invasion assay (7, 8, 88, 89) to measure the invasive ability of first-trimester trophoblast cells in culture. In both assays trophoblast cells were prelabeled with *$^{125}$I-deoxyuridine* ($^{125}$I-dUR) or $^{3}$H-TdR and placed on the invasion substrate. In the amnion invasion assay, the invasion at a particular time point was then measured as the percent of radioactivity retained in the body of the amniotic membrane resulting from the radiolabeled cells in transit. In the matrigel invasion assay, this was measured as the percent of radioactivity accumulating in a lower well because of transgression of the matrigel barrier by the radioactive cells.

Using both assays we have shown that first-trimester human trophoblast cells in culture are highly invasive cells (6–8, 97) and that their invasive ability is comparable to highly invasive malignant cells inclusive of choriocarcinomas (97, 99). Other investigators have also demonstrated the invasive ability of first-trimester human trophoblast cells isolated by different techniques (100–103). These results show that the uterine microenvironment is not required for their invasive function. Indeed, it had earlier been shown that mouse ectoplacental cones transplanted in ectopic sites—for example, under the kidney capsule, the spleen, or the testis—rapidly invaded these tissues (104–107).

What are the mechanisms underlying trophoblast invasion? The ability of cells to invade a tissue barrier, such as basement membrane or ECM, is a multistep process whereby cells must (i) initially attach to the substrate, (ii) detach from the substrate and be capable of migration, and (iii) degrade the matrix by producing matrix-degrading enzymes in order to migrate through the lysed matrix. Basement membranes commonly consist of type IV collagen, laminin, fibronectin, entactin, and heparan sulfate proteoglycan (108). Binding to laminin may be an important step for trophoblast invasion of a basement membrane. Our laboratory has shown that trophoblast cells in culture produce laminin (97). Loke et al. (109) noted that trophoblast cells adhere to laminin-coated substrates better than to plastic surfaces. Damsky et al. (110) reported that the $\alpha1/\beta1$ integrin, the laminin receptor, is expressed by the invasive intermediate trophoblast cells embedded in the uterine wall. It has also been reported that when normal and malignant trophoblast cells bind to laminin, there is a stimulation of type IV collagenase activity (111) that is required for the degradation of type IV collagen.

Interaction with fibronectin may also be important for trophoblast invasion of the ECM. Damsky et al. (110) have demonstrated that the invasive intermediate trophoblast cells distant from the villi express the integrin $\alpha5/\beta1$ that binds fibronectin. Using immunoelectron microscopy, we have recently demonstrated that migrant trophoblast cells in primary explant cultures of first-trimester human chorionic villi produce and secrete fibronectin, as well as oncofetal fibronectin (112). This observation is in agreement with in situ immunolabeling of extravillous trophoblast and its surrounding ECM for fibronectin and oncofetal fibronectin (113, 114). These results suggest that trophoblast cell production and binding to fibronectin may be important for their migratory or invasive behavior.

We have recently shown that invasive first-trimester trophoblast cells, similar to metastatic cancer cells, express complex-type Asn-linked oligosaccharides on their cell surface that are essential for their migratory ability (115, 116). Alteration of the structure of these oligosaccharides by culturing cells in the presence of swainsonine (which alters the glycosylation process) abrogated their ability to detach from a basement membrane substrate and, hence, their invasive capability (115, 116).

A variety of matrix-degrading enzymes have been identified in the body. Of these, two major classes have been shown to be important for tumor and trophoblast cell invasiveness: (i) metalloproteases—in particular, type IV collagenases; and (ii) serine proteases—in particular, uPA. Blocking metalloprotease activity with either a chemical inhibitor (e.g., 1,10-phenanthroline) or a natural tissue inhibitor of metalloproteases (TIMP-1) abrogated trophoblast invasiveness. Blocking of plasminogen activation with aprotinin (Trasylol) or anti-uPA antibody also abrogated their invasive ability (97). It appears that activation of collagenase is largely plasmin dependent in the case of both trophoblast (97) and tumor cells (96, 97).

Experiments using primary cultures of enzyme-dispersed trophoblast have suggested that the 92-kd type IV collagenase is essential for first-trimester trophoblast invasion (103). However, we have noted that the 72-kd type IV collagenase is the major metalloprotease produced by first-trimester trophoblast cells propagated in culture (6). While these differences may have resulted from the employment of different trophoblast cell subpopulations, recent experiments testing the anti-invasive function of antisense oligonucleotides targeted against various domains of these collagenase types indicate that both forms of type IV collagenases may be important for trophoblast invasion (117).

First-trimester trophoblast cells in culture produce uPA (97) and express high-affinity receptors for uPA that remain saturated by the endogenous uPA (85). Since the binding of uPA to its receptor does not block the catalytic function of uPA, it is possible that this provides a mechanism for the trophoblast to express an "invasive front" on the cell membrane (5).

There exist natural inhibitors for both metalloproteases and serine proteases. *Tissue inhibitors of metalloproteases* (TIMPs) are molecules produced by most cells that are capable of producing metalloproteases (118). They can inactivate metalloproteases by binding at a equimolar ratio. Two forms of TIMPs have been described: TIMP-1 (119) and TIMP-2 (120). We have found that both TIMP-1 and TIMP-2 mRNA are expressed by first-trimester trophoblast cells (6, 89). Thus, trophoblast type IV collagenase activity represents a balance between the production of type IV collagenase and TIMPs. The uPA can be inactivated by its natural inhibitors *plasminogen activator inhibitor I* (PAI-I) and PAI-II. While both inhibitors have been localized immunocytochemically to the trophoblast cell membrane, PAI-I has been selectively localized on the invasive trophoblast in situ (121). Thus, trophoblast uPA activity must result from a balance between uPA production and PAI production by these cells.

What controls trophoblast invasion of the uterus in situ? We found that primary cultures of term trophoblast cells isolated by enzyme dispersion, followed by density gradient purification (122), had significant invasive ability, indicating that their invasion was not lost in a preprogrammed manner (7). These findings are in agreement with Kliman and Feinberg (100) and suggest that trophoblast invasion must be controlled by the uterine microenvironment. Indeed, Kirby had shown in the mouse model that trophoblast invasion following blastocyst transfer was higher in the nondecidualized uterus than in the decidualized uterus, indicating that decidua may be responsible for the control of invasion (107).

We tested whether a decidua-derived soluble factor(s) influenced trophoblast invasiveness in vitro (6). Cell-free conditioned media from first-trimester human decidua were found to abrogate trophoblast invasiveness, and this effect was reversed partially or totally in the presence of neutralizing antibodies against TIMP-1 or TGFβ, indicating that both molecules produced by the decidua had anti-invasive effects (6). These antibodies, when added alone, stimulated trophoblast invasion beyond control levels, indicating that trophoblast cells themselves make these molecules.

Chemically pure TGFβ also blocked trophoblast invasion and reduced the type IV collagenase activity of the trophoblast cells in vitro (6). Since this activity represents a balance between the production of type IV collagenase and TIMPs, a Northern analysis of mRNA for these molecules was carried out in the presence of TGFβ or its neutralizing antibody. TGFβ caused a significant increase in TIMP-1 mRNA, while the antibody caused a reduction of the expression (6). We further showed that human decidual cell culture supernatants had high levels of TGFβ secreted in an inactive form and possibly activated in the invasion assays by trophoblast-derived enzymes, such as plasmin (6, 19). Thus, locally produced TGFβ (primarily by the decidua and to a minor extent by the trophoblast) appears to provide the primary mode of invasion control in vivo by up-regulating TIMP-1 in the trophoblast.

Are trophoblast proliferation and invasion linked biological events? This question was raised because both the proliferative and invasive functions were inhibited by TGFβ. Since all EGF receptor ligands stimulated trophoblast growth, we tested whether EGF and TGFα also promoted trophoblast invasion in vitro. We found that neither of these molecules had any effect on first-trimester trophoblast invasion when tested at doses that caused significant stimulation of proliferation (89). Northern analysis of 72-kd type IV collagenase and TIMP-1 mRNA revealed that both mRNAs were up-regulated to a similar extent by EGF or TGFα (89). Thus, there was no net shift in the balance between the production of type IV collagenase and its inhibitor, TIMP-1, that would affect invasion. These results further support the view that trophoblast proliferation and invasion are functions that are not necessarily linked.

Is the invasive property of trophoblast cells dependent on an autocrine growth factor(s)? Because of the interesting observation that IGF-II mRNA is selectively expressed by the invasive intermediate trophoblast cells in situ throughout gestation (Han, personal communication) and that IGFBP-1 mRNA is selectively expressed by the decidual cells adjacent to the intermediate trophoblast (Han, personal communication), we investigated the role of these molecules on trophoblast invasion (90). Matrigel invasion by first-trimester human trophoblast cells was stimulated in a dose-dependent manner with increasing doses of IGF-II when the invasion assay was carried out in low-serum (1% FCS) medium (90). This stimulation was further enhanced in a synergistic manner in the presence of IGFBP-1, even at minute (0.1 nM) doses (Lysiak, Lala, unpublished observations). This synergy may have resulted from an increased affinity or stability of IGF-II binding to its receptor when complexed with IGFBP-1. The latter is a likely possibility since IGFBP-1 contains an RGD sequence (58) that can bind to RGD binding sites on the cell surface. Such binding sites are known to be a part of cell surface integrin molecules.

The precise mechanisms by which IGF-II stimulates first-trimester trophoblast invasion are currently under study. The finding that IGF-II stimulates trophoblast invasion without influencing proliferation further substantiates the view that the two processess are not linked. Figure 5.2 presents a schematic diagram of proposed autocrine and paracrine mechanisms of regulation of trophoblast invasion and proliferation by locally derived growth factors.

## Summary

Trophoblast proliferation, differentiation, and invasion are key events during normal human placental growth and development. Our studies suggest that locally produced growth factors regulate trophoblast growth and invasion in a positive as well as a negative manner by autocrine and/or paracrine pathways. The EGF receptor ligands stimulate trophoblast growth in both autocrine and paracrine modes without affecting invasion. TGFβ produced by the decidua—and to a minor extent by the trophoblast—has a dual role: down-regulation of trophoblast growth and invasion. IGF-II produced by the invasive intermediate trophoblast cells stimulates their invasiveness without affecting growth. The mechanisms of the invasion-regulating functions of these growth factors may be multiple. Of these, up- or down-regulation of the genes for the matrix-degrading enzymes or their natural inhibitors appears to be important. The differential effects of certain growth factors on trophoblast proliferation and invasive ability suggest that the two processes are not linked. Thus, normal placental growth and development must depend on the temporal and spatial regulation of the local production of growth factors in the placenta and the decidua, on expression of growth factor receptors on specific cellular subsets, and on cellular responses resulting from the interactions of growth factors and their receptors. Dysregulation of one or more of these events may be relevant to pathological conditions of abnormal placental growth and invasion; for example, choriocarcinoma and preeclampsia.

## *References*

1. Hertig AT, Rock J. Two human ova of the previllous stage, having a developmental age of about seven and nine days respectively. Contrib Embryol Carnegie Inst 1945;31:65–84.
2. Boyd JD, Hamilton WJ. Development and structure of the human placenta from the end of the third month of gestation. J Obstet Gynaecol Br Commonw 1967;74:161–226.
3. Loke YW. Experimenting with human extravillous trophoblast: a personal view. Am J Reprod Immunol 1990;24:21–8.
4. Lala PK. Similarities between immunoregulation in pregnancy and in malignancy: the role of prostaglandin E2. Am J Reprod Immunol 1990;20:147–52.
5. Lala PK, Graham CH. Mechanisms of trophoblast invasiveness and their control: the role of proteases and protease inhibitors. Cancer Metastasis Rev 1990;9:369–79.
6. Graham CH, Lala PK. Mechanism of control of trophoblast invasion in situ. J Cell Physiol 1991;148:228–34.
7. Graham CH, Lala PK. Mechanisms of placental invasion of the uterus and their control. Biochem Cell Biol 1992;70:867–74.
8. Graham CH, McCrae KR, Lala PK. Molecular mechanisms controlling trophoblast invasion of the uterus. Troph Res 1993;7:237–50.
9. Kearns M, Lala PK. Characterization of hematogenous cellular constituents of the murine decidua: a surface marker study. J Reprod Immunol 1985; 8:213–34.
10. Zhinken LN, Samoskina NA. DNA synthesis and cell proliferation during the formation of deciduomata in mice. J Embryol Exp Morphol 1967; 17:593–605.
11. Galassi L. Autoradiographic study of the decidual cell reaction in the rat. Dev Biol 1968;17:75–84.
12. Das RM, Martin L. Uterine DNA synthesis and cell proliferation during early decidualization induced by oil in mice. J Reprod Fertil 1978;53:125–8.

13. Bell SC. Decidualization and associated cell types: implications for the role of the placenta bed in the materno-fetal immunological relationship. J Reprod Immunol 1983;5:185–94.
14. Kearns M, Lala PK. Bone marrow origin of decidual cell precursors in the pseudopregnant mouse uterus. J Exp Med 1982;155:1537–54.
15. Johnson S, Graham CH, Lysiak JJ, Lala PK. Hemopoietic origin of certain decidual cell precursors in murine pregnancy. Am J Anat 1989;185(1): 9–18.
16. Lysiak JJ, Lala PK. In situ localization and characterization of bone marrow-derived cells in the decidua of normal murine pregnancy. Biol Reprod 1992;47:603–13.
17. Riddick DH, Kusmik WF. Decidua: a possible source of amniotic fluid prolactin. Am J Obstet Gynecol 1976;127:187–90.
18. Kubota T, Kumasaka T, Yaoi Y, Suzuki A, Saito M. Study on immunoreactive prolactin of decidua in early pregnancy. Acta Endocrinol (Copenh) 1981;960:2580–640.
19. Graham CH, Lysiak JJ, McCrae KR, Lala PK. Localization of transforming growth factor-β at the human fetal-maternal interface: role in trophoblast growth and differentiation. Biol Reprod 1992;46:561–72.
20. Lysiak JJ, Han VKM, Lala PK. Localization of transforming growth factor-α (TGF-α) in the human placenta and decidua: role in trophoblast growth. Biol Reprod 1993.
21. Parhar RS, Kennedy TG, Lala PK. Suppression of lymphocyte alloreactivity by early gestational human decidua, I. Characterization of suppressor cells and suppressor molecules. Cell Immunol 1988;116:392–410.
22. Lala PK, Kennedy TG, Parhar RS. Suppression of lymphocyte alloreactivity by early gestational human decidua, II. Characterization of suppressor mechanisms. Cell Immunol 1988;116:411–22.
23. Lala PK, Scodras JM, Graham CH, Lysiak JJ, Parhar RS. Activation of maternal killer cells in the pregnant uterus with chronic indomethacin therapy, IL-2 therapy or a combination therapy is associated with embryonic demise. Cell Immunol 1990;127:368–81.
24. Scodras JM, Parhar RS, Kennedy TG, Lala PK. Prostaglandin-mediated inactivation of natural killer cells in the murine decidua. Cell Immunol 1990;127:352–67.
25. Parhar RS, Yagel S, Lala PK. PGE2-mediated immunosuppression by first trimester human decidual cells blocks activation of maternal leukocytes in the decidua with potential anti-trophoblast activity. Cell Immunol 1989; 120:61–74.
26. Tawfik OW, Hunt JS, Wood GW. Implication of prostaglandin E in soluble factor-mediated immune suppression by murine decidual cells. Am J Reprod Immunol Microbiol 1986;12:111–7.
27. Mathews CJ, Searle RF. The role of prostaglandins in the immunosuppressive effects of supernatants from adherent cells of murine decidual tissue. J Reprod Immunol 1987;12:109–24.
28. Clark DA, Flanders KC, Banwatt D, et al. Murine pregnancy decidua produces a unique immunosuppressive molecule related to transforming growth factor β-2. J Immunol 1990;144:3008–14.

29. Carpenter G, Wahl MI. The epidermal growth factor family. In: Sporn MB, Roberts AB, eds. Peptide growth factors and their receptors I. New York: Springer-Verlag, 1991:69–171.
30. Cohen S. Isolation of a mouse submaxillary gland protein accelerating incisor eruption and eyelid opening in the new-born animal. J Biol Chem 1962;237:1555–62.
31. Bell GI, Fong NM, Stempie NM, et al. Human epidermal growth factor precursor: cDNA sequence, expression in vitro and gene organization. Nucleic Acids Res 1986;14:8427–46.
32. Teixido J, Wong ST, Lee DC, Massague J. Generation of transforming growth factor-α from the cell surface by an O-glycosylation-independent multistep process. J Biol Chem 1990;265:6410–5.
33. Bringman TS, Lindquist PB, Derynck R. Different transforming growth factor-α species are derived from a glycosylated and palmitoylated transmembrane precursor. Cell 1987;48:429–40.
34. Brachmann R, Lindquist PB, Nagashima N, et al. Transmembrane TGF-α precursors activate EGF/TGF-α receptors. Cell 1989;56:691–700.
35. Lee DC, Rockford R, Todaro GJ, Villarreal LP. Developmental expression of rat transforming growth factor-α mRNA. Mol Cell Biol 1985;5:3644–6.
36. Wong ST, Winchell LF, McCune BK, et al. The TGF-α precursor expressed on the cell surface binds to the EGF receptor on adjacent cells, leading to signal transduction. Cell 1989;56:495–506.
37. Massague J. The transforming growth factor-β family. Annu Rev Cell Biol 1990;6:597–641.
38. Shoyab M, McDonald VL, Bradley JG, Todaro GJ. Amphiregulin: a bifunctional growth-modulating glycoprotein produced by the phorbol 12-myristate 13-acetate-treated human breast adenocarcinoma cell line MCF-7. Proc Natl Acad Sci USA 1988;85:6528–32.
39. Plowman GD, Green JM, McDonald VL, et al. The amphiregulin gene encodes a novel epidermal growth factor-related protein with tumor-inhibitory activity. Mol Cell Biol 1990;10(5):1969–81.
40. Shoyab M, Plowman GD, McDonald VL, Bradley JG, Todaro GJ. Structure and function of human amphiregulin: a member of the epidermal growth factor family. Science 1989;243:1074–6.
41. Johnson GR, Saeki T, Auersperg N, et al. Response to and expression of amphiregulin by ovarian carcinoma and normal ovarian surface epithelial cells: nuclear localization of endogenous amphiregulin. Biochem Biophys Res Commun 1991;180:481–8.
42. Johnson GR, Saeki T, Gordon AW, Shoyab M, Salomon DS, Stromberg K. Autocrine action of amphiregulin in a colon carcinoma cell line and immunocytochemical localization of amphiregulin in human colon. J Cell Biol 1992;118:741–51.
43. Derynck R, Jarret JA, Chen EY, et al. Human transforming growth factor-β complementary DNA sequence and expression in normal and transformed cells. Nature 1985;316:701–5.
44. Pircher R, Jullien P, Lawrence DA. β-Transforming growth factor is stored in human blood platelets as a latent high molecular weight complex. Biochem Biophys Res Commun 1986;136:30–7.

45. Miyazono K, Heldin CH. Interaction between TGF-β1 and carbohydrate structures in its precursor renders TGF-β1 latent. Nature 1989;388:158–60.
46. Miyazono K, Yuki K, Takaku F, et al. Latent forms of TGF-beta: structure and biology. Ann NY Acad Sci 1990;593:51–8.
47. Wakefield LM, Smith DM, Flanders KC, Sporn MB. Latent transforming growth factor-β from human platelets. J Biol Chem 1988;263:7646–54.
48. Lyons RM, Keski-Oja J, Moses HL. Proteolytic activation of latent transforming growth factor-β from fibroblast-conditioned medium. J Cell Biol 1988;106:1659–65.
49. Wakefield LM, Smith DM, Masui T, Harris CC, Sporn MB. Distribution and modulation of the cellular receptor for transforming growth factor-beta. J Cell Biol 1987;105:965–75.
50. O'Connor-McCourt MD, Wakefield LM. Latent transforming growth factor-β in serum. J Biol Chem 1987;262:14090–9.
51. Yamaguchi Y, Mann DM, Ruoslahti E. Negative regulation of transforming growth factor-β by the proteoglycan decorin. Nature 1990;346:281–4.
52. Border WA, Noble NA, Yamamoto T, et al. Natural inhibitor of transforming growth factor-β protects against scarring in experimental kidney disease. Nature 1992;360:361–4.
53. Cheifetz S, Bassols A, Stanley K, Ohta M, Greenberger J, Massague J. Heterodimeric transforming growth factor-β. Biological properties and interaction with three types of cell surface receptors. J Biol Chem 1988; 263:10783–9.
54. Jennings JC, Mohan S, Linkhart TA, Widstorm R, Baylink DJ. Comparison of the biological activities of TGF-β1 and TGF-β2: differential activity in endothelial cells. J Cell Physiol 1988;137:167–72.
55. Rechler MM, Nissley SP. Insulin-like growth factors. In: Sporn MB, Roberts AB, eds. Peptide growth factors and their receptors I. New York: Springer-Verlag, 1991:263–368.
56. Kiess W, Blickenstaff GD, Sklar MM, Thomas CL, Nissley SP, Sahagian GG. Biochemical evidence that the type II insulin-like growth factor is identical to the cation-independent mannose 6-phosphate receptor. J Biol Chem 1988;263:9339–44.
57. Mathieu M, Rochefort H, Barenton B, Prebois C, Vignon F. Interactions of cathepsin-D and insulin-like growth factor-II (IGF-II) on the IGF-II-mannose-6-phosphate receptor in human breast cancer cells and possible consequences on mitogenic activity of IGF-II. Mol Endocrinol 1990;4: 1327–35.
58. Shimasaki S, Ling N. Identification and molecular characterization of insulin-like growth factor binding proteins (IGFBP-1, -2, -3, -4, -5 and -6). Prog Growth Factor Res 1991;3:243–66.
59. DeMellow JSM, Baxter RC. Growth hormone-dependent insulin-like growth factor (IGF) binding protein both inhibits and potentiates IGF-I-stimulated DNA synthesis in human skin fibroblasts. Biochem Biophys Res Commun 1988;156:199–204.
60. Clemmons DR. Insulin-like growth factor binding proteins' roles in regulating IGF physiology. J Dev Physiol 1991;15:105–10.
61. McCusker RH, Clemmons DR. The insulin-like growth factor binding proteins: structure and biological functions. In: Schofield P, ed. The insulin-

like growth factors: structure and biological functions. Oxford: Oxford University Press 1992:110–50.
62. Stromberg K, Pigott DA, Ranchalis JE, Twardzik DR. Human term placenta contains transforming growth factors. Biochem Biophys Res Commun 1982; 106:354–61.
63. Haining REB, Schofield JP, Jones DSC, Rajput-Williams J, Smith SK. Identification of mRNA for epidermal growth factor and transforming growth factor-α in low copy number in human endometrium and decidua using reverse transcriptase-polymerase chain reaction. J Mol Endocrinol 1991; 6:207–14.
64. Lai WH, Guyda H. Characterization and regulation of epidermal growth factor receptors in human placental cell cultures. J Clin Endocrinol Metab 1984;58:344–52.
65. Chen CF, Kurachi H, Fujita Y, Terakawa N, Miyake A, Tanizawa. Changes in epidermal growth factor receptor and its messenger ribonucleic acid levels in human placenta and isolated trophoblast cells during pregnancy. J Clin Endocrinol Metab 1988;67(6):1171–7.
66. Muhlhauser J, Crescimanno C, Kaufmann P, Hofler H, Zaccheo D, Castellucci M. Differentiation and proliferation patterns in human trophoblast revealed by c-erbB-2 oncogene product and EGF-R. J Histochem Cytochem 1993;41(2):165–73.
67. Hofmann GE, Scott RT Jr, Bergh PA, Deligdisch L. Immunohistochemical localization of epidermal growth factor in human endometrium, decidua, and placenta. J Clin Endocrinol Metab 1991;73(4):882–7.
68. Bissonnette F, Cook C, Geoghegan T, et al. Transforming growth factor-α and epidermal growth factor messenger ribonucleic acid and protein levels in human placentas from early, mid, and late gestation. Am J Obstet Gynecol 1992;166:192–9.
69. Filla MS, Zhang CX, Kaul KL. A potential transforming growth factor α/epidermal growth factor receptor autocrine circuit in placental cytotrophoblasts. Cell Growth Differ 1993;4:387–93.
70. Han VK, Hunter ES, Pratt RM, Zendegui JG, Lee DC. Expression of rat transforming growth factor α mRNA during development occurs predominantly in the maternal decidua. Mol Cell Biol 1987;7:2335–43.
71. Bonvissuto AC, Lala PK, Kennedy TG, Nygard K, Lee DC, Han VKM. Induction of transforming growth factor-α expression in rat decidua is independent of the conceptus. Biol Reprod 1992;46:607–16.
72. Lysiak JJ, Johnson GR, Lala PK. Localization and function of amphiregulin in the human placenta [Abstract]. Proc Can Fed Biol Soc 1993;36:60.
73. Frolick CA, Dart LL, Meyers CA, Smith DM, Sporn MB. Purification and initial characterization of a type β transforming growth factor from human placenta. Proc Natl Acad Sci USA 1983;80:3676–80.
74. Dungy LJ, Siddiqi TA, Khan S. Transforming growth factor-β expression during placental development. Am J Obstet Gynecol 1991;4(1):853–7.
75. Lysiak JJ, Graham C, Riley S, Johnson G, Lala PK. Localization of transforming growth factor β (TGFβ) and amphiregulin in the human placenta and decidua throughout gestation [Abstract]. Am J Reprod Immunol 1992; 27(1/2):46.

76. Mitchell EJ, Fitz-Gibbon L, O'Connor-McCourt MD. Subtypes of betaglycan and of type I and type II transforming growth factor-β (TGF-β) receptors with different affinities for TGF-β1 and TGF-β2 are exhibited by human placental trophoblast cells. J Cell Physiol 1992;150:334–43.
77. Lysiak JJ, Pringle GA, Lala PK. Immunolocalization of TGF-β and decorin (its natural inhibitor) in the human placenta and decidua. Placenta 1993.
78. Hill DJ, Clemmons DR, Riley SC, Bassett N, Challis JRG. Immunohistochemical localization of insulin-like growth factors (IGFs) and IGF binding proteins-1, -2, and -3 in human placenta and fetal membranes. Placenta 1993;14:1–12.
79. Brice AL, Cheetham JE, Bolton VN, Hill NCW, Schofield PN. Temporal changes in the expression of the insulin-like growth factor II gene associated with tissue maturation in the human fetus. Development 1989;106:543–54.
80. Rutanen EM, Pekonen F, Makinen T. Soluble 34K binding protein inhibits the binding of insulin-like growth factor I to its cell receptors in human secretory phase endometrium: evidence for autocrine/paracrine regulation of growth factor action. J Clin Endocrinol Metab 1988;66:173–80.
81. Rutanen EM, Partanen S, Pekonen F. Decidual transformation of human extrauterine mesenchymal cells is associated with the appearance of insulin-like growth factor-binding protein-1. J Clin Endocrinol Metab 1991;72(1): 27–31.
82. Waites GT, James RFL, Bell SC. Human pregnancy-associated endometrial α-globin, an insulin-like growth factor-binding protein: immunohistological localization in the decidua and placenta during pregnancy employing monoclonal antibodies. J Endocrinol 1989;120:351–7.
83. Yagel S, Casper RF, Powel W, Parhar RS, Lala PK. Characterization of pure human first trimester cytotrophoblast cells in long term culture: growth pattern, markers, and hormone production. Am J Obstet Gynecol 1989; 160:938–45.
84. Irving JA, Lysiak JJ, Han VKM, Lala PK. Properties of trophoblast cells growing out of first trimester human chorionic villus explants prior to their propagation. Placenta 1993.
85. Zini JM, Murray SC, Graham CH, et al. Identification and characterization of urokinase receptors expressed by human trophoblasts. Blood 1992;79: 2917–29.
86. Graham CH, Lysiak JJ, Irving JA, et al. Characteristics of first trimester normal human trophoblast cells propagated in culture. Placenta 1993.
87. Shorter SC, Starkey PM, Ferry BL, Sargent IL, Redman CWG. Isolation of cell island cytotrophoblast from first trimester human placenta and characterization of NDOG5, a monoclonal antibody specific for a human trophoblast subpopulation. Placenta 1991;12(4):434–5.
88. Graham CH, Hawley TS, Hawley RG, et al. Establishment and characterization of first trimester human trophoblast cells with extended lifespan. Exp Cell Res 1993;206:204–11.
89. Lysiak JJ, Connelly IH, Khoo NKS, Stetler-Stevenson W, Lala PK. Role of transforming growth factor-α (TGF-α) and epidermal growth factor (EGF) on proliferation and invasion by first trimester human trophoblast. Troph Res 1993.

90. Lysiak JJ, Han VKM, Lala PK. Role of insulin-like growth factor (IGF)-II on human first trimester trophoblast cell growth and invasion. Placenta 1993.
91. Hart IR, Fidler IJ. An in vitro assay for tumour cell invasion. Cancer Res 1978;38:3218–24.
92. Chambers AF, Shafir R, Ling V. A model system for studying metastasis using the embryonic chick. Cancer Res 1982;42:4018–25.
93. Mareel M, Klint J, Meyvisch C. Methods of study of the invasion of malignant C3H mouse fibroblasts into embryonic chick heart in vitro. Virchows Arch [B] 1979;30:95–111.
94. Starkey JR, Hosick HL, Stanford DR, Liggitt HD. Interaction of metastatic tumour cells with bovine lens capsule basement membrane. Cancer Res 1984;44:1585–94.
95. Liotta LA, Lee CW, Moraski DJ. New method for preparing large surfaces of intact human basement membrane for tumour invasion studies. Cancer Lett 1980;11:141–52.
96. Mignatti P, Robbins E, Rifkin DB. Tumour invasion through the human amniotic membrane: requirement for a proteinase cascade. Cell 1986;47: 487–98.
97. Yagel S, Parhar RS, Jeffrey JJ, Lala PK. Normal nonmetastatic human trophoblast cells share in vitro invasive properties of malignant cells. J Cell Physiol 1988;136:455–62.
98. Repesh LA. A new in vitro assay for quantitating tumour cell invasion. Invasion Metastasis 1989;9:192–208.
99. Connelly I, Lysiak JJ, Khoo N, Graham CH, Lala PK. Differential regulation of normal trophoblast and choriocarcinoma cell proliferation and invasiveness by transforming growth factors [Abstract]. Am J Reprod Immunol 1992;27(1/2):47.
100. Kliman HJ, Feinberg RF. Human trophoblast-extracellular matrix (ECM) interactons in vitro: ECM thickness modulates morphology and proteolytic activity. Proc Natl Acad Sci USA 1990;87:3057–61.
101. Fisher SJ, Leitch MS, Kantor MS, Basbaum CB, Kramer RH. Degradation of extracellular matrix by trophoblastic cells of first trimester human placentas. J Cell Biochem 1985;27:31–41.
102. Fisher SJ, Cui T, Zhang L, et al. Adhesive and degradative properties of human placental cytotrophoblast cells in vitro. J Cell Biol 1989;109:891–902.
103. Librach CL, Werb Z, Fitzgerald ML, et al. 92-kD type IV collagenase mediates invasion of human cytotrophoblasts. J Cell Biol 1991;113:437–49.
104. Kirby DRS. The development of mouse eggs beneath the kidney capsule. Nature 1960;187:707–8.
105. Kirby DRS. The development of mouse blastocysts transplanted to the spleen. J Reprod Fertil 1963;5:1–12.
106. Kirby DRS. The development of the mouse blastocyst transplanted to the cryptorchid and scrotal testis. J Anat 1963;97:119–30.
107. Kirby DRS. The "invasiveness" of the trophoblast. In: Park WW, ed. The early conceptus, normal and abnormal. Edinburgh: University of St. Andrews Press, 1965:68–74.
108. Liotta LA, Rao CN, Wewer UM. Biochemical interactions of tumour cells with the basement membrane. Annu Rev Biochem 1986;55:1037–57.

109. Loke YW, Gardner L, Grabowska A. Isolation of human extravillous trophoblast cells by attachment to laminin-coated magnetic beads. Placenta 1989;10:407–15.
110. Damsky CH, Fitzgerald ML, Fisher SJ. Distribution patterns of extracellular matrix components and adhesion receptors are intricately modulated during first trimester cytotrophoblast differentiation along the invasive pathway, in vivo. J Clin Invest 1992;89:210–22.
111. Emonard H, Christiane Y, Smet M, Grimaud JA, Foidart JM. Type IV and interstitial collagenolytic activities in normal and malignant trophoblast cells are specifically regulated by the extracellular matrix. Invasion Metastasis 1990;10:170–7.
112. Lysiak JJ, Hearn SA, Lala PK. Immuno-electron microscopic analysis of trophoblast cells in explant cultures of human first trimester chorionic villi [Abstract]. Proc Can Fed Biol Soc 1993;36:57.
113. Feinberg RF, Kliman HJ, Lockwood CJ. Is oncofetal fibronectin a trophoblast glue for human implantation. Am J Pathol 1991;138(3):537–43.
114. Hearn S, Walton J, Chapman W. Evidence that fibronectin in human placentas is derived from intermediate trophoblasts [Abstract]. Lab Invest 1992;66(1):64a.
115. Yagel S, Feinmesser R, Waghorne C, Lala PK, Breitman ML, Dennis JW. Evidence that β1–6 branched asn-linked ogliosaccharides on metastatic tumour cells facilitate invasion of basement membranes. Int J Cancer 1990; 44:685–90.
116. Yagel S, Kerbel RS, Lala PK, Elder-Gara T, Dennis JW. Basement membrane invasion by first trimester human trophoblast: requirement for branched complex-type asn-linked ogliosaccharides. Clin Exp Metastasis 1990; 8:305–17.
117. Connelly IH, Lala PK. Effects of antisense oligonucleotides targeted against metalloproteinases on matrigel invasion by normal and malignant trophoblasts [Abstract]. Proc Can Fed Biol Soc 1993;36:139.
118. Welgus HG, Campbell EJ, Bar-Shavit Z, Senior RM, Teitelbaum SC. Human alveolar macrophages produce a fibroblast-like collagenase and collagenase inhibitor. J Clin Invest 1985;76:219–24.
119. Welgus HG, Stricklin GP. Human skin fibroblast collagenase inhibitor. J Biol Chem 1983;258:12259–64.
120. Stetler-Stevenson WG, Krutzach HL, Liotta LA. Tissue inhibitor of metalloproteinases (TIMP-2). J Biol Chem 1989;264:17372–8.
121. Feinberg RF, Kao L, Haimowitz JE, et al. Plasminogen activator inhibitor types 1 and 2 in human trophoblasts PAI-1 is an immunocytochemical marker of invading trophoblasts. Lab Invest 1989;61:20–6.
122. Kliman HJ, Nestler JE, Sermasi E, Sanger JM, Strauss JF III. Purification, characterization, and in vitro differentiation of cytotrophoblasts from human term placentae. Endocrinology 1986;118(4):1567–82.

# 6

# Uterine Epithelial GM-CSF and Its Interlocutory Role During Early Pregnancy in the Mouse

SARAH A. ROBERTSON, ANNA C. SEAMARK, AND ROBERT F. SEAMARK

It is now recognized that the murine uterine epithelium has an extraordinary capacity to synthesize and release a range of lymphohemopoietic cytokines. Recent studies have shown that these cytokines are produced in a precisely orchestrated sequence during the reproductive cycle and are important components of the intercellular signaling language mediating in the actions of steroid hormones in reproductive tissues (1). Cytokines produced during the dramatic remodeling events of early pregnancy are postulated to have key roles in communication between resident cells and trafficking leukocytes in the endometrium and the developing conceptus. In this chapter we review our findings regarding the production and roles of *granulocyte macrophage colony stimulating factor* (GM-CSF) during the preimplantation period of murine pregnancy. We provide evidence that GM-CSF originating within the epithelium acts to coordinate local leukocyte participation, both in tissue remodeling and in modulating the local immune milieu to accommodate invasion by semiallogeneic fetal tissue. Nonhemopoietic cells, including the developing embryo, may provide additional targets for the action of this pleiotrophic cytokine.

## Epithelial Cell-Derived Cytokines in the Preimplantation Uterus

Endometrial cells harvested from cycling or pregnant uteri are a potent source of GM-CSF, releasing up to 10-fold more bioactivity on a cellular basis than conventional sources of this cytokine, including activated macrophages and lymphocytes (2). Epithelial cells have been identified as the source of this activity in monoclonal antibody-facilitated cell isolation studies, and GM-CSF mRNA localizes predominantly to luminal, and to a lesser degree glandular, epithelial cells in vivo (2, 3). The capacity of epithelial cells to release GM-CSF fluctuates through the cycle and is maximal at estrus (3). Steroid replacement experiments in ovariectomized mice and the use of steroid agonists have demonstrated that GM-CSF production is dependent on estrogen, but is antagonized moderately by progesterone (3). This contrasts with other epithelial cytokines, including *colony stimulating factor I* (CSF-I), *leukemia inhibitory factor* (LIF), and *tumor necrosis factor α* (TNFα), that are up-regulated in response to progesterone as placentation is initiated (1). The mechanism by which steroids influence GM-CSF synthesis remains unclear, but the failure of epithelial cells from ovariectomized mice to respond to estrogen in vitro suggests that it may depend upon the intermediate paracrine or juxtacrine activities of endometrial stromal cells.

Within hours after mating, GM-CSF mRNA expression by epithelial cells (4 and Robertson, unpublished observations) and the GM-CSF content of uterine luminal fluid increase more than 20-fold as a consequence of an interaction between the ejaculate and epithelial cells (2, 5). The active factor in semen is not associated with the sperm, but appears to be associated with a high-molecular weight, non-MHC component of seminal plasma originating within the seminal vesicle (2, 6). Our current aim is to ascertain the significance of this burst of cytokine activity. In recent experiments we have investigated the expression of GM-CSF receptor in the female reproductive tract, and our data suggest that GM-CSF has the potential to exert both uterotrophic and embryotrophic actions.

## GM-CSF Receptor Complex

GM-CSF is characteristically described as a regulator of macrophage and/or granulocyte differentiation and function. In hemopoietic cells the bioactivity of GM-CSF is mediated following binding of the protein to its membrane-bound receptor (7). The murine GM-CSF receptor, like that of the human, is a heterodimer composed of subunits belonging to the cytokine receptor superfamily (8, 9). The *α-subunit* (GM-CSF-R) of the GM-CSF receptor binds GM-CSF with low affinity, and the *β-subunit* (designated AIC2B in the mouse) does not bind GM-CSF by itself, but, together with the α-subunit, forms a high-affinity ligand binding complex. Association between the α- and β-receptor molecules is necessary for receptor activation and growth signal transduction (8, 10). The mechanism of signal transduction remains unclear since neither the human nor the mouse α- or β-subunit molecules have such obvious signal transducing elements as tyrosine kinase or G-protein interaction domains, but critical residues for inducing phosphorylation or cell proliferation have been identified in the cytoplasmic domain of the human β-subunit (11).

## GM-CSF Receptor Expression in the Endometrium During Early Pregnancy

### *RT-PCR Analysis of GM-CSF Receptor Expression*

In recent experiments *reverse transcriptase-polymerase chain reaction* (RT-PCR) analysis has been used to determine whether mRNAs encoding the α-subunit (GM-CSF-R) and β-subunit (AIC2B) of the GM-CSF receptor are expressed in the uterus at estrus and during early pregnancy. RNA was prepared by guanidine isothiocyanate extraction, followed by DNAse treatment, from pools of 4 uteri from estrous mice and day 1, day 2, day 3, and day 4 pregnant mice. *Complementary DNA* (cDNA) was prepared by reverse transcription using oligo-dT primers. Uterine tissue was found to contain GM-CSF-R mRNA at estrus and on each of the first 4 days of pregnancy (Fig. 6.1). The 235-bp product was indistinguishable from that obtained by RT-PCR of murine myeloid cell (FD 5/12) cDNA under the same reaction conditions. GM-CSF-R expression appeared to be greater on days 1 and 2 following mating than in estrous or implantation stage uteri. The AIC2B mRNA was also found to be present in the uterus on each of the days examined. These results show that cells in estrous and early pregnant uteri express mRNAs for both the α-subunit and β-subunit of the GM-CSF receptor and so potentially express the high-affinity GM-CSF receptor complex that is characteristic of myeloid leukocytes in other tissues.

A second reaction product of 450 bp was also generated by RT-PCR from each of the uterine cDNAs. The identity of this band is not clear, but a band of similar size was sometimes generated from cDNAs prepared from liver and ovary (data not shown). There was an inverse relationship between the density of the 235- and 450-bp bands. It is possible that this band is the product of an alternatively spliced GM-CSF-R transcript, and it will be of interest to determine whether the product of this is related to unusual GM-CSF-R isoforms that have been described in the human placenta. However, the primers used in these experiments span a cDNA sequence corresponding to a membrane-proximal region of the extracellular domain of the GM-CSF-R protein that is 5′ of the position equivalent to the splice point for the soluble and α2 forms of the human receptor (see below) and so would not be expected to span regions that would be different in murine counterparts of the known human isoforms.

### *Distribution of GM-CSF Receptor-Bearing Cells in the Uterus on Day 1 of Pregnancy*

GM-CSF-responsive leukocytes, including macrophages, neutrophils, and eosinophils, comprise major populations in the rodent uterus, and fluctuations in their numbers parallel circulating estrogen levels (13). Indeed,

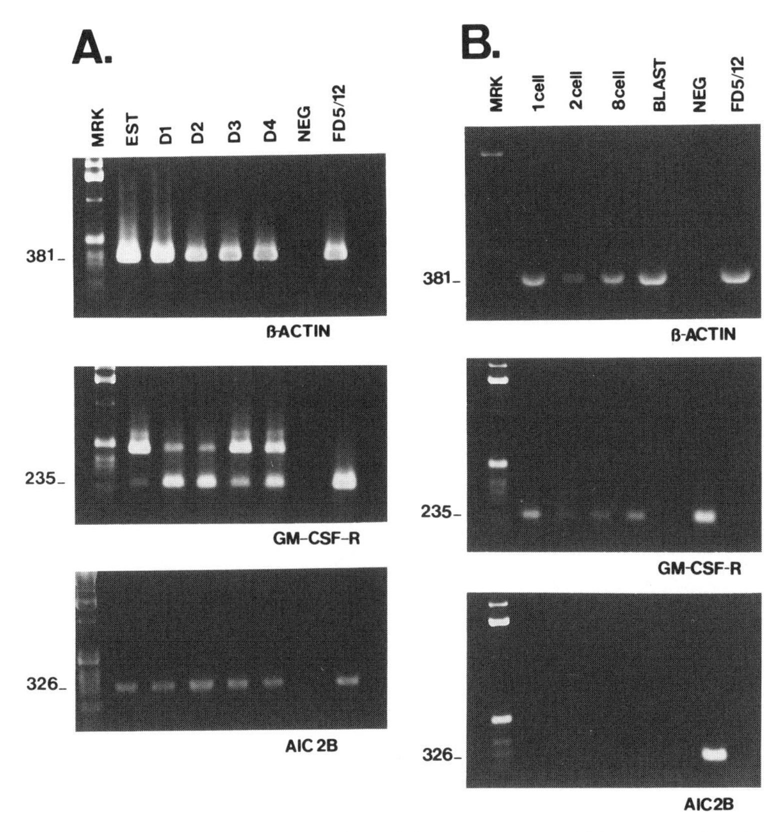


FIGURE 6.1. RT-PCR analysis of GM-CSF receptor mRNA expression in (*A*) estrous and preimplantation pregnant uterus and (*B*) preimplantation embryos. RNA was isolated from 1-cell, 2-cell, 8-cell, and blastocyst stage (BLAST) embryos according to the method described in reference 52 and from estrous, day 1 through day 4 pregnant uterus, and murine myeloid cells (FD5/12) by guanidine thiocyanate extraction. First-strand cDNA was reverse-transcribed from DNAse-treated mRNA using oligo-dT primers, then amplified with primers for β-actin, GM-CSF-R (the α-chain of the GM-CSF receptor), and AIC2B (the β-chain of the GM-CSF receptor). GM-CSF-R primers corresponded to bp positions 717–738 and 930–951 of the GM-CSF-R cDNA sequence, as described in reference 9, and the AIC2B primers used were those described in reference 53 to discriminate between the closely related AIC2A and AIC2B mRNAs.

there is a striking concurrence between the trafficking and functional activation of these cells and the levels of GM-CSF in the cycling and early pregnant uterus, with both peaking at estrus and again during the 24-h period following mating (13, 14). During the 24-h period after mating, macrophages and granulocytes become concentrated in the subepithelial endometrial stroma, and this is accompanied by a transient increase in their expression of *interleukin-1* (IL-1) and TNFα (14, 15). A similar inflammatory response is elicited by mating with vasectomized males, but not by cervical stimulation of estrous mice (15).

The finding that expression of mRNA for the high-affinity GM-CSF receptor is maximal on day 1 and day 2 of pregnancy (see above) when macrophage and granulocyte numbers peak is consistent with these cells being the predominant source. This proposal is also supported by the localization of GM-CSF receptor-bearing cells in sections of day 1 uterus. The $^{125}$I-GM-CSF was found to label cells identified as leukocytes on the basis of their distribution throughout the endometrium and myometrium and predominant localization to subepithelial areas of the endometrial stroma (12).

However, these findings do not preclude the possibility that nonhemopoietic cells in the uterus also respond to GM-CSF. For example, it has been claimed that secretion of PGE and $PGF_{2\alpha}$ by stromal fibroblasts from pregnant bovine endometrium in vitro is reduced moderately by GM-CSF (16), although the contribution to this effect of contaminating leukocytes remains to be determined. Alteration in the local vascular architecture is one of the earliest uterine responses to the embryo at implantation, and GM-CSF has been reported to induce migration and proliferation of endothelial cells (17).

## Effect of GM-CSF on Leukocyte Trafficking into the Endometrium

To investigate whether GM-CSF can alter the distribution or the numbers of leukocytes in the endometrium, leukocyte populations were examined in the uteri of ovariectomized mice following local exposure to recombinant GM-CSF. This experimental system was chosen since uteri in ovariectomized mice are relatively deficient in GM-CSF (3) and contain fewer macrophages and granulocytes than the uteri of intact mice (13 and Robertson, unpublished observations). Groups of 4 ovariectomized mice were anesthetized and, through a dorsal incision, were given a unilateral intraluminal injection of either 40, 200, or 1000 U of recombinant GM-CSF in 50 μL of 1% BSA or carrier alone. Sixteen hours later the mice were sacrificed, and their uteri were removed. The right and left uterine horns were processed and analyzed separately. The numbers and the distribution of uterine leukocytes were assessed immunohistochemically with mAbs against CD45/LCA, F4/80 antigen, Mac-1, and Ia antigen. Eosinophils were assessed on the basis of their endogenous peroxidase activity.

In each of two experiments, intraluminal administration of GM-CSF dramatically altered the numbers of leukocytes in endometrial tissues.

TABLE 6.1. Effect of intraluminal GM-CSF on the number of eosinophils in the endometrium of ovariectomized mice.

| | Treatment | | | |
|---|---|---|---|---|
| Mouse No. | Carrier | 40 U | 200 U | 1000 U |
| Experiment 1 | | | | |
| 1 | 14 ± 5 | | 344 ± 80 | <1 |
| 2 | 34 ± 4 | | 214 ± 11 | 90 ± 7 |
| 3 | 28 ± 13 | | 249 ± 62 | 3 ± 1 |
| 4 | | | 222 ± 56 | 4 ± 1 |
| Experiment 2 | | | | |
| 1 | <1 | 2 ± 1 | <1 | 3 ± 2 |
| 2 | <1 | 105 ± 16 | 170 ± 82 | 292 ± 14 |
| 3 | <1 | 4 ± 3 | <1 | <1 |
| 4 | <1 | <1 | 170 ± 16 | 257 ± 9 |

*Note:* Recombinant GM-CSF (40, 200, or 1000 U) in 50 µL of 1% BSA or 1% BSA alone was injected into the left uterine horn of 4 mice in each experimental group. Two sections from each uterine horn were stained with DAB and $H_2O_2$. The numbers of endogenous peroxidase-positive cells within 3 grid areas of 0.09 $mm^2$ were counted and pooled. There was no significant difference between the numbers of eosinophils in tissue from the left and right uterine horns. Data are expressed as the mean ± SD number of eosinophils (per 0.27 $mm^2$) in each of 4 (2 each from the left and right horn) sections per uterus.

Eosinophils appeared to be the subpopulation influenced to the greatest degree. However, not all mice responded, and treated mice appeared to fall into two categories: (i) those with uteri that contained very dense populations of eosinophils (responders) and (ii) those with uteri that were indistinguishable from control uteri (nonresponders). In total, 1 of 4 mice that received 40 U, 6 of 8 mice that received 200 U, and 3 of 8 mice that received 1000 U of GM-CSF were found to have between 4-fold and 200-fold increases in the number of eosinophils within the endometrial stroma and were classified as responders (Table 6.1).

In the first experiment the effect was dependent on the dose of GM-CSF. Mice that received 200 U of GM-CSF were found to have significantly greater numbers of endometrial eosinophils than mice that received carrier alone or mice that received 1000 U of GM-CSF (both $P = 0.0001$). However, in a second experiment only 5 of 12 mice responded to cytokine treatment, reducing the significance of this effect ($P = 0.167$). Eosinophils were distributed throughout the endometrial stroma in GM-CSF-treated mice and in some uteri were concentrated in areas adjacent to the luminal surface (Fig. 6.2).

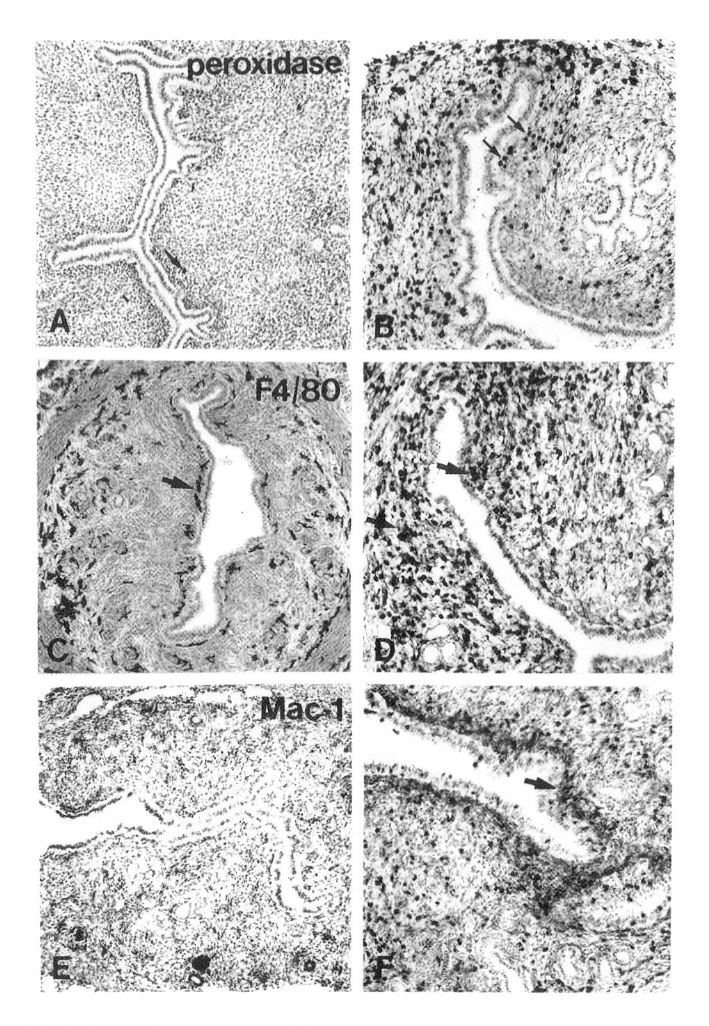


FIGURE 6.2. Effect of exogenous GM-CSF on the numbers of leukocytes in the uteri of ovariectomized mice. Sections of uterus are from responder mice that had received intraluminal injections of 200 U of GM-CSF (right-hand panel: *B*, *D*, and *F*), or from mice that received carrier (left-hand panel: *A*, *C*, and *E*). The sections were stained with DAB and $H_2O_2$ to detect eosinophils (*A* and *B*) or incubated with mAbs specific for F4/80 (*C* and *D*) or Mac-1 (*E* and *F*). Endogenous peroxidase-positive cells are evident as small, granular, dark-stained cells (small arrows) in all sections in addition to larger mAb-reactive cells (large arrows in sections *C* to *F*).

GM-CSF was also found to influence the numbers of other leukocytes in the uteri of ovariectomized mice, as determined by the use of specific *monoclonal antibodies* (mAbs) against various leukocyte subpopulations (Table 6.2). Uteri from carrier-treated ovariectomized mice contained

TABLE 6.2. Effect of intraluminal GM-CSF on the numbers of uterine cells bearing leukocyte, macrophage, and dendritic cell-specific markers.

| | | Treatment | | |
|---|---|---|---|---|
| mAb | Specificity | Carrier | 200 U | 2000 U |
| LCA | All leukocytes | ++ | ++++ | ++ |
| F4/80 | Macrophages | + | +++ | + |
| Mac-1 | Macrophages/neutrophils | + | +++ | + |
| Ia | Macrophages/dendritic cells | ++ | +++ | ++ |

*Note:* Sections were prepared from GM-CSF and carrier-treated mice as described in Table 6.1 and incubated with mAbs against LCA (CD45), F4/80, Mac-1 (CD11b/CD18), and Ia. The numbers of cells reactive with each antibody were estimated, and qualitative scores are given for sections of uteri from mice in experiment 1.

moderate numbers of cells that reacted strongly with mAbs against F4/80 (Fig. 6.2) and Ia (not shown) and weakly with mAbs against Mac-1 (CD11b/CD18). These cells were generally dendritic in shape and were presumed to be predominantly dendritic cells and/or macrophages. Intraluminal instillation of GM-CSF clearly increased the numbers of F4/80$^+$ and Mac-1$^+$ cells, as well as increasing the intensity of Mac-1 expression. While the numbers of Ia$^+$ cells increased, the difference was less striking than for the other mAbs. Uteri from mice classified as responder mice for eosinophils were found generally to contain a greater number of cells that expressed F4/80, Mac-1, and Ia than control mice. The F4/80$^+$ and Ia$^+$ cells were distributed throughout the stroma, but were most concentrated in areas underlying the epithelium in both treated and untreated mice. Small round cells that were Mac-1$^+$ but F4/80$^-$ (and so presumed to be neutrophils) were found between epithelial cells and appeared to be in the process of trafficking into the luminal cavity in GM-CSF-treated mice.

That epithelial GM-CSF may directly influence the recruitment of leukocytes into endometrial tissue is also indicated by the previous finding that GM-CSF is chemotactic in vitro for both monocytes and granulocytes (18). GM-CSF is proposed to be important in trapping these cells within inflammatory sites in vivo by enhancing Mac-1 expression and reducing expression of the MEL-14 homologue LAM-1 (19). Local endothelial cell/leukocyte interactions may also be indirectly altered by GM-CSF; for example, through enhanced TNFα release from macrophages.

## GM-CSF Receptor Expression in the Conceptus

Placental trophoblast cells have biological responses in vitro to many of the cytokines that are released by uterine cells during pregnancy and in some instances have been shown to express the appropriate receptors.

GM-CSF receptor (α-subunit) mRNA is expressed in human placenta (20), but expression of the GM-CSF receptor in the rodent placenta has not been examined. Primary cultures of cells from day 12 mouse placenta and cytokeratin-positive placental cell lines have been found to proliferate in response to GM-CSF (21). Purified natural GM-CSF also stimulated the incorporation of $^{3}$H-thymidine into trophoblast cells harvested as outgrowths from day 8–9 ectoplacental cones or by enzymatic disruption of day 12 placenta (22). Human choriocarcinoma cell lines JAR, JEG, and BeWo secrete GM-CSF, and an autocrine action for this factor is indicated by the finding that an anti-GM-CSF antibody blocks their proliferation (23). GM-CSF and CSF-I have been shown to induce the differentiation of human cytotrophoblast cells into syncytium and to stimulate release of placental lactogen and chorionic gonadotropin in vitro (23).

There is also evidence that the lymphohemopoietic cytokines are among those growth factors synthesized by uterine cells that can influence the growth and development of preembryos (24). GM-CSF has been reported to have both positive and negative effects on the development of embryos in culture. The effects of this cytokine may be dependent on the developmental stage of the embryo and the influence of other growth factors. GM-CSF has been shown to inhibit the development of 2-cell embryos into morulae (25) and the attachment of blastocysts to fibronectin-coated culture dishes (26), whereas we have found that GM-CSF promotes the development of periimplantation murine embryos cultured in fetal calf serum, particularly through enhancing the rate of blastocyst implantation (27).

We have reported previously that murine blastocysts specifically bind $^{125}$I-GM-CSF (12), but whether this interaction was of high or low affinity was not determined. To investigate whether murine embryos express mRNA for either subunit of the GM-CSF receptor, mRNA was harvested from embryos harvested on each of the first 4 days after fertilization and subjected to RT-PCR analysis. One-cell, 2-cell, 8-cell, and blastocyst stage embryos were found to express GM-CSF-R mRNA. The cDNAs from each embryonic stage generated a GM-CSF-R amplicon of the expected size (235 bp) (Fig. 6.2). The reaction product from 2-cell stage cDNA yielded an additional, minor band of approximately 450 bp. The identity of this band is unknown, but it is the same size as a band that is a more abundant product of uterine cDNA (see above). Messenger RNA for the β-subunit of the GM-CSF receptor was not detected in embryos. AIC2B primers failed to amplify product from any of the 4 embryo cDNAs under reaction conditions that yielded substantial product from FD5/12 cDNA (Fig. 6.2). To increase the sensitivity of the analysis, $^{32}$P-dATP was included in the reaction mix, and gels were autoradiographed after capillary transfer to nylon filters. No reaction product was detected when filters were exposed for up to 2 weeks (data not shown).

This result is not surprising in view of the lack of expression of the β-subunit of the high-affinity GM-CSF receptor in all other GM-CSF-responsive, nonhemopoietic cells. The mechanism by which nonhemopoietic cells, including endothelial cells and melanoma cells (28, 29), and embryonic or placental trophoblast respond to GM-CSF therefore remains unclear. Whether the α-subunit can initiate certain types of signaling in the absence of the β-subunit, perhaps by associating with other signal-transducing proteins, remains to be fully investigated. There is evidence that additional GM-CSF receptor systems may exist, and, interestingly, these data have been generated in human conceptus-derived tissues. In the human, alternatively spliced mRNAs give rise to at least another two isoforms of the α-subunit of the GM-CSF receptor in addition to the α1-subunit described above. One of the additional isoforms, designated *α2*, is expressed in human bone marrow and placenta and has a longer cytoplasmic domain than the α1 molecule (30). This domain is rich in serine residues, a feature that is typical of regions critical for signal transduction for other receptors of the hemopoietic receptor superfamily.

In addition, a soluble form of the GM-CSF receptor α-subunit has been found to be released from the JAR, BeWo, and JEG-3 human choriocarcinoma cells (31, 32). This receptor is transcribed from an alternatively spliced α-subunit mRNA (31) that is similar to an mRNA found in human placental tissue (33). The soluble receptor may modulate the local response to GM-CSF by competing with membrane-bound receptor for available ligand, and soluble GM-CSF receptor-ligand complexes may transmit signal after association with membrane-bound β-subunit, as has been documented for the *interleukin-6* (IL-6) receptor system (34). Furthermore, in view of the recent report that GM-CSF has a nucleotide binding site (35), it will be of interest to determine whether GM-CSF has an alternative mode of action within the target cell nucleus, as has been described for the structurally related protein prolactin (36).

## IL-6 in the Preimplantation Uterus

The production of IL-6 by epithelial cells parallels in many respects the production of GM-CSF. IL-6, like GM-CSF, has been identified as a product of the uterine epithelium in the estrous cycle and during early pregnancy, when seminal factors evoke a 200-fold increase in the IL-6 content of uterine luminal fluid (2). This peak in bioactivity is associated with a transient elevation in IL-6 mRNA (6). In contrast to GM-CSF, IL-6 is constitutively released in vitro by epithelial cells from ovariectomized and nonpregnant mice, regardless of the steroid status of the animal from which they were harvested (3, 37), although the vectorial secretion of IL-6 from uterine epithelial cells has been found to be inhibited by physiological levels of estrogen and progesterone (37). The apparent steroid independence of IL-6 secretion in vitro may not, however, accurately reflect the physiological situation. In a recent study neither IL-6 bioactivity nor mRNA could be detected in homogenates of uterine tissue from ovariectomized mice unless induced by the synergistic action of progesterone and estrogen (38).

Like GM-CSF, IL-6 can alter the differentiation state and function of macrophages and granulocytes, as well as act on a broader range of lymphocytes and nonhemopoietic cells (39), although a role for IL-6 in influencing leukocyte dynamics in the preimplantation uterus remains to be demonstrated. A further action of this cytokine in promoting edema in the vicinity of the implanting embryo is suggested by the finding that IL-6 can markedly increase the permeability of endothelial cell monolayers in vitro (40).

The growth and function of blastocyst and placental trophoblast cells may also be regulated by IL-6. The attachment and trophoblast outgrowth of murine blastocysts in vitro is reported to be inhibited by this cytokine (37), but the expression of IL-6 receptors by preimplantation embryos remains to be investigated. Human placental trophoblast cells synthesize IL-6 and IL-6 receptors, and an IL-6-mediated autocrine regulation of hCG synthesis has been proposed (41). Whether a parallel situation exists in rodents is not yet clear.

## Epithelial Cytokines As Primary Interlocutors in the Inflammatory Response to Mating

These data strongly implicate GM-CSF and possibly IL-6 as primary initiators of the sequence of cellular events consequent to mating. Our current working hypothesis (illustrated in Fig. 6.3) is that estrogen primes epithelial cells to release GM-CSF and IL-6 in response to a specific factor(s) in seminal plasma. This precipitates the inflammatory cascade that results in a dramatic infiltration and activation of macrophages and granulocytes within the underlying uterine stroma (14, 15). Macrophages are extraordinarily versatile cells with a broad spectrum of activities, including the secretion of a vast array of cytokines and other mediators with pleiotrophic activities, in a microenvironment-specific manner (42). Neutrophils and eosinophils are potential sources of factors, including plasminogen activator, collagenase, histamine, PAF, and prostaglandins, that have been implicated in implantation and induction of the decidual response. These leukocytes are therefore prime candidates as major effector cells in tissue remodeling in the preimplantation uterus. Epithelial GM-CSF may also act, possibly in concert with other cytokines such as TGFβ (43), to modulate the immunoaccessory functions of macrophages (42, 44, 45) and thus maintain a local immunological environment that is less hostile to the alloantigenically foreign conceptus.

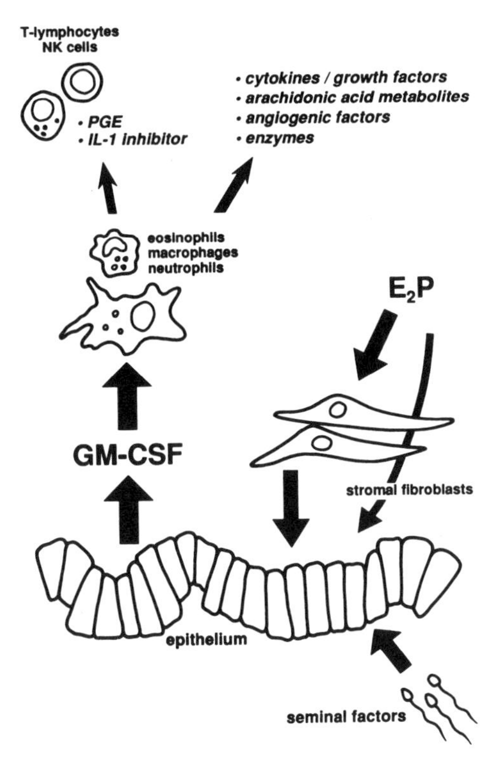


FIGURE 6.3. A model for the role of GM-CSF in establishing receptivity in the preimplantation uterus. Seminal factors stimulate enhanced release of GM-CSF from estrogen-primed epithelial cells. GM-CSF initiates an inflammatory cascade, during which large numbers of macrophages and granulocytes infiltrate the endometrium and become activated. GM-CSF and other epithelial cell cytokines subsequently coordinate the secretory profile of these cells during the days leading up to and during implantation of the blastocyst. Growth factors, enzymes, and bioactive lipids are macrophage products that are potentially important both in tissue remodeling and in interacting with local lymphocyte populations to help establish an immunologically receptive uterine environment.

The role of GM-CSF in the subsequent cascade of events leading to successful placentation needs to be considered in the context of the changing profile of other epithelial cytokines and environmental determinants, but the importance of GM-CSF to the ongoing success of pregnancy is highlighted by the dramatic consequences of administering relatively small amounts of this cytokine to pregnant mice. The high rate of early embryo malformation and loss that occurs spontaneously in the CBA/J × DBA/2 mating combination is reversed following daily systemic administration of rGM-CSF during the preimplantation period (46). In midgestation exogenous GM-CSF protects against IFN$\gamma$- or TNF$\alpha$-induced fetal resorption and increases fetal and placental weights (47). The doses used in these studies were far too small to have direct effects on placental cell growth, and the authors hypothesize a mechanism involving a cytokine cascade. Support for the role of seminal plasma in potentiating this cascade comes from studies showing that reduced success rates are typical of pregnancies initiated in the absence of exposure to seminal factors (48, 49) and that presensitization of the uterus to seminal factors in previous estrous cycles contributes to pregnancy success in rodents (50) and humans (51).

## Summary

Recognition of the uterine epithelium as a potent source of lymphohemopoietic cytokines has enhanced our understanding of the function of this tissue to include a role analogous to that of the epithelia in the skin and other mucosae, which act to coordinate the recruitment and behavior of local leukocytes during inflammatory responses. The present study shows that GM-CSF emanating from the uterine epithelium has a potentially important role in recruiting macrophages and granulocytes into the endometrium. We propose that the seminal factor-evoked release of GM-CSF from estrogen-primed epithelial cells at mating triggers a sequence of cytokine- and leukocyte-mediated events underpinning the establishment and viability of a pregnancy. This includes the participation of GM-CSF-responsive leukocytes in tissue remodeling and in establishing an immunologically receptive environment conducive to embryo implantation. In addition, we have provided evidence that GM-CSF is embryotrophic and, together with other epithelial cytokines, would contribute to the role of the uterus in modulating embryo growth and development. Together, these data suggest that during early pregnancy epithelial GM-CSF is an important mediator in the communication interlinking resident uterine cells with the often neglected but major populations of uterine leukocytes and the conceptus.

## *References*


1. Robertson SA, Brannstrom M, Seamark RF. Cytokines in rodent reproduction and the cytokine-endocrine interaction. Curr Opin Immunol 1992;4: 585–90.
2. Robertson SA, Mayrhofer G, Seamark RF. Uterine epithelial cells synthesize granulocyte-macrophage colony-stimulating factor (GM-CSF) and interleukin 6 (IL-6) in pregnant and non-pregnant mice. Biol Reprod 1992;46: 1069–79.

3. Robertson SA, Mayrhofer G, Seamark RF. Ovarian steroid hormones regulate granulocyte-macrophage colony-stimulating factor (GM-CSF) production by uterine epithelial cells in the mouse. Biol Reprod 1993.
4. Sanford TR, De M, Wood GW. Expression of colony-stimulating factors and inflammatory cytokines in the uterus of CD1 mice during days 1 to 3 of pregnancy. J Reprod Fertil 1992;94:213–20.
5. Robertson SA, Seamark RF. Granulocyte macrophage colony stimulating factor (GM-CSF) in the murine reproductive tract: stimulation by seminal factors. Reprod Fertil Dev 1990;2:359–686.
6. Robertson SA, Seamark RF. Uterine granulocyte-macrophage colony stimulating factor in early pregnancy: cellular origin and potential regulators. In: Mowbray JF, Chaouat G, eds. Molecular and cellular biology of the feto-maternal relationship, Colloque INSERM 212. Paris: Editions John Libbey Eurotext, 1991:113–21.
7. Nicola N, Metcalf D. Subunit promiscuity among hemopoietic growth factor receptors. Cell 1991;67:1–4.
8. Gorman DM, Itoh N, Kitamura T, et al. Cloning and expression of a gene encoding an interleukin 3 receptor-like protein: identification of another member of the cytokine receptor gene family. Proc Natl Acad Sci USA 1990;87:5459–63.
9. Park LS, Martin U, Sorensen R, et al. Cloning of the low-affinity murine granulocyte-macrophage colony-stimulating factor receptor and reconstitution of a high-affinity receptor complex. Proc Natl Acad Sci USA 1992;89: 4295–9.
10. Kitamura T, Hayashida K, Sakamaki K, Yokota T, Arai K, Miyajima A. Reconstitution of functional receptors for human granulocyte/macrophage colony-stimulating factor (GM-CSF): evidence that the protein encoded by the AIC2B cDNA is a subunit of the murine GM-CSF receptor. Proc Natl Acad Sci USA 1991;88:5082–6.
11. Sakamaki K, Miyajima I, Kitamura T, Miyajima A. Critical cytoplasmic domains of the common beta subunit of the human GM-CSF, IL-3 and IL-5 receptors for growth signal transduction and tyrosine phosphorylation. EMBO J 1992;11:3541–9.
12. Robertson SA, Mayrhofer G, Seamark RF. Epithelial cell cytokine synthesis in the pre-implantation rodent uterus. In: Proc 5th Int Cong Reprod Immunol. New York: Raven Press, 1993.
13. De M, Wood GW. Influence of oestrogen and progesterone on macrophage distribution in the mouse uterus. J Endocrinol 1990;126:417–24.
14. De M, Choudhuri R, Wood GW. Determination of the number and distribution of macrophages, lymphocytes, and granulocytes in the mouse uterus from mating through implantation. J Leukoc Biol 1991;50:252–62.
15. McMaster MT, Newton RC, Dey SK, Andrews GK. Activation and distribution of inflammatory cells in the mouse uterus during the preimplantation period. J Immunol 1992;148:1699–705.
16. Betts JG, Hansen PJ. Regulation of prostaglandin secretion from epithelial and stromal cells of the bovine endometrium by interleukin-1 beta, interleukin-2, granulocyte-macrophage colony stimulating factor and tumor necrosis factor-alpha. Life Sci 1992;51:1171–6.

17. Bussolino F, Wang JM, Defilippi P, et al. Granulocyte- and granulocyte-macrophage-colony stimulating factors induce human endothelial cells to migrate and proliferate. Nature 1989;337:471–3.
18. Wang JM, Colella S, Allavena P, Mantovani A. Chemotactic activity of human recombinant granulocyte-macrophage colony-stimulating factor. Immunology 1987;60:439–44.
19. Kishimoto TK, Jutila MA, Lakey Berg E, Butcher EC. Neutrophil Mac-1 and Mel-14 adhesion proteins inversely regulated by chemotactic factors. Science 1989;245:1238–41.
20. Gearing DP, King JA, Gough NM, Nicola NA. Expression cloning of a receptor for human granulocyte-macrophage colony-stimulating factor. EMBO J 1989;8:3667–76.
21. Athanassakis I, Bleackley RC, Guilbert L, Paetkau V, Barr PJ, Wegmann TG. The immunostimulatory effect of T cells and T cell lymphokines on murine fetally derived placental cells. J Immunol 1987;138:37–44.
22. Armstrong DT, Chaouat G. Effects of lymphokines and immune complexes on murine placental cell growth in vitro. Biol Reprod 1989;40:466–74.
23. Garcia-Lloret M, Guilbert L, Morrish PW. Functional expression of CSF-1 receptors on normal human trophoblast [Abstract]. In: Cocard L, Alsat E, Challier J-C, Chaouat G, Malaesine A, eds. Placental communications: biochemical, morphological and cellular aspects, Colloque INSERM 199. Paris: Editions John Libbey Eurotext, 1990:135.
24. Pampfer S, Arceci RJ, Pollard JW. Rolc of colony stimulating factor-1 (CSF-1) and other lympho-hematopoietic growth factors in mouse pre-implantation development. Bioessays 1991;13:535–40.
25. Hill JA, Haimovici F, Anderson DJ. Products of activated lymphocytes and macrophages inhibit mouse embryo development in vitro. J Immunol 1987; 139:2250–4.
26. Haimovici F, Hill JA, Anderson DJ. The effects of soluble products of activated lymphocytes and macrophages on blastocyst implantation events in vitro. Biol Reprod 1991;44:69–75.
27. Robertson SA, Lavranos TC, Seamark RF. In vitro models of the maternal fetal interface. In: Wegmann TG, Nisbett-Brown E, Gill TG, eds. The molecular and cellular immunobiology of the maternal-fetal interface. New York: Oxford University Press, 1991:191–206.
28. Baldwin GC, Golde DW, Widhopf GF, Economou J, Gasson JC. Identification and characterization of a low-affinity granulocyte-macrophage colony-stimulating factor receptor on primary and cultured human melanoma cells. Blood 1991;78:609–15.
29. Bussolino F, Wang JM, Turrini F, et al. Stimulation of the $Na^+/H^+$ exchanger in human endothelial cells activated by granulocyte- and granulocyte-macrophage-colony-stimulating factor: evidence for a role in proliferation and migration. J Biol Chem 1989;264:18284–7.
30. Crosier KE, Wong GG, Mathey-Prevot B, Nathan DG, Sieff C. A functional isoform of the human granulocyte/macrophage colony stimulating factor receptor has an unusual cytoplasmic domain. Proc Natl Acad Sci USA 1991;88: 7744–8.
31. Raines MA, Liu L, Quan SG, Joe V, DiPersio JF, Golde DW. Identification and molecular cloning of a soluble human granulocyte-macrophage colony-stimulating factor receptor. Proc Natl Acad Sci USA 1991;88:8203–7.

32. Sasaki K, Chiba S, Mano H, Yazaki Y, Hirai H. Identification of a soluble GM-CSF binding protein in the supernatant of a human choriocarcinoma cell line. Biochem Biophys Res Commun 1992;183:252–7.
33. Ashworth A, Kraft A. Cloning of a potentially soluble receptor for human GM-CSF. Nucleic Acids Res 1990;18:71–8.
34. Hibi M, Murakami M, Saito M, Hirano T, Taga T, Kishimoto T. Molecular cloning and expression of an IL-6 signal transducer, gp130. Cell 1990;63: 1149–57.
35. Doukas MA, Chavan AJ, Gass C, Boone T, Haley BE. Identification and characterization of a nucleotide binding site on recombinant murine granulocyte/macrophage-colony stimulating factor. Bioconjug Chem 1992;3: 484–92.
36. Clevenger CV, Altmann SW, Prystowsky MB. Requirement of nuclear prolactin for interleukin-2-stimulated proliferation of T-lymphocytes. Science 1991:77–9.
37. Jacobs AL, Sehgal PB, Julian J, Carson DD. Secretion and hormonal regulation of interleukin-6 production by mouse uterine stromal and polarized epithelial cells cultured in vitro. Endocrinology 1992;131:1037–46.
38. De M, Sanford TR, Wood GW. Interleukin-1, interleukin-6, and tumor necrosis factor alpha are produced in the mouse uterus during the estrous cycle and are induced by estrogen and progesterone. Dev Biol 1992;151: 297–305.
39. Le JM, Vilcek J. Interleukin 6: a multifunctional cytokine regulating immune reactions and the acute phase protein response. Lab Invest 1989;61:588–602.
40. Maruo N, Morita I, Shirao M, Murota S. IL-6 increases endothelial permeability in vitro. Endocrinology 1992;131:710–4.
41. Nishino E, Matsuzaki N, Masuhiro K, et al. Trophoblast derived interleukin-6 (IL-6) regulates human chorionic gonadotrophin release through IL-6 receptor on human trophoblasts. J Clin Endocrinol Metab 1990;71:436–41.
42. Rappolee DA, Werb Z. Macrophage secretions: a functional perspective. Bull Inst Pasteur 1989;87:361–94.
43. Tsunawaki S, Sporn M, Ding A, Nathan C. Deactivation of macrophages by transforming growth factor-β. Nature 1988;334:260–2.
44. Mazzei GJ, Bernasconi LM, Lewis C, Mermod J-J, Kindler V, Shaw AR. Human granulocyte-macrophage colony-stimulating factor plus phorbol myristate acetate stimulate a promyelocytic cell line to produce an IL-1 inhibitor. J Immunol 1991;145:585–91.
45. Heidenreich S, Gong J-H, Schmidt A, Nain M, Gemsa D. Macrophage activation by granulocyte/macrophage colony-stimulating factor: priming for enhanced release of tumor necrosis factor-α and prostaglandin $E_2$. J Immunol 1989;143:1198–205.
46. Tartakovsky B, Ben-Yair E. Cytokines modulate preimplantation development and pregnancy. Dev Biol 1991;146:345–52.
47. Chaouat G, Menu E, Clark DA, Dy M, Minkowski M, Wegmann TG. Control of fetal survival in CBA × DBA/2 mice by lymphokine therapy. J Reprod Fertil 1990;89:447–58.
48. Bellinge BS, Copeland CM, Thomas TD, Mazzucchelli RE, O'Neil G, Cohen MJ. The influence of patient insemination on the implantation rate in an in vitro fertilization and embryo transfer program. Fertil Steril 1986;46: 2523–6.

49. Pang SF, Chow PH, Wong TM. The role of the seminal vesicle, coagulating glands and prostate glands on the fertility and fecundity of mice. J Reprod Fertil 1979;56:129–32.
50. Beer AE, Billingham RE. Host responses to intra-uterine tissue, cellular and fetal allografts. J Reprod Fertil Suppl 1974;21:59–89.
51. Klonoff-Cohen HS, Savitz DA, Celafo RC, McCann MF. An epidemiologic study of contraception and preeclampsia. JAMA 1989;262:3143–7.
52. Arcellana-Panlilio MY, Schultz GA. Guide to techniques in mouse development; analysis of messenger RNA in early mouse embryos. Methods Enzymol 1993.
53. Fung MC, Mak NK, Leung KN, Hapel AJ. Distinguishing between mouse IL-3 and IL-3 receptor-like (IL-5/GM-CSF receptor converter) mRNAs using the polymerase chain reaction method. J Immunol Methods 1992;149:97–103.

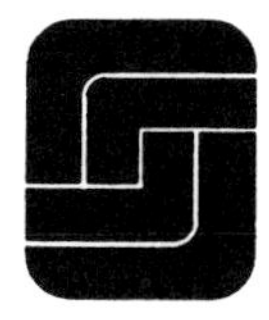

# 7

# Tilt Toward $T_{H2}$ in Successful Pregnancy

THOMAS G. WEGMANN, HUI LIN, JANE YUI, MARIA GARCIA-LLORET, TIM MOSMANN, AND LARRY GUILBERT

The hypothesis explored in this chapter is that the immune system of the pregnant female is shifted toward antibody-mediated immunity regulated by $T_{H2}$ helper cells as a result of spontaneous cytokine production by placental-decidual tissues. This alteration of cytokine balance in turn has pronounced consequences for maternal health and fetal survival. We have recently reviewed clinical and experimental evidence in support of this point of view (1) and briefly summarize it here.

## Autoimmune and Alloimmune Responses in Pregnancy

It has long been apparent that pregnant humans generally obtain relief from the inflammatory autoimmune disease rheumatoid arthritis during pregnancy. There are also reports that indicate that the antibody-mediated autoimmune disease systemic lupus erythematosis is exacerbated by pregnancy. Furthermore, women exposed to the intracellular pathogens leading to tuberculosis, leprosy, coccidioidomycosis, toxoplasmosis, malaria, and the *human immunodeficiency virus* (HIV) are reported to be at greater risk if they are pregnant (reviewed in 1). These clinical observations could be explained by postulating that $T_{H2}$-type immune responses predominate over those mediated by $T_{H1}$ (i.e., cell-mediated immunity) in pregnancy.

There is corroborative and indeed less controversial evidence for this postulate in murine gestation. Pregnant mice are known to be more susceptible to the intracellular pathogens *Listeria* and *Toxoplasma* than nonpregnant mice. They are also resistent to the induction of delayed-type hypersensitivity, one manifestation of cell-mediated immunity, and show diminished *natural killer* (NK) cell responsiveness when exposed to *Corynebacterium*, which stimulates a vigorous NK response in non-

TABLE 7.1. $T_{H2}/T_{H1}$ cytokine ratios in cell supernatants.

| | Unstimulated day 12 decidual cells | Con A-stimulated spleen cells |
|---|---|---|
| IL-10/IFNγ | 10 | 0.02 |
| IL-4/IFNγ | 600 | 3.00 |

*Source:* Reprinted with permission from Wegmann, Lin, Guilbert, and Mosmann (1).

pregnant mice. On the other hand, $T_{H2}$-type antibody-mediated immunity is enhanced during murine pregnancy, as reflected in the elevation of the plaque-forming cell response to sheep red blood cells and in the enhancement of the IgG1 versus IgG2 isotype (reviewed in 1).

## $T_{H1}$- and $T_{H2}$-Related Cytokines in Pregnant Mice

These prior observations led us to examine the immune and reproductive tissues of pregnant mice for the secretion of relevant $T_{H1}$ and $T_{H2}$ cytokines using double monoclonal antibody sandwich *enzyme-linked immunoassays* (ELISA). To summarize our unpublished findings to date, *interferon γ* (IFNγ) is present at low levels in the placenta during the first trimester. It almost disappears during the second trimester and is undetectable during the third trimester. *Interleukin-4*, *-5*, and *-10* (IL-4, IL-5, and IL-10), however, are constitutively produced by placental cells throughout gestation. The presence of IL-10 function was ascertained by testing the ability of placental supernatants to block the IL-2-induced induction of IFNγ by normal spleen cells and the prevention of that activity by treatment of the supernatants with anti-IL-10 monoclonal antibody. No constitutive cytokine production is found in draining lymph nodes from the uterus, the mesenteric lymph nodes, or the spleen. Indeed, if spleen cells are stimulated in vitro with Con A, they produce a cytokine profile that is far more skewed toward the $T_{H1}$ pattern (Table 7.1).

## IL-10 in Mouse Pregnancy

We have also begun to localize the placental production of IL-10 mRNA by in situ hybridization. To date, we have observed that IL-10 mRNA is expressed in the maternal decidua on the antimesometrial side of the uterus on day 6 of murine pregnancy. We are currently investigating its localization in the other stages of gestation and will eventually extend it to other cytokines. Kourilsky and his colleagues at the Institut Pasteur have corroborated our results by quantitative *polymerase chain reaction* (PCR) (personal communication).

In preliminary studies we have identified IL-10 mRNA in cultures of purified human trophoblast by PCR. We are in the process of developing quantitative PCR techniques for this and the other $T_{H1}/T_{H2}$ cytokines to evaluate their status in these cultures and in other components of the human placenta. Bruce Smith and his colleagues in Philadelphia have localized IL-4, -5, and -6 and IFNγ in trophoblasts within intact human placentas by in situ hybridization and immunohistochemistry (personal communication). IFNγ and *tumor necrosis factor α* (TNFα) have been reported to be expressed in the human placenta (2, 3; Chapter 10, this volume). We are now collaborating with B. Smith to localize IL-10 mRNA expression in the human placenta by in situ hybridization and are examining the $T_{H1}/T_{H2}$ cytokine profiles in placental cells and tissues. Although we postulate that the human will show a bias toward $T_{H2}$ cytokine expression, except perhaps at parturition (for reasons reviewed in 1), this point remains to be established experimentally.

## Inflammation-Associated Cytokines in Pregnancy

We know from a variety of different experiments that some cytokines are harmful for the continuance of pregnancy while others are beneficial. For example, the inflammatory cytokines IL-2, IFNγ, and TNFα can enhance murine fetal resorption when it is of low frequency or exacerbate it when it is of high frequency (4). In collaboration with Raj Raghupathy at the National Institute of Immunology and Gerard Chaouat at Hopital Clamart in Paris, we have recently observed that the $T_{H1}$ cytokines IL-2, IFNγ, and TNFα are elevated in the resorbing mouse combination CBA × DBA/2 when compared to CBA × BALB pregnancies, which show a low level of resorption. When the resorption is reversed by IFNα or alloimmunization, the $T_{H2}$ cytokines once again predominate, thus showing a correlation with prevention of abortion (unpublished observations). In addition, recombinant IL-10, prepared by John Elliott of the University of Alberta, can directly prevent abortion in CBA × DBA/2 matings when injected interperitoneally during pregnancy (unreported observations). Thus, fairly direct evidence exists for the postulate that $T_{H2}$ cytokines with anti-inflammatory effects can contribute to successful pregnancy when injected in vivo, presumably by dampening down the inflammatory cytokines that are known to be harmful to pregnancy (unpublished observations).

How do these harmful cytokines do their damage? We have recently obtained evidence, as yet unpublished, for a direct effect of these inflammatory cytokines on placental trophoblasts from human term placentas. By preparing pure trophoblasts using glass bead columns with a negative selection protocol based on CD9 antibodies, we have been able to culture pure trophoblast monolayers in culture that are capable of long-term survival without fibroblast overgrowth. These cultures are >99.9% epithelial and <0.1% mesenchymal. Thus, they are virtually pure populations of term villous trophoblast. When these cultures are exposed to physiologically relevant levels of TNFα (100 pg to 100 ng/mL), up to 40% of the DNA content is lost via apoptotic cell death, as evidenced by the appearance of fragmented DNA and a characteristic morphology. Furthermore, TNFα induces the formation of sporadic holes in the trophoblast monolayer. IFNγ shows little direct trophoblast killing, but strongly synergizes with TNFα to kill up to 70% of the cells. Furthermore, this cytokine-induced cell death is antagonized by GM-CSF in a dose-dependent fashion. We are currently exploring the mechanisms of both TNFα- and IFNγ-induced apoptosis and the GM-CSF-induced resistance to apoptotic death.

These observations have implications for the in utero transmission of infectious diseases, such as AIDS. Previous reports indicate that the appearance of focal placental inflammation (placental villitis) correlates positively with the incidence of vertical transmission of HIV (5). We postulate that maternal TNFα and/or IFNγ, up-regulated in the host response, allow the placental barrier formed by syncytiotrophoblast to be breached. The resulting focal inflammatory site lacks the protective trophoblastic epithelium, making it more likely that the infectious agent would be transmitted from the maternal to the fetal circulation.

## Summary

Overall, this complex network of anti-inflammatory cytokines dampening the effects of inflammatory cytokines at the fetomaternal interface has profound implications for fertility, spontaneous abortion, and possibly even parturition. However, we have only uncovered the tip of this iceberg; thus, we know very little about the interregulation of these two groups of cytokines in various normal and pathological states of murine pregnancy and even less about them in the human situation. No doubt a clearer picture will unfold as we further explore this regulatory network of antagonistic cytokines at the fetomaternal interface.

## *References*

1. Wegmann TG, Lin H, Guilbert L, Mosmann TR. Bidirectional cytokine interactions in the maternal-fetal relationship: is successful pregnancy a TH2 phenomenon? Immunol Today 1993;14:353–6.
2. Chen H-L, Yang Y, Hu X-L, Yelavarthi KK, Fishback JL, Hunt JS. Tumor necrosis factor alpha mRNA and protein are present in human placental and

uterine cells at early and late stages of gestation. Am J Pathol 1991;139: 327–37.
3. Bulmer JN, Morrison L, Johnson PM, Meager A. Immunohistochemical localization of interferons in human placental tissues in normal, ectopic and molar pregnancy. Am J Reprod Immunol 1990;22:109–16.
4. Chaouat G, Menu E, Clark D, Dy M, Minkowski M, Wegmann TG. Control of fetal survival in CBA × DBA/2 mice by lymphokine therapy. J Reprod Fertil 1990;89:447–58.
5. Chandwani S, Greco MA, Mittal K, Antoine C, Krasinski K, Borkowsky W. Pathology and human immunodeficiency virus expression in placentas of seropositive women. J Infect Dis 1991;163:1134–8.

# 8

# Use of the Osteopetrotic Mouse for Studying Macrophages in the Reproductive Tract

PAULA E. COHEN AND JEFFREY W. POLLARD

Macrophages are found at all levels of the reproductive tract in both males and females. They represent a constitutive cellular component of some tissues, such as the interstitium of the testis, while in other tissues their numbers may fluctuate at specific times, as is the case for the uterus. The increase in macrophage numbers in the uterus and other tissues—and their subsequent activation—is known to be regulated by the steroid hormones, estrogen and progesterone (1), and by polypeptide growth factors. These growth factors include *colony stimulating factor I* (CSF-I), also known as *macrophage colony stimulating factor* (M-CSF), *granulocyte macrophage colony stimulating factor* (GM-CSF), *transforming growth factor β1* (TGFβ1), *tumor necrosis factor α* (TNFα), and *interferon γ* (IFNγ). Moreover, there is significant evidence, in the uterus at least, to suggest that production of these growth factors in target tissues is steroid hormone regulated. Thus, it appears that steroid-induced events in the reproductive tract may be mediated by macrophages via their steroid-induced chemoattractants and growth factors.

The classical functions of macrophages may be divided broadly into three categories: (i) immunoregulation, in particular their antigen-presenting functions; (ii) phagocytosis and tissue remodeling; and (iii) synthesis of prostaglandins and polypeptide growth factors, including CSF-I, TGFβ1, TNFα, *interleukin-1* (IL-1), and *interleukin-6* (IL-6). In the mouse uterus, for example, macrophages produce and secrete large amounts of *prostaglandin* $E_2$ ($PGE_2$) (2), are able to phagocytose bacteria (3), and will express *major histocompatibility complex* (MHC) class II molecules and present antigens when activated (4). Moreover, in recent years it has become clear that macrophages may also exhibit tissue-specific properties in addition to those outlined above, depending on the tissue in which they reside. It is these additional properties of macrophages in the reproductive tract that are the focus for this chapter.

The existence of macrophages in reproductive tissues is well documented, and numerous roles have been suggested for them; however, little is known of their tissue-specific regulation. In the testis, for example, while the presence of macrophages and their close association with Leydig cells has been described in detail, it is still not known how and why these cells are present.

The *osteopetrotic* (*op*/*op*) mouse provides an excellent model in which to study these complexities since it suffers from a severe depletion in macrophages due to an inactivating mutation in the gene for the mononuclear phagocytic growth factor CSF-I (5). As a further consequence of this mutation, the *op*/*op* mouse has very few osteoclasts, which results in the absence of teeth and the bone-restructuring disorder known as *osteopetrosis*. Studies in this laboratory have demonstrated that mice homozygous for the recessive *op* mutation have reduced fertility in terms of successful pregnancies and numbers of live offspring (5). Furthermore, the absence of CSF-I appears to affect both male and female fertility in these mice since *op*/*op* × *op*/*op* crosses yield far fewer pregnancies than when +/*op* males are crossed with *op*/*op* females. This suggests that macrophages and macrophage-derived signals are vital to reproductive functioning in both males and females. The aim of this chapter is to provide evidence that macrophages are required at all levels of the reproductive tract and to demonstrate, with reference to the macrophage-deficient osteopetrotic mouse, the importance of these macrophage-associated events in the functioning of the reproductive system.

## Macrophages in the Gonads

### *Ovary*

Follicular development requires complex interactions between the granulosa and thecal cells surrounding the developing oocyte. In addition, significant histological changes take place within the developing follicle and in the subsequent formation of the corpus luteum that require rapid tissue remodeling, angiogenesis, and leukocyte invasion. The control mechanisms involved are both autocrine and paracrine in nature and may involve gonadotropins, ovarian steroids, and certain growth factors. Numerous cytokines have been identified within the ovary, including *fibroblast growth factor* (FGF), *epidermal growth factor* (EGF), *insulin-like growth factor I* (IGF-I), *platelet-derived growth factor* (PDGF), GM-CSF (6), IL-6 (6), CSF-I (7), TNFα (2), TGFβ1, IFNγ, and IL-1 (reviewed in 8).

These growth factors, together with their ovary-specific actions, are listed in Table 8.1. While in situ and immunohistochemical methods have

TABLE 8.1. Cytokine and growth factor synthesis in the ovary.

| | Follicle layers (mRNA present) | | | Receptors (to cytokines) | | | Ovarian functions | | | |
|---|---|---|---|---|---|---|---|---|---|---|
| Cytokine | GC | TC | CL | GC | TC | CL | GC prolif. | Foll. growth | $P_4$ output | LH-R and FSH-R |
| EGF | + | + | + | + | (+) | + | ↑ | ↑ | ↑ | ↑ |
| FGF | (+) | + | ? | + | + | ? | ↑ | ↑ | ↑↓ | ? |
| TGFα | + | + | + | + | ? | ? | ↑ | ↑ | ? | ? |
| TGFβ | + | + | + | + | ? | ? | ? | ↑ | ↑ | ? |
| TNFα | + | + | + | + | + | + | ↑ | ↑ | ↑↓ | ↓ |
| IFNγ | + | ? | ? | ? | ? | ? | – | – | ↓ | ↓ |
| IL-1 | – | + | ? | + | + | ? | ↑ | ↑ | ↓ | ↓ |
| CSF-I | ? | ? | ? | ? | ? | ? | ? | ? | ? | ? |
| GM-CSF | +? | +? | + | ? | ? | ? | ? | ? | ? | ? |

*Note:* Localization of growth factor mRNA to specific layers within the ovary does not necessarily indicate that the ovarian cells within that layer are the source of the mRNA. Instead, it may be that macrophages present within the cell layer may be responsible for the transcription of the growth factor gene. (GC = granulosa cell; TC = theca cell; CL = corpus luteum; + indicates that the mRNA or receptor for the growth factor is present; (+) indicates that the mRNA or receptor may be present under certain circumstances or only in certain species; – indicates that no mRNA or receptor is present; ? indicates that the presence of mRNA or receptor for the growth factor has not yet been determined.) The effect of these growth factors on various ovarian parameters, namely granulosa cell proliferation, follicular growth, progesterone ($P_4$) output, and LH and FSH receptor numbers (LH-R and FSH-R) are indicated by arrows: ↑ = positive effect and ↓ = negative effect. Data taken from references given in the text.

shown that virtually all cell types in the ovary are capable of cytokine production, it is probable that macrophages are responsible for the majority of ovarian cytokine production. The mRNA for TNFα, for example, has been localized to both theca and granulosa cells of the mouse ovary, but the protein is not observed in these cell layers unless macrophagelike cells are also present, indicating that macrophages may be the primary source for TNFα (9). Similarly, TGFα, TGFβ, and IFNγ are all present during follicular development, but their sites of production remain uncertain.

Macrophages are distributed throughout the ovary of rats (10–12), mice (13), and humans (14). They constitute a major cellular component of the interstitial tissue between developing follicles and are present in the thecal cell layers and occasionally in among the granulosa cells (11, 13). Their numbers are elevated prior to ovulation throughout the theca and medullary regions of the ovary (12). The macrophages are excluded from the developing follicle itself, but are able to infiltrate the newly formed corpus luteum following ovulation in mice (13) and rats (15). Luteolysis is associated with a rapid accumulation of macrophages within the corpus luteum, and this also occurs in follicles undergoing atresia. In the human ovary macrophages increase in number during follicular development and ovulation (16) and are found in significant numbers in human follicular fluid (17, 18). Macrophages are almost entirely absent from the ovaries of *op/op* females, but can be restored partially by continuous treatment with CSF-I from birth (Pollard, Dominguez, Chisholm, Stanley, unpublished observations). This suggests an important role for CSF-I in the regulation of ovarian macrophages.

Macrophages are thought to play a role in regulation of steroidogenesis in the ovary since splenic macrophages significantly enhance progesterone secretion from cultured rat granulosa cells in combination with prolactin (19). Furthermore, macrophages are known to stimulate granulosa cell proliferation (11) and the accumulation of LH receptors within cultured granulosa cells (20). Macrophage-conditioned media also stimulate progesterone production by porcine granulosa cells in vitro (20), suggesting that macrophage-derived secretory products are involved. These effects may be mediated by macrophage-derived IL-1 or TGFβ since both of these cytokines also stimulate the proliferation of granulosa cells in vitro (21, 22).

However, the role of cytokines in vivo remains unclear since they appear to have different effects on steroidogenesis depending on the cell type and endocrine status of the ovary at the time of investigation. For example, IL-1, which is produced primarily in the thecal layers (23), stimulates progesterone production in preovulatory rat follicles and in cultured thecal cells (24), but inhibits androgen production by these cells (25). Furthermore, IL-1 inhibits both basal and FSH-induced progesterone secretion by cultured granulosa cells taken from immature or developing follicles (21), but appears to lose this ability once the granulosa cells become luteinized (21, 26). Therefore, IL-1 appears to affect steroidogenesis in both thecal and granulosa cells, and both of these actions are thought to be direct, rather than through cell-to-cell mediators (25). Levels of IL-1 increase by about 5-fold just prior to ovulation (23), and there are suggestions that it may be involved in the process of ovulation by sufficiently increasing progesterone levels to induce follicle rupture (24, 27).

The role(s) of TNFα in ovarian steroidogenesis is also unclear since this cytokine appears to have a variety of actions, much like IL-1. In granulosa cell cultures from immature rat follicles, TNFα inhibits FSH-stimulated progesterone secretion (28, 29), but increases progesterone output in granulosa cells from proestrous rats and human preovulatory follicles (30, 31). Following luteinization, however, progesterone secretion appears to become insensitive to TNFα, while estrogen output is reduced by TNFα at this time (26). TNFα is thought to have direct effects on theca, granulosa, and luteal cells since its receptors are distributed throughout these cell types (29, 32).

The presence of such large numbers of macrophages within the ovary suggests that these cells may also play a role in follicular development.

This is supported by the fact that the proportion of macrophages associated with developing follicles is higher than that associated with mature follicles (11). In rats ovulation is associated with a dramatic rise in ovarian cytokines, including IL-1, IL-6, GM-CSF, and TNFα (6). In addition, CSF-I treatment significantly increases the number of ovulated oocytes per cycle, and ovulation is inhibited by antibodies to CSF-I (33). Furthermore, in situ hybridization analysis of *CSF-I receptor* (CSF-I-R) mRNA expression in the mouse ovary has shown that the receptor is located in the oocyte from stage III of development, as well as on ovarian macrophages. It persists in the oocyte right through until fertilization, when the message is degraded to undetectable levels (34, 35). Thus, in addition to its role in regulating ovarian macrophage populations, CSF-I may also be involved in the induction of ovulation through the interaction of maternal CSF-I and its receptors present on the oocyte. These reports are further supported by the observation that *op*/*op* mice have irregular estrous cycles and, therefore, do not ovulate as frequently as do wild-type or +/*op* females (Cohen, Pollard, unpublished observations).

CSF-I mRNA is located in isolated mouse cumulus cells (35), and both CSF-I and the CSF-I-R mRNA are present among human granulosa cell populations (36). However, any endogenous CSF-I involved in ovulation is likely to be of macrophage origin or concentrated from the serum since there is no evidence to suggest that either the granulosa or theca cells produce CSF-I.

Recent evidence has pointed toward a role for developing follicles in the control of theca and granulosa cell function. Microsurgical removal of oocytes from FSH-stimulated follicles in vitro results in a 17- to 36-fold increase in progesterone secretion (37) that is reduced in the presence of oocyte-conditioned media. Similarly, follicles lacking oocytes produced 30% less estradiol, but 2-fold more estradiol in the presence of oocyte-conditioned media (37), suggesting the presence of a soluble steroidogenic stimulus. The involvement of TNFα in these processes is suggested by the observation that the TNFα gene is transcribed in the mouse oocyte (9) and that transcription is initiated in a stage-specific manner. The TNFα transcript and the protein itself are present in the follicle from developmental stage IV (the stage at which 2 layers of granulosa cells surround the oocyte), but the protein levels decline as the preovulatory stage approaches.

In summary, it is clear that macrophages and their secretory products are vital to both follicular development and steroidogenesis. However, the source and nature of the chemotactic stimuli responsible for inducing macrophage infiltration into the ovary remain unclear. Possible sources for these factors are the theca or granulosa cells or the macrophages themselves, with the macrophages being the major contributor. If CSF-I is involved in macrophage recruitment and activation in the ovary, then it is likely that the ovary of the *op*/*op* mouse will differ dramatically from that of normal mice in terms of histology, steroidogenic capability, and follicular development. Indeed, preliminary studies in our laboratory have indicated that the absence of macrophages in the ovaries of these mice disrupts ovulation, either by altering steroidogenesis, or by interfering with follicular development, or both. These studies indicate that CSF-I plays an important role in the control of ovarian function, either by a direct action on ovarian cells or by its ability to induce macrophage recruitment, proliferation, and cell spreading in the ovary. A model for the role of macrophages in the ovary is given in Figure 8.1.

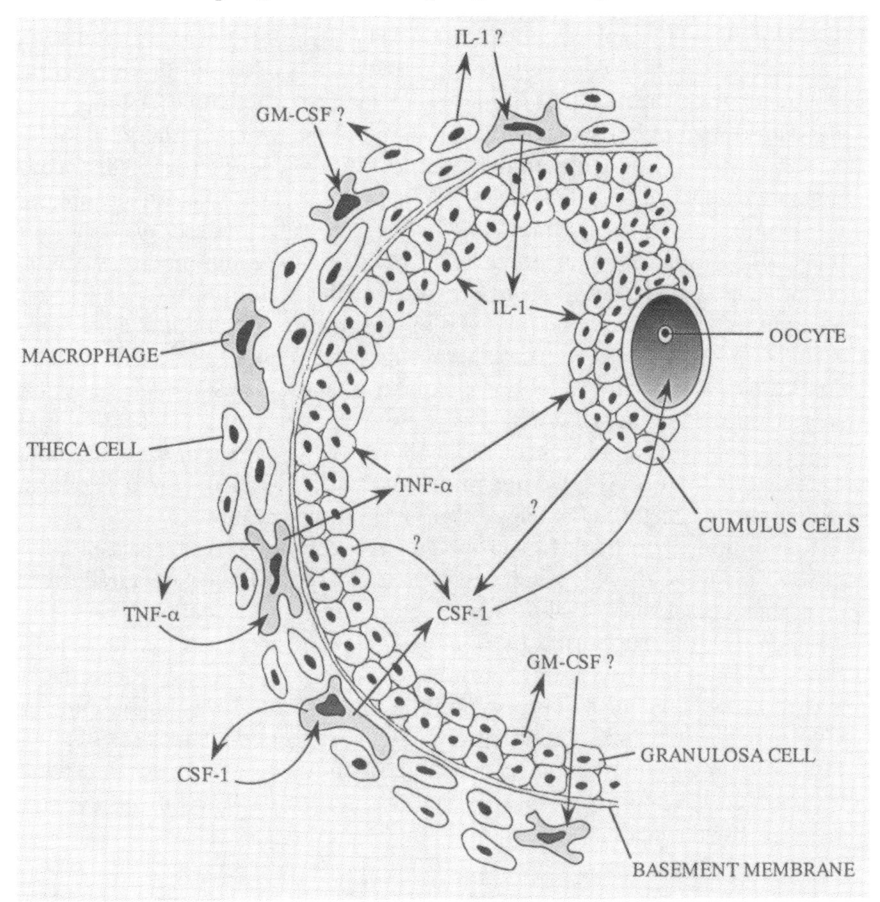


FIGURE 8.1. A model for the role of macrophages in the ovary. A diagrammatic representation of an antral follicle with surrounding theca-interstitial cells. Activated macrophages present in the thecal layer (and possibly in the granulosa cell layer) secrete CSF-I, TNFα, and IL-1, all of which induce the recruitment, proliferation, and activation of other macrophages. These factors may have additional effects on granulosa cell function and on the developing follicle. In addition, GM-CSF, IL-1, and TNFα may be synthesized by the thecal cells, and CSF-I may be synthesized by granulosa cells.

## *Testis*

In the rat testis macrophages constitute ~25% of interstitial cells (38, 39), with no macrophage populations being observed inside the seminiferous tubules. These testicular macrophages are arranged in cluster formation around the neighboring Leydig cells such that the microvilli of the Leydig cells fit into coated vesicles on the macrophage surface (38, 40, 41). Macrophages first appear in the rat testis at day 19 of gestation and increase 15-fold during the first 50 days of postnatal life (42). Furthermore, the testicular macrophages undergo a 2-fold increase in size between days 13 and 47 postpartum. The distinct digitations between macrophages and Leydig cells form between days 20 and 30 postpartum (43), just prior to the surge of androgen secretory activity of Leydig cells associated with puberty. In culture, the macrophages maintain their ability to adhere to Leydig cells, thus forming distinct rosettes (44). This property is Leydig cell specific since no adherence occurs when macrophages are cocultured with spermatozoa or Sertoli cells. These observations suggest that macrophages are able to adhere to specific binding sites on the Leydig cell surface that are not present in the intratubular cells. It is possible that CSF-I may be involved in this process, such that cell surface-bound CSF-I on the Leydig cell membrane would bind to CSF-I-R located on the surface of the neighboring macrophages. These suggestions are supported by the fact that CSF-I is present in the adult testis (7) and that macrophages are significantly reduced in the testis of *op/op* mice and are partially restored by daily injections of CSF-I from birth (Pollard, Dominguez, Chisholm, Stanley, unpublished observations).

Regulation of testicular macrophages is likely to come from a number of sources, none of which is necessarily exclusive: (i) macrophage-derived cytokines, (ii) Leydig cell-derived factors, and (iii) gonadotropins. This last possibility is supported by numerous lines of evidence, including the observations that testicular macrophages from adult rat testes have high-affinity, low-capacity receptors for FSH and are FSH responsive both in vivo and in vitro (45). Furthermore, hCG treatment of rats results in a rapid rise in the number of testicular macrophages, suggesting that these cells are LH responsive (46). Vasectomy, which causes a drop in circulating testosterone and a rise in both FSH and LH, is associated with Leydig cell and macrophage hypertrophy in the testis (47). Thus, gonadotropins may have a direct effect on testicular macrophages via their specific receptors located on the surface of the macrophages or may exert their effects indirectly by altering local production of CSF-I by testicular cells.

As in females, there is evidence to suggest that macrophages are involved in the control of gonadal steroidogenesis in males. Macrophage-conditioned media are known to increase testosterone secretion by Leydig cells in vitro (45). Initial studies on male *op/op* mice have indicated that their serum testosterone levels are significantly diminished as compared with heterozygote littermates (+/*op*). However, these observations are complicated by the fact that macrophages inhibit LH-stimulated testosterone output from Leydig cells in culture (48), but do not inhibit LH binding. This inhibition is not thought to be induced by IL-1 (48), suggesting that other macrophage factors are involved. However, IL-1 mRNA has been detected in testicular macrophages (49), and the protein itself appears to have differing effects on Leydig cell function in culture (50–52). Clearly, the role(s) of IL-1 in the testis in vivo has yet to be clarified.

Conditioned media from testicular macrophages contain high levels of TNFα activity (43), and TNFα is capable of suppressing both basal and 8-Br-cAMP-stimulated testosterone production in vitro (52–54). Moreover, treatment of Leydig cells with TNFα causes a significant and dose-dependent decrease in the mRNA and protein for the steroidogenic enzymes, $P450_{17\alpha}$ and $P450_{scc}$ (54). However, the mechanisms by which testicular macrophages are activated into producing TNFα in vivo remain unclear since both FSH and testosterone have no effect on testicular TNFα (43). Furthermore, there is some debate as to whether lipopolysaccharide is able to increase testicular TNFα output (43, 55).

Testicular macrophages are not present in the compartment in which spermatogenesis takes place. It is possible that macrophage secretory products are capable of traversing the tubular basement membrane to enter the seminiferous tubules and thereby influence spermatogenesis or Sertoli cell function. However, as yet, no evidence exists to suggest that either TNFα or IL-1 is able to cross into the seminiferous tubules, nor are data available to suggest that they have any effect on spermatogenesis. Interestingly, a recent study has shown that Sertoli cells possess receptors for TNFα and that two different TNFα messages are present in spermatid cells (56). One of these, a 1.9-kb transcript, is found only in pachytene spermatids and is induced in vitro by lipopolysaccharide. The 2.8-kb message, on the other hand, is found only in round spermatids and is insensitive to lipopolysaccharide (56).

Studies performed in this laboratory on the mating success of *op*/*op* mice indicated that there is some sort of male factor infertility in these animals as a consequence of the depleted macrophages. These studies showed that while *op*/*op* females mated with +/*op* males yielded a 59% successful pregnancy rate, *op*/*op* females mated with *op*/*op* males produced less than a 1% success rate (5). However, matings of +/*op* females with *op*/*op* males produce a success rate of 93%, a figure that is similar to heterozygote matings, suggesting that any male fertility disorders may be overcome in nonmutant females. If such fertility problems arise at the testicular level, then one would expect a much lower success rate for +/*op* female by *op*/*op* male crosses since spermatogenesis could not be repaired by mating with heterozygote females. Thus, it is more likely that the problems arise at the sperm maturation stage in the epididymis, but can be corrected once the sperm reach the female tract. This would also suggest that the reduced testosterone levels in *op/op* mice do not affect spermatogenesis or Sertoli cell function, but may influence the male accessory glands.

In summary, macrophages constitute a major proportion of the interstitial space of the testis, and as such, they are well placed to play a crucial role in the control of Leydig, and possibly Sertoli, cell function.

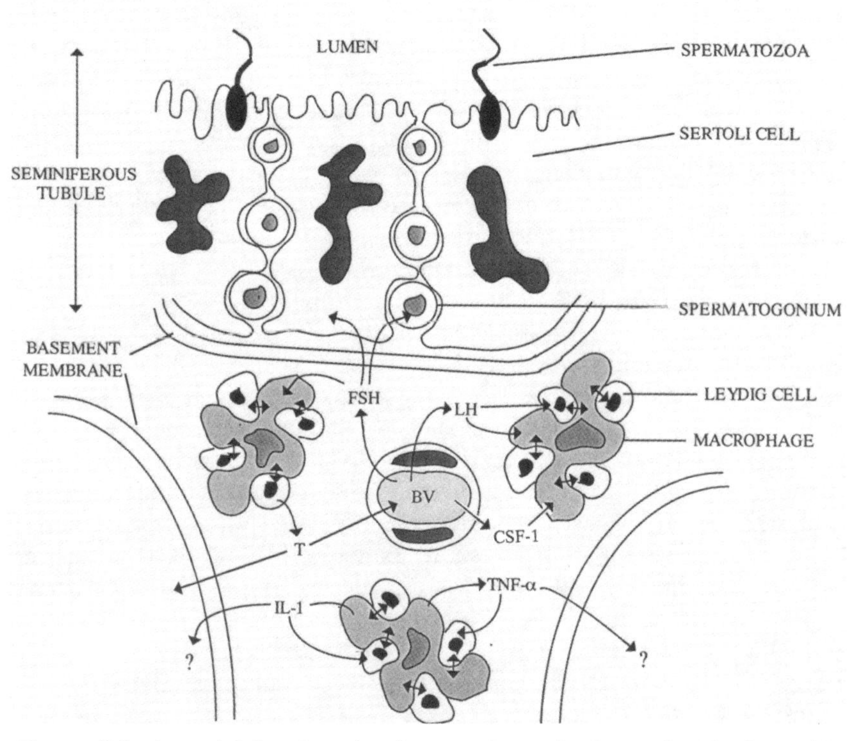


FIGURE 8.2. A model for the role of macrophages in the testis. The interstitial space of the testis contains Leydig cells, macrophages, and blood vessels (BV), while the seminiferous tubule contains Sertoli cells, spermatogonia, and spermatozoa. Circulating CSF-I is targeted to the macrophages of the interstitial space, along with follicle stimulating hormone (FSH) and luteinizing hormone (LH). FSH and LH also stimulate the testicular cells to produce testosterone (T) (Leydig cells) and to support spermatogenesis (Sertoli cells). The T is released into the circulation and also stimulates spermatogenesis. Macrophages, in response to CSF-I, LH, and FSH, produce additional cytokines, such as IL-1 and TNFα, that may act on Sertoli cells and Leydig cells. Finally, the close association between Leydig cells and macrophages may be mediated by CSF-I binding to its receptor, as illustrated by 2-way arrows between the macrophages and Leydig cells.

Testicular macrophages are unusual in that they are responsive to gonadotropins and have intrinsic binding sites for Leydig cells. In addition, they are able to influence testosterone production in some way, either directly through surface-bound molecules or through secretory products (Fig. 8.2).

## Influence of Macrophages on Gamete Transport and Development

Very little information is available concerning the influence of macrophage and macrophage secretory products on gamete transport and development. As mentioned above, studies on the *op/op* mouse have demonstrated a possible male fertility defect at the level of the epididymis or beyond. This defect may arise as a consequence of the absence of CSF-I or the depletion in resident macrophages.

Macrophages are present at all levels of the male reproductive tract, but are highest in the caput epididymis and vas deferens. In *op/op* males macrophages are significantly reduced in all areas of the reproductive tract, as well as in the prostate and seminal vesicles, and this may help to explain the fertility defects described above. Interestingly, reconstitution of these animals from birth by daily injections of human recombinant CSF-I fully restores caput epididymis macrophage numbers, but only partially restores macrophages to all other parts of the reproductive tract (Pollard, Dominguez, Chisholm, Stanley, unpublished observations), suggesting a need for locally produced CSF-I in these latter sites.

Macrophages are present in large numbers in seminal fluid, arriving there either from the accessory glands or by migration through the wall of the epididymis or vas deferens. They are capable of engulfing whole sperm heads and even sperm with multiple heads or tails. Macrophages in human semen are often found to contain sperm fragments (57), suggesting a role for macrophages in the phagocytosis of sperm. Furthermore, increased macrophage numbers are associated with high-quality human sperm and are independent of sperm count, suggesting that they are involved in phagocytosis of abnormal or dysfunctional sperm (57). Indeed, the discrepancy between sperm production and the number of sperm ejaculated would certainly agree with this observation (58). Sperm are also phagocytosed by leukocytes in the cervix of the female tract (59).

There have been few studies on the role of cytokines on sperm maturation in vivo. In vitro, at least, it seems that only IFNγ and TNFα have any appreciable effect on sperm maturation parameters, having negative effects on sperm motility and sperm penetration (60). Whether these cytokines affect extratesticular sperm maturation in vivo remains to be seen.

Following ovulation in females, the oocyte remains in the oviduct in the time leading up to, and immediately following, fertilization. Therefore, the majority of its maturation takes place within the oviductal environment. Under such circumstances it is likely that macrophages and/or macrophage products present in the oviduct would exert significant effects over the oocyte that could influence both its viability and development. Macrophages are present in the oviduct throughout the cycle and in early pregnancy and have been shown to actively transcribe the TNFα gene (61). Furthermore, the expression of this gene in oviductal macrophages is not dependent on CSF-I since TNFα is also expressed in the oviduct of *op/op* mice (61).

## Macrophages in the Uterus

In virgin mice macrophages are distributed throughout both the endometrium and myometrium (4), as well as in the mesometrial triangle (5). Their distribution within the endometrium is thought to be controlled by the ovarian steroids (62), such that distinct distributional changes occur throughout the course of the estrous cycle as the steroid levels fluctuate. The majority of endometrial macrophages are present within stroma and, to a lesser extent, close to the luminal epithelium (4). Mating and pregnancy are associated with a dramatic rise in macrophage numbers within the endometrium and myometrium, with the majority of macrophages being localized to nondecidual tissue in both mice (5) and rats (63) and in the mesometrial triangle in mice (5). Interestingly, there is a complete absence of macrophages in the primary decidual zone. In humans endometrial macrophage numbers appear to be fairly stable during the menstrual cycle (64), but are abundant throughout the decidua from early on in the first trimester of pregnancy (65). High macrophage numbers persist throughout pregnancy in both mice and humans.

Macrophages are thought to be involved in a wide number of uterine processes, involving both their classical immune functions and their secretory properties. They are activated by a number of stimulants, including bacterial-derived endotoxins, IFNs, and other cytokines. In their activated state they secrete large amounts of cytokines that may have wide-ranging effects on neighboring cell types and may take part in numerous steroid-induced processes. However, macrophages are best known for their phagocytic properties, and uterine macrophages are able to phagocytose invading microorganisms, such as *Listeria monocytogenes* (3). During pregnancy in the mouse, uterine macrophages secrete $PGE_2$, which inhibits the proliferation of lymphocytes in vitro (2). In this way the macrophages may reduce the possibility of fetal rejection by reducing the local population of cytotoxic T cells. In addition, macrophages may release

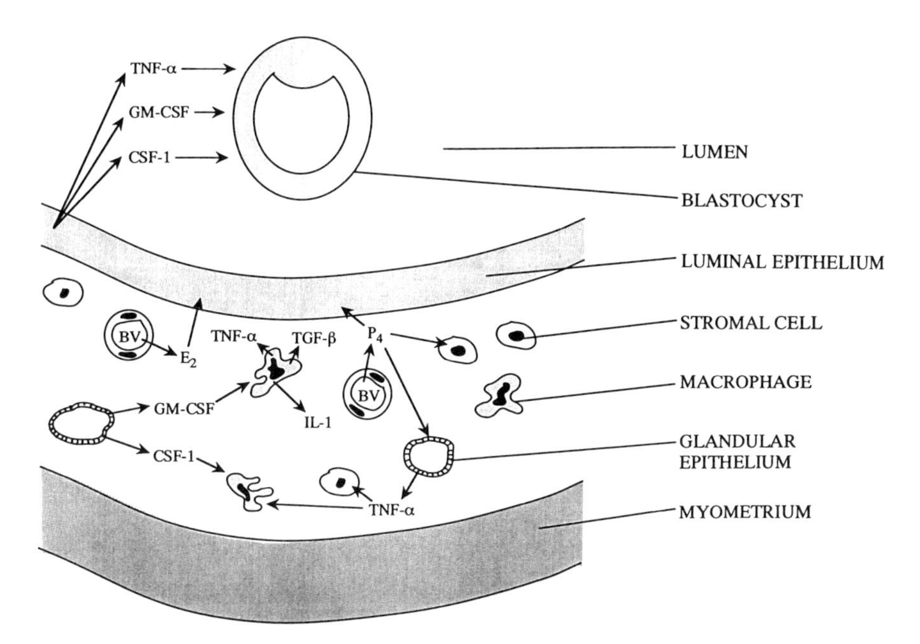


FIGURE 8.3. A model for the role of macrophages in the uterus during implantation. The uterine epithelium, in response to circulating estradiol ($E_2$) and progesterone ($P_4$), secretes TNFα, GM-CSF, and CSF-I, which are targeted to the blastocyst and to the macrophages present within the endometrial stroma. The macrophages in turn release various cytokines, including CSF-I, TNFα, TGFβ, and IL-1, all of which are targeted to nearby stromal cells. These stromal cells are then able to undergo decidualization in response to the implanting blastocyst.

additional immunosuppressive agents, such as TNFα (66) and TGFβ1 (67).

The invasion of the uterine tissue by macrophages during pregnancy is under the control of the ovarian steroids (9, 62, 68), either directly or indirectly, via chemoattractants. (See Fig. 8.3 for summary.) Potential chemotactic agents in the uterus include CSF-I, GM-CSF, TNFα, and TGFβ1. These are ideal candidates since they share the following fundamental properties: (i) They are known to be steroid induced (69–73); (ii) they are produced by specific cells within the endometrium (7, 66, 67, 74–76); (iii) their levels rise during pregnancy (7, 61, 67, 72, 75); and (iv) these rises are in line with the increase in uterine macrophage numbers.

The distribution of chemotactic cytokines within the endometrium also shows distinct similarities; each may or may not be expressed in the nonpregnant state, but the major region of expression following mating is consistently the uterine luminal and glandular epithelium (61, 67, 72, 77). During pregnancy GM-CSF is additionally expressed in the non-decidualized endometrium (72), while the mRNA for TNFα and TGFβ1 are located in the primary decidual zone (61, 67). In addition, both the CSF-I and TNFα genes are actively transcribed in human placental tissue, while TNFα is also transcribed in the mouse placenta (75, 78).

CSF-I is produced by the luminal epithelial cells of the endometrium (69). By implantation the uterine levels of CSF-I are 5 times higher than nonpregnant levels, and by term the levels are 1000-fold higher than in the nonpregnant state (77). The CSF-I is probably required for the recruitment of macrophages into the uterus and is likely to be assisted in the early preimplantation stage by GM-CSF (79). Following implantation, however, CSF-I may be the major chemotactic stimulus for macrophages entering the uterus and also stimulates their survival, proliferation, and spreading within the endometrium. In *op/op* females macrophages are present in reasonable numbers from day 7 of pregnancy, but they are abnormal in shape and completely disappear by day 14 (5). This suggests that factors other than CSF-I are capable of inducing macrophage recruitment, but that CSF-I is required for their viability and survival in the uterus. One likely candidate is TNFα, whose expression in the uterine epithelium, decidua, and trophoblast is similar in *op/op* and +*/op* animals (61) and is therefore independent of CSF-I levels and macrophage numbers. The absence of functional macrophages during pregnancy will reduce the uterine production of cytokines that may ultimately be the cause of the pregnancy failures seen in these mice.

## Summary

Macrophages appear to play a large role in the control of reproductive functioning, having diverse roles at all levels of the reproductive tract in both males and females. Aside from the traditional immunological roles assigned to macrophages in the past, macrophages of the reproductive tract appear to play an important role in cell-to-cell interactions within each system. Moreover, it is apparent that the macrophages are able to take on extremely specialized roles depending on the tissue into which they are recruited. In the testis, for example, the macrophages acquire the ability to respond to gonadotropins and subsequently to influence steroidogenesis in neighboring Leydig cells.

It is interesting to note that there are numerous similarities between macrophage distribution and functions in the ovary and testis. For example, the macrophages are unable to gain access to the compartment most proximate to the developing gamete in both male and female gonads, although they are able to infiltrate more distal areas, such as the granulosa and thecal layers of the developing follicle and the Leydig/interstitial layers of the testis. Thus, macrophages cannot come into direct cell-to-cell contact with the gametes until they enter the oviduct or epididymis, where more direct interactions may be possible. Nevertheless, macrophages are able to influence gamete maturation prior to fertilization: In the ovary macrophages are able to influence follicular development indirectly via the cytokines, and it is possible that macrophage-derived cytokines are also able to cross into the seminiferous tubules of the testis. Despite the tight blood-testis barrier that exists between the interstitial compartment and the interior of the seminiferous tubules, testosterone and other substances are able to enter the Sertoli cells, suggesting that cytokines, too, may cross over.

The role of macrophages in reproductive functioning beyond the gonads varies considerably between males and females. Macrophages in the male reproductive tract appear to have a more traditionally immune role and to be responsible for phagocytosis of abnormal sperm. In the oviduct the effects of macrophages on oocyte transport remain uncertain, but the macrophages in this region are known to transcribe TNFα. Finally, macrophages are thought to play vital and diverse roles in the uterus during pregnancy that include the release of cytokines known to assist in cell signaling between the different levels of the endometrium, between the uterus and blastocyst, and between the maternal and fetal components of the placenta.

The *op/op* mouse has been particularly useful in elucidating macrophage regulation in the male and female reproductive tracts. Illustrative of the complexities involved in this regulation is the observation that macrophages, while essentially absent from nonpregnant uteri, are present in early pregnancy in *op/op* females, but disappear from about day 14 onward. This would imply that there is a factor(s) present in the uterus that can compensate for the lack of CSF-I, but that it can only do so during early gestation. Thus, while it appears that the absence of macrophages in the latter part of pregnancy does not prevent continuation to term, the reduced litter sizes in *op/op* mice suggest that there may be some influence of fetal survival once the pregnancy is established. Similar studies to these using other strains of mice with mutations affecting macrophage populations, such as a null mutation for GM-CSF or the tissue-specific ablation of macrophage populations (80), will further delineate the regulation and function of various populations of macrophages in the reproductive tract.

*Acknowledgments.* This work was supported by grants from the American Cancer Society (DB-28) and The Chiron Corp., CA. J.W. Pollard is a Monique Weill-Caulier Scholar.

## *References*

1. Finn CA, Pope M. Control of leukocyte infiltration into the decidualized uterus. J Endocrinol 1986;110:93–6.

2. Tawfik OW, Hunt JS, Wood GW. Implication of prostaglandin E2 in soluble factor-mediated immune supression by murine decidual cells. Am J Reprod Immunol Microbiol 1986;12:111–7.
3. Redline RW, Lu CY. Specific defects in the anti-listerial immune response in discrete regions of the murine uterus and placenta account for susceptibility to infection. J Immunol 1988;140:3947–55.
4. Hunt JS, Manning LS, Mitchell D, Selanders JR, Wood GW. Localization and characterization of macrophages in the mouse uterus. J Leuk Biol 1985;38:255–65.
5. Pollard JW, Hunt JS, Wiktor-Jedrzejczak W, Stanley ER. A pregnancy defect in the osteopetrotic (*op/op*) mouse demonstrates the requirement for CSF-1 in female fertility. Dev Biol 1991;148:273–83.
6. Brännstrom M, Norman RJ, Seamark RF, Robertson SA. The rat ovary produces cytokines during ovulation. Biol Reprod 1994;50:88–94.
7. Bartocci A, Pollard JW, Stanley ER. Regulation of colony-stimulating factor-1 during pregnancy. J Exp Med 1986;164:956–61.
8. Carson RS, Zhang Z, Hutchinson LA, Herington LA, Findlay JK. Growth factors in ovarian function. J Reprod Fertil 1989;85:735–46.
9. Chen H-L, Marcinkiewicz JL, Sancho-Tello M, Hunt JS, Terranova PF. Tumor necrosis factor alpha gene expression in mouse oocytes and follicular cells. Biol Reprod 1993;48:707–14.
10. Bulmer D. The histochemistry of ovarian macrophages in the rat. J Anat 1968;98:313–9.
11. Fukumatsu Y, Katabushi H, Naito M, Takeya M, Takahashi K, Okamura H. Effect of macrophages on proliferation of granulosa cells in the ovary of rats. J Reprod Fertil 1992;96:241–9.
12. Brännstrom M, Mayrhofer G, Robertson SA. Localization of leukocyte subsets in the rat ovary during the periovulatory period. Biol Reprod 1993; 48:277–86.
13. Hume DA, Halpin D, Charlton H, Gordon S. The mononuclear phagocyte system of the mouse defined by immunohistochemical localization of antigen F4/80: macrophages of endocrine organs. Proc Natl Acad Sci USA 1984; 81:4174–7.
14. Wang LJ, Pascoe V, Petrucco OM, Norman RJ. Distribution of leukocyte subpopulations in the human corpus luteum. Hum Reprod 1992;7:197–202.
15. Brännstrom M, Giesecke L, Moore IC, van den Heuval CJ, Robertson SA. Leukocyte subpopulations in the rat corpus luteum during pregnancy and pseudopregnancy. Biol Reprod 1994 (in press).
16. Katabuchi H, Fukumatsu Y, Okamura H. Immunohistochemical and morphological observations of macrophages in the human ovary. In: Hirshfield AN, ed. Growth factors and the ovary. New York: Plenum Press, 1989: 409–13.
17. Khan SA, Schmidt K, Hallin P, Pauli R, De Geyter CH, Nieschlag E. Human testis and ovarian follicular fluid contain high amounts of interleukin-1-like factor(s). Mol Cell Endocrinol 1988;58:221–30.
18. Loukides JA, Loy RA, Edwards R, Honig J, Visintin I, Polan ML. Human follicular fluid contains tissue macrophages. J Clin Endocrinol Metab 1990; 71:1363–7.

19. Yamanouchi K, Matsuyama S, Nishihara M, Shiota K, Tachi C, Takahashi M. Splenic macrophages enhance prolactin-induced progestin secretion from mature rat granulosa cells in vitro. Biol Reprod 1992;46:1109–13.
20. Chen TT, Lane TA, Doody MC, Caudle MR. The effect of peritoneal macrophage-derived factor(s) on ovarian progesterone secretion and LH receptors: the role of calcium. Am J Reprod Immunol 1992;28:43–50.
21. Fukuoka M, Mori T, Taii S, Yasuda K. Interleukin-1 stimulates growth and inhibits progesterone secretion in cultures of porcine granulosa cells. Endocrinology 1989;124:884–90.
22. Bendell JJ, Dorrington J. Estradiol-17β stimulates DNA synthesis in rat granulosa cells: action mediated by transforming growth factor-b. Endocrinology 1991;128:2663–5.
23. Hurwitz A, Richiarelli E, Botero L, Rohan RM, Hernandez ER, Adashi EY. Endocrine and autocrine-mediated regulation of rat ovarian (theca-interstitial) interleukin-1B gene expression: gonadotrophin-dependent preovulatory aquisition. Endocrinology 1991;129:3427–9.
24. Brännstrom M, Wang L, Norman RJ. Effects of cytokines on prostaglandin production and steroidogenesis of incubated preovulatory follicles of the rat. Biol Reprod 1993;48:165–71.
25. Hurwitz A, Payne DW, Hackman JN, et al. Cytokine-mediated regulation of ovarian function: interleukin-1 inhibits gonadotrophin-induced androgen biosynthesis. Endocrinology 1991;129:1250–6.
26. Fukuoka M, Yasuda K, Emi N, et al. Cytokine-modulation of progesterone and estradiol secretion in cultures of luteinized human granulosa cells. J Clin Endocrinol Metab 1992;75:254–8.
27. Brännstrom M, Jansen PO. Progesterone is a mediator in the ovulatory process in the in vitro perfused ovary. Biol Reprod 1989;40:1170–7.
28. Adashi EY, Resnick CE, Crofts CS, Payne DW. Tumor necrosis factor alpha inhibits gonadotrophin hormonal action in non-transformed ovarian granulosa cells: a modulatory non-cytotoxic property. J Biol Chem 1993;264: 11591–7.
29. Veldhuis JD, Ganney JC, Urban RJ, Demers LM, Aggarwal BB. Ovarian actions of tumor necrosis factor-alpha (TNF-α): pleiotropic effects of TNF-α on differentiated functions of un-transformed swine granulosa cells. Endocrinology 1991;129:641–8.
30. Roby KF, Terranova PF. Tumor necrosis factor alpha alters follicular steroidogenesis in vitro. Endocrinology 1988;123:2952–4.
31. Yan Z, Hunter V, Weed J, Hutchison S, Lyles R, Terranova P. Tumor necrosis factor-α alters steroidogenesis and stimulates proliferation of human ovarian granulosa cells in vitro. Fertil Steril 1993;59:332–8.
32. Andreani CL, Payne DW, Packman JN, Resnick CE, Hurwitz A, Adashi EY. Cytokine-mediated regulation of ovarian function: tumor necrosis factor alpha inhibits gonadotrophin-supported ovarian androgen biosynthesis. J Biol Chem 1991;266:6761–6.
33. Nishimura K, Tanaka N, Fukumatsu Y, Matsura K, Okamura H. The effect of macrophage colony-stimulating factor (M-CSF) on folliculogenesis and ovulation in the gonadotrophin-treated immature rat [Abstract]. Biol Reprod 1993;48:64.

34. Arceci RJ, Pampfer S, Pollar JW. Role and expression of colony-stimulating factor-1 and steel factor receptors and their ligands during pregnancy in the mouse. Reprod Fertil Dev 1992;4:619–32.
35. Arceci RJ, Pampfer S, Pollard JW. Expression of CSF-1/c-fms and SF/c-kit mRNA during preimplantation mouse development. Dev Biol 1992;151:1–8.
36. Witt BR, Barad DH, Cohen BL, Pollard JW. Colony stimulating factor-1 in human follicular fluid [Abstract]. Soc Gynecol Invest 1993;40:P289.
37. Vanderhyden BC, Cohen JN, Morley P. Mouse oocytes regulate granulosa cell steroidogenesis. Endocrinology 1993;133:423–6.
38. Miller SC. Localization of plutonium-241 in the testis: an interspecies comparison using light and electron microscopy. Int J Radiat Biol 1982; 41:633–43.
39. Niemi M, Sharpe RM, Brown RA. Macrophages in the interstitial tissue of the rat testis. Cell Tissue Res 1986;243:337–44.
40. Wing JY, Lin HS. The fine structure of testicular interstitial cells in the adult golden hamster with special reference to seasonal changes. Cell Tissue Res 1977;183:385–93.
41. Miller SC, Bowman BM, Rowland HG. Structure, cytochemistry, endocytotic activity and immunoglobulin (Fc) receptors of rat testicular interstitial-tissue macrophages. Am J Anat 1983;168:1–13.
42. Hutson JC. Changes in the concentration and size of testicular macrophages during development. Biol Reprod 1990;43:885–90.
43. Hutson JC. Development of cytoplasmic digitations between Leydig cells and testicular macrophages of the rat. Cell Tissue Res 1992;267:385–9.
44. Rivenson A, Ohmori T, Hamazaki M. Cell surface recognition: spontaneous identification of mouse Leydig cells by lymphocytes, macrophages and eosinophils. Cell Mol Biol 1981;27:49–56.
45. Yee YB, Hutson JC. Effects of testicular macrophage-conditioned medium on Leydig cells in culture. Endocrinology 1985;116:2682–4.
46. Raburn DJ, Coquelin A, Hutson JC. Human chorionic gonadotrophin increases the concentration of macrophages in neonatal rat tissues. Biol Reprod 1991;45:172–7.
47. Geierhaas B, Bornstein SR, Jarry H, Scherbaum WA, Herrmann M, Pfeiffer EF. Morphological and hormonal changes following vasectomy in rats, suggesting a functional role for Leydig cell-associated macrophages. Horm Metab Res 1991;23:373–8.
48. Sun X-R, Hedger MP, Risbridger GP. The effect of testicular macrophages and interleukin-1 on testosterone production by purified adult rat Leydig cells cultured under in vitro maintenance conditions. Endocrinology 1993;132: 186–92.
49. Hales DB. Interleukin-1 inhibits Leydig cell steroidogenesis primarily by decreasing 17α-hydroxylase/C17–20 lyase cytochrome P450 expression. Endocrinology 1992;131:2165–72.
50. Calkins JH, Sigel MN, Nankin HR, Lin T. Interleukin-1 inhibits Leydig cell steroidogenesis in primary culture. Endocrinology 1988;123:1605–10.
51. Verhoeven G, Cailleau J, Van Damme J, Billiau A. Interleukin-1 stimulates steroidogenesis in cultured rat Leydig cells. Mol Cell Endocrinol 1988; 57:51–60.

52. Calkins JH, Guo H, Sigel MN, Lin T. Differential effects of recombinant interleukin-1 alpha and beta on Leydig cell function. Biochem Biophys Res Commun 1990;167:548–53.
53. Warren DW, Pasupuleti V, Lu Y, Platter BW, Horton R. Tumor necrosis factor and interleukin-1 stimulate testosterone secretion in adult male rat Leydig cells in vitro. J Androl 1990;11:353–60.
54. Xiong Y, Buchanan Hales D. The role of tumor necrosis factor-alpha in the regulation of mouse Leydig cell steroidogenesis. Endocrinology 1993; 132:2438–44.
55. Xiong Y, Buchanan Hales D. Expression, regulation and production of tumor necrosis factor-α in mouse testicular interstitial macrophages in vitro. Endocrinology 1993;133:2568–73.
56. De SK, Chen H-L, Pace JL, Hunt JS, Terranova PF, Enders GC. Expression of tumor necrosis factor alpha in mouse spermatogenic cells. Endocrinology 1993;133:389–96.
57. Tomlinson MJ, White A, Barratt CLR, Boulton AE, Cooke ID. The removal of morphologically abnormal sperm forms by phagocytes: a positive role for seminal leukocytes? Hum Reprod 1992;7:517–22.
58. Barratt CLR, Cohen J. Quantifiction of sperm disposal and phagocytic cells in the tract of short and long vasectomized mice. J Reprod Fertil 1987; 81:377–84.
59. Pandya IJ, Cohen J. The leukocytic reaction of the human uterine cervix to spermatozoa. Fertil Steril 1985;43:417–21.
60. Hill JA, Cohen J, Anderson DJ. The effect of lymphokines and monokines on human sperm fertilizing ability in the zona-free hamster egg penetration test. Am J Obstet Gynecol 1989;160:1154–9.
61. Hunt JS, Chen H-L, Hu X-L, Pollard JW. Normal distribution of tumor necrosis factor-alpha messenger ribonucleic acid and protein in the uteri, placentas and embryos of osteopetrotic (*op*/*op*) mice lacking colony stimulating-factor 1. Biol Reprod 1993;49:441–52.
62. De M, Wood GW. Influence of oestrogen and progesterone on macrophage distribution in the mouse uterus. J Endocrinol 1990;126:417–24.
63. Tachi C, Tachi S, Knysxynshi A, Lindner HR. Possible involvement of macrophages in embryo-maternal relationships during ovum implantation in the rat. J Exp Zool 1981;217:81–92.
64. King A, Wellings V, Gardner L, Loke YW. Immunocytochemical characterization of the unusual large granular lymphocytes in human endometrium throughout the menstrual cycle. Hum Immunol 1989;24:195–205.
65. Bulmer JN, Morrison L, Smith JC. Expression of class II MHC gene products by macrophages in human uteroplacental tissue. Immunology 1988;63:707–14.
66. Yelavarthi KK, Chen HL, Yang YP, Cowley BD, Fishback JL, Hunt JS. Tumor necrosis factor-α mRNA and protein in rat uterine and placental cells. J Immunol 1991;146:3840–8.
67. Tamada H, McMaster MT, Flanders KC, Andrews GK, Dey SK. Cell type-specific expression of transforming growth factor-β1 in the mouse uterus during the periimplantation period. Mol Endocrinol 1990;4:965–72.
68. Choudhuri R, Wood GW. Leukocyte distribution in the pseudopregnant mouse uterus. Am J Reprod Immunol 1992;27:69–76.

69. Pollard JW, Bartocci A, Arceci R, Orlofsky A, Ladner MB, Stanley ER. Apparent role of macrophage growth factor, CSF-1, in placental development. Nature 1993;330:484–6.
70. Das SK, Flanders KC, Andrews GK, Dey SK. Expression of transforming growth factor-β isoforms (β2 and β3) in the mouse uterus: analysis of the preimplantation period and effects of ovarian steroids. Endocrinology 1992; 130:3459–66.
71. De M, Sanford TH, Wood GW. Detection of interleukin-1, interleukin-6 and tumor necrosis factor-α in the uterus during the second half of pregnancy in the mouse. Endocrinology 1992;131:14–20.
72. Robertson SA, Seamark RF. Uterine epithelial cells synthesize granulocyte-macrophage colony-stimulating factor and interleukin-6 in pregnant and non-pregnant mice. Biol Reprod 1992;46:1069–79.
73. Wood GW, De M, Sanford TH, Choudhuri R. Macrophage colony-stimulating factor controls macrophage recruitment to the cycling mouse uterus. Dev Biol 1992;152:336–43.
74. Robertson SA, Seamark RF. Granulocyte-macrophage colony-stimulating factor (GM-CSF) in the murine reproductive tract: stimulation by seminal factors. Reprod Fertil Dev 1990;2:359–69.
75. Chen HL, Yang YP, Hu X-L, Yelavarthi KK, Fishback JL, Hunt JS. Tumor necrosis factor alpha mRNA and protein are present in human placental and uterine cells and early and late stages of gestation. Am J Pathol 1991;139: 327–35.
76. Rotello RJ, Leberman RC, Purchio AF, Gerschenson LE. Co-ordinated regulation of apoptosis and cell proliferation by transforming growth factor-β1 in cultured uterine epithelial cells. Proc Natl Acad Sci USA 1991;88: 3412–5.
77. Arceci RJ, Shanahan F, Stanley ER, Pollard JW. Temporal expression and location of colony-stimulating factor 1 (CSF-1) and its receptor in the female reproductive tract are consistent with CSF-1-regulated placental development. Proc Natl Acad Sci USA 1989;86:8818–22.
78. Daiter E, Pampfer S, Yeung YG, Barad D, Stanley ER, Pollard JW. Expression of colony-stimulating factor 1 in the human uterus and placenta. J Clin Endocrinol Metab 1992;74:850–8.
79. Robertson SA, Seamark RF. Granulocyte-macrophage colony stimulating factor (GM-CSF) in the murine reproductive tract: stimulation by seminal factors. Reprod Fertil Dev 1993;2:359–68.
80. Lang RA, Bishop JM. Macrophages are required for cell death and tissue remodeling in the developing mouse eye. Cell 1993;74:453–62.

# Part III

## Growth Factor Networks in Pregnancy Loss and Cancer

# 9

# Novel Transforming Growth Factor Betas (TGFβ2) in Pregnancy and Cancer

DAVID A. CLARK, KATHLEEN C. FLANDERS, GILL VINCE, PHYLLIS STARKEY, HAL HIRTE, JUSTIN MANUEL, JENNIFER UNDERWOOD, AND JAMES MOWBRAY

A number of cytokines may be present at the fetomaternal interface (1), and these may determine whether the implanted mammalian conceptus will succeed or fail. Certain cytokines have the potential to compromise the pregnancy and cause abortion: *tumor necrosis factor α* (TNFα), *interleukin-1* (IL-1), and *interferon γ* (IFNγ). This may involve activation of *natural killer* (NK) lineage cells into *lymphokine-activated killer* (LAK) cells that have the capacity to damage fetal trophoblast cells (2–7) that lie at the fetomaternal interface and are crucial for placentation and embryo survival (8). Other cytokines have been postulated to be favorable to survival of the pregnancy. Three of these—*interleukin-10* (IL-10), *granulocyte-macrophage colony stimulating factor* (GM-CSF), and *transforming growth factor β* (TGFβ)—appear to counter those cytokine-dependent processes that are antagonistic to pregnancy. IL-10 does not block generation of LAK cells in response to *interleukin-2* (IL-2) (9–13), but can inhibit release of IL-2, TNFα, and related cytokines that are abortogenic. IL-10-deficient mice give birth to babies that are smaller than normal, but apparently do not have a higher resorption (abortion) rate (14). This may be due to the well-known phenomenon of redundancy, whereby several different cytokines produce the same effects, and this preserves a degree of protection when one cytokine is missing. GM-CSF, which is produced by fetal trophoblast (15, 16), has been shown to inhibit the generation of antitrophoblast LAK-like cells in vivo (17 and Clark, Chaouat, Mogil, Wegmann, manuscript submitted). Further, a potent immunosuppressive molecule closely related to TGFβ2 that blocks LAK generation and macrophage activation and cytokine release is also present at the fetomaternal interface (7, 18–20). These TGFβ2-like molecules appear to be released by bone marrow-derived *natural suppressor cells* and to have unusual molecular properties (20, 21).

This chapter summarizes current data concerning these TGFβ-like molecules and the cells that release them and shows that similar factors may be produced in association with malignant neoplasms. Hence, cytokines that promote the success of pregnancy may similarly promote the "success" of cancer.

## TGFβs in Murine Pregnancy

Approximately 4 days after implantation in DBA/2-mated C3H/HeJ mice, supernatants of mechanically disaggregated decidua cultured for 48 h contain a potent immunosuppressive activity, as determined in a *cytotoxic T lymphocyte* (CTL) response assay, on IL-2-driven LAK response assays, and on IFN-driven NK response (18). The suppressive activity is reversed by neutralizing antibody against TGFβ and, specifically, by antibody recognizing TGFβ2 and -β3 (19, 20). Molecular separation studies using sephadex have indicated that most of the activity was associated with 80- to 100-kd molecules (22), and aqueous phase sieving HPLC separation supported this finding, although most of the activity was "lost" on the column (19). Using 0.1–0.05 M KCl and acetic acid (pH 2.5–3.0 HPLC) buffer prevented sticking and brought out a prominent spike of suppressive activity coseparating with cytochrome C (19). Purified TGFβ standards (both β1 and β2) eluted at slightly higher molecular weights, and by Western blotting and PAGE-electroelution, it has been confirmed that the TGFβ2-like activity is slightly smaller than "authentic" TGFβ2 (20). Nevertheless, these molecules function as TGFβs in promoting colony formation by NRK cells in the presence of EGF in vitro—the "gold standard" for TGFβ activity (20, 23).

More recently, it has been possible to obtain supernatants in serum-free medium for direct Western blotting without a preliminary purification by HPLC (7 and Clark, Flanders, Hirte, Dasch, manuscript submitted). Suppressive activity is present at time zero in wash medium without a need for incubation at 37°C, and activity increases with incubation to peak at 18–24 h. A doublet slightly higher in molecular weight (28,000–32,000 estimated) is present at zero time, and with incubation the previously described lower-molecular weight species of molecules appears. The latter appears to be due to the activity of decidua-associated cells, as no transition in band size occurs if the zero-time supernatant is incubated at 37°C in the absence of cells, and there is no increase in suppressive activity (Clark, Flanders, Hirte, Dasch, manuscript submitted). A transition can be effected by incubation of the time-zero supernatant at low pH, but not by incubation at pH $> 7$ with N-glycanases that cleave N-linked carbohydrate.

Why these murine decidua-derived factors should have unusual molecular weights is unclear. Recombinant TGFβ2 with a 25,000 molecular weight injected into the peritoneal cavity of mice induces a proinflammatory state with marked fibrosis (thought to be mediated by type I TGFβ receptors), as well as systemic immunosuppression (thought to be mediated by type II receptors) (24 and unpublished observations). Fibrosis and monocyte recruitment are not evident in decidua (25), so if the TGFβ2-like factor in decidua is active, it may bind selectively to type II receptors.

The cells responsible for suppression in mechanically disaggregated decidua appear to be small lymphocytic cells of the *null* phenotype with cytoplasmic granules and lack the asialo-GM1 marker of NK cells. Two markers that are present are FcR-II and an epitope found on bone marrow-derived natural suppressor cells, IE5:B5.1 (21, 26, 27). The cells appear to hybridize selectively in situ with the pGEM-G1G2 simian TGFβ2 probe (25). The sensitivity of this detection method is uncertain, but considerable specificity appears to exist, as other cells in the decidua, placenta, and embryo that may contain TGFβ2 mRNA (and do contain TGFβ2 by immunohistochemical staining) are not labeled (25). Release of the factor appears to require activation of the null cells by soluble factors from trophoblast (22, 28), and decidual cell suspensions exposed to tryptic proteases no longer release the factor (19). Immunohistochemical staining of sections of decidua shows foci of increased staining when the ability to release the suppressor factor in vitro is acquired (25).

A deficiency of suppressor cell activity and of release of the TGFβ2-like suppressor activity has been associated with trophoblast failure and spontaneous abortion/resorption (28–32). The abortion rate in the DBA/2J-mated CBA/J female mouse can be boosted significantly by injecting 50–100 μg of neutralizing antibody specific for TGFβ2 and -β3 (7, 21); nonneutralizing antibody is ineffective (7). It is uncertain if these effects are local or systemic (7). There is suppressor cell activity in the uterine venous blood of pregnant mice (33), and collagen-induced arthritis in mice remits at about the time the cells in decidua acquire the ability to release TGFβ2-like factors (34). Other evidence for systemic immunoregulatory effects has been discussed elsewhere (6, 35). IL-10 is another immunosuppressive molecule that has been found in the murine placenta/uterus, and a contributory role of locally produced IL-10 in these systemic effects is possible (36). IL-10 is apparently present in the murine ectoplacental cone area (day 6.5), which is much earlier than the time of appearance of TGFβ2-like factors (25, 36).

The suppressor cells and soluble suppressor factor have been isolated from the decidua of every strain combination of inbred and outbred mouse studied to date, including SCID and SCID-Beige mice (20). Recent HPLC studies of supernatants from pregnant rats also show a suppressive activity peak similar to the murine TGFβ2-like factor (Lea,

Clark, unpublished observations). Therefore, the phenomenon would seem to have some generality. As activity is present in immunodeficient and syngeneically mated SCID mice and in syngeneically mated immunocompetent mice (20, 22), neither an immune response nor a *major histocompatibility complex* (MHC) disparity between embryo and mother appears obligatory for activation of suppression. Immune system stimulation by allogeneic cells may, however, boost suppressive activity and reduce the rate of spontaneous abortion (31, 37). The mechanism of this boosting effect has not been determined.

## TGFβs in Human Pregnancy

Based on studies in mice, it was logical to look for suppressor cells and TGFβ2-like suppressor factors in human pregnancy decidua. Initial studies showed that small-sized lymphocytic suppressor cells could be detected in first-trimester pregnancy decidua, and with missed abortion (trophoblast death as reflected in the disappearance of β-hCG), suppression was lost (38–40).

Analysis of the soluble suppressive activity from whole, nondisaggregated decidua biopsies from the placental attachment site indicated that most of the activity inhibiting proliferation of Con A-stimulated lymphocytes was not TGFβ (21, 40, 41; manuscript submitted). By HPLC these non-TGFβ factors are either large (>60 kd) or small (<1300 d). Of interest are 3 peaks of activity in the 20- to 45-kd range. These 3 peaks can also be obtained by culturing the lymphocytes isolated from decidua and by separately culturing in vitro the $CD56^+$ granulated lymphocytes in contrast to the $CD56^-$ lymphocytes (40, 42; manuscript in preparation). A subset of these $CD56^+$ cells stain with anti-TGFβ2/β3 antibody, and the expected 3 peaks are released by the $CD56^+$, but not by the $CD56^-$ cells (manuscript in preparation). Similar peaks have been generated by culturing human luteal phase lymphocytes with trophoblast membrane vesicles in vitro (43). Of interest, the 3 peaks in supernatants from whole decidua from successful pregnancies are deficient in a pool of supernatants from biopsies from patients with incipient spontaneous abortion, most of whom have had recurrent unexplained losses (40). Analysis of individual biopsies using the simian probe for TGFβ2 suggests that only a subset of recurrent aborters may be deficient and that most sporadic (nonrecurrent) aborters have cell numbers indistinguishable from those found in decidua from normal healthy pregnancies presenting for legal termination (40 and manuscript submitted).

The human system shows some similarities and differences in comparison with the murine system. The magnitude of suppressive activity purified by HPLC is significantly less in the human (although perhaps not on a per-uterus basis). The human $CD56^+$ cells that stain for TGFβ2 do not appear to be $FcR^+$, and the suppressive peaks in humans differ in molecular weight from those in the mouse, particularly the higher-molecular weight species. Preliminary studies by Western blotting support this thesis. Antibody neutralization studies indicate that the high-molecular weight peak reacts with anti-TGFβ2/β3 and not anti-TGFβ (40 and manuscript submitted).

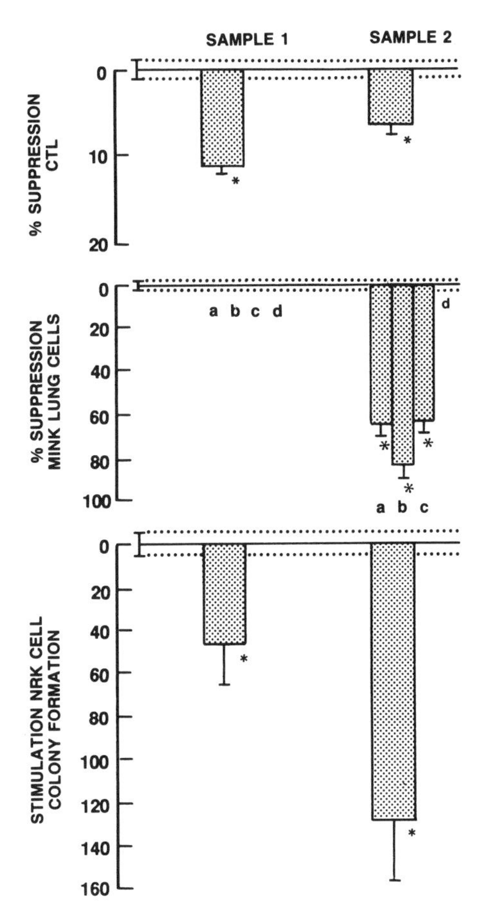


FIGURE 9.1. Comparison of immunologic (CTL supression) and nonimmunologic (mink lung growth suppression and NRK fibroblast colony) assays of TGFβ2 in decidual mononuclear supernatant. The top panel is an MLC-CTL assay 1/10 test dilution. The middle panel is a mink lung growth suppression assay 1/4 test dilution. (a = untreated supernatants; b = acidified to pH 2.5 to activate latent TGFß and neutralized; c = heated to 80ºC for 5 min to activate latent TGFß; d = heated plus 10 μg/mL anti-TGFβ antibody.) The bottom panel is an NRK fibroblast colony formation assay 1/4 test dilution. The parallel dotted lines at the top of each assay represent medium control ± 1 SEM, 4–8 replicates. The CTL suppression assay and mink lung assay tested supernatants run with 4 replicates; the NRK assay was run with 8 replicates. All error bars are ± 1 SEM. An asterisk (*) denotes significant activity by Student's t-test.

It has been suggested by Graham and Lala that the TGFβ released from decidua is in an inactive rather than an active form (44). This was based on using the mink lung growth inhibition assay, wherein it was necessary to heat the decidual supernatants to 80°C to activate latent TGFβ. We had found some latent TGFβ in murine zero-time supernatants and wanted to check for latent activity in the human supernatants that, surprisingly, were biologically active in stimulating NRK cell growth without prior heating. Figure 9.1 shows data obtained using two decidual lymphocyte-derived supernatants that were suppressive (albeit at a lower level than in the mouse). The middle panel of the figure shows that one supernatant was inactive against mink lung cells in spite of activation by heating or acidification. The second supernatant was active without treatment (a small amount of latent factor may have been present, as indicated by the effect of acid treatment), and specific antibody neutralization confirmed the TGFβ nature of the activity. The lower panel shows that both samples were active in the NRK assay. (Activation treatment produced no additional effects.) We have noted that the mink lung assay may have some problems when used to test crude biological samples for TGFβ activity. There may be growth factors present that counteract TGFβ, and heating may reveal suppression by destroying the growth factors rather than by activating latent TGFβ. Further, the mink lung assay may detect suppression in the absence of demonstrable TGFβ, as shown by studies on cloned human decidual leukocytes (45, 46).

The presence of TGFβ2-like molecules arising from freshly isolated granulated lymphoid cells in decidua (similar to the mouse) is of considerable interest. The significance of the anomalous molecular weights in the human remains to be determined.

## TGFβs in Cancer

Natural suppressor cell activation has also been associated with cancer (47), and many types of malignant cells and cell lines produce TGFβ (23, 48). Indeed, TGFβ2 was initially purified from a prostatic carcinoma (49). Glioblastoma cells secrete a TGFβ2-like factor, and a variety of molecular weights, including 25,000 and 80,000–120,000 species with bioactivity, have been described (50, 51).

In studies we have done using ascites from ovarian cancer patients, a 25,000 classical TGFβ1 has been isolated by acid HPLC separation (52). However, under neutral conditions, a 40,000 species may also be detected. The latter stimulates NRK cell growth, may or may not possess immunosuppressive activity, stimulates ovarian cancer cell lines to grow, is not neutralized by anti-TGFβ antibody but reacts on Western blots, and vanishes when separations are done under low pH to bring out the 25,000 factor (which is antibody neutralizable) (53).

These data suggest that there may be TGFβ-related factors with different spectra of bioactivities at anomalous molecular weights, but that the factors can be changed into other molecular weight species with different spectra of activities. Of particular interest have been studies on TGFβ human breast cancer (54), wherein TGFβ1-like molecules with molecular weights of 40,000 and 30,000–32,000 (a doublet) have been described (55). Simple enumeration and quantitation of TGFβ positivity of breast cancer biopsies give little prognostic information (56), perhaps due to the need to know exactly which molecules are present and which receptors are borne by the target cells (tumor and host). Indeed, the presence of nonclassical types of TGFβ may explain the negative result in a recent report of attempts to detect antibody-neutralizable TGFβ released by human breast cancer cells (57).

## Summary

The definition of TGFβ-like molecules of anomalous molecular weight has opened up new opportunities to understand important biological events related to the TGFβ family of cytokines. Further, if anomalous functional properties are confirmed by analysis of purified factors of this type, there will likely be expanded possibilities for diagnosis and treatment of human diseases.

## *References*

1. Clark DA. Cytokines, decidua and early pregnancy. In: Milligan SR, ed. Oxf Rev Reprod Biol 1993:83–111.
2. Chaoaut G, Menu E, Clark DA, Minkowsky M, Dy M, Wegmann TG. Control of fetal survival in CBA × DBA/2 mice by lymphokine therapy. J Reprod Fertil 1990;89:447–58.
3. Gendron RL, Nestel FP, Lapp WS, Baines MG. Lipopolysaccharide-induced fetal resorption in mice is associated with the intrauterine production of tumour necrosis factor-alpha. J Reprod Fertil 1990;90:395–402.
4. Silen M, Firpo A, Morgello S, Lowry SF, Francus T. Interleukin-1 alpha and tumor necrosis factor alpha cause placental injury in the rat. Am J Pathol 1989;135:239–44.
5. Ecker JL, Laufer MR, Hill JA. Measurement of embryotoxic factors is predictive of pregnancy outcome in women with a history of recurrent abortion. Obstet Gynecol 1993;81:1–4.
6. Clark DA. Controversies in reproductive immunology. Boca Raton, FL: CRC Press, 1991:215–47.
7. Clark DA, Lea RG, Flanders KC, Banwatt D, Chaoaut G. Role of a unique species of TGF-β in preventing rejection of the conceptus during pregnancy. In: Gergely J, Benczúr M, Erdei N, et al., eds. Progress in immunology VIII. Budapest: Springer-Verlag, 1992:841–7.

8. Rossant J, Mauro VM, Croy BA. Importance of trophoblast genotype for survival of interspecific murine chimeras. J Embryol Exp Morphol 1982;69: 141–9.
9. Fiorentino DF, Zlotnik A, Mosmann TR, Howard M, O'Garra A. Il-10 inhibits cytokine production by activated macrophages. J Immunol 1991;147: 3815–22.
10. Schandené L, Gerard C, Crusiaux A, Abramowicz D, Velu T, Goldman M. Interleukin-10 inhibits OKT3-induced cytokine release: in vitro comparison with pentoxifylline. Transplant Proc 1993;25:55–6.
11. Spagnoli GC, Juretic A, Schult-Thater E, et al. On the relative roles of interleukin-2 and interleukin-10 in the generation of lymphokine-activated killer cell activity. Cell Immunol 1993;146:391–405.
12. Wegmann TG, Guilbert LJ, Mosmann TR, Lin H. The role of placental IL-10 in maternal-fetal immune interactions. In: Gergely J, Benczúr M, Erdei N, et al., eds. Progress in immunology VIII. Budapest: Springer-Verlag, 1992: 857–9.
13. Gazzinelli RT, Oswald IP, James Sl, Sher A. IL-10 inhibits parasite killing and nitrogen oxide production by IFN-$\gamma$-activated macrophages. J Immunol 1992;148:1792–6.
14. Kuhn R, Rajewsky K, Muller W. IL-4 and IL-10 deficient mice [Abstract]. 8th Int Cong Immunology, Budapest, Hungary, August 23–28, 1992:203.
15. Crainie M, Guilbert L, Wegmann TG. Expression of novel cytokine transcripts in the murine placenta. Biol Reprod 1990;43:999–1005.
16. Kanzaki H, Crainie M, Lin H, Yui J, Guilbert LJ, Wegmann TG. The in-situ expression of granulocyte-macrophage colony-stimulating factor (GM-CSF) mRNA at the maternal-fetal interface. Growth Factors 1991;5:69–74.
17. Clark DA, Lea RG, Chaouat G, Pearce M, Abrams J. $CD8^+$ suppressor cells and GM-CSF in the CBA-DBA/2 murine recurrent spontaneous abortion model [Abstract]. EOS 1992;12:82.
18. Clark DA, Lea RG, Denburg J, et al. Transforming growth factor-beta related suppressor factor in mammalian pregnancy decidua: homologies between the mouse and human in successful pregnancy and in recurrent unexplained abortion. In: Chaouat G, Mowbray JF, eds. Materno-fetal relationship: molecular and cellular biology. Colloque INSERM 1991;212:171–9.
19. Clark DA, Falbo M, Rowley RB, Banwatt D, Stedronska-Clark J. Active suppression of host-versus-graft reaction in pregnant mice, IX. Soluble suppressor activity obtained from allopregnant mouse decidua that blocks the response to interleukin 2 is related to TGF-$\beta$. J Immunol 1988;141:3833–40.
20. Clark DA, Flanders KC, Banwatt D, et al. Murine pregnancy decidua produces a unique immunosuppressive molecule related to transforming growth factor $\beta$-2. J Immunol 1990;144:3008–14.
21. Clark DA, Lea RG, Podor T, Daya S, Banwatt D, Harley C. Cytokines determining the success or failure of pregnancy. Ann NY Acad Sci 1991;626: 524–36.
22. Clark DA, Chaput A, Walker C, Rosenthal KL. Active suppression of host-versus-graft reaction in pregnant mice, VI. Soluble suppressor activity obtained from decidua of allopregnant mice blocks the response to IL-2. J Immunol 1985;134:1659–64.

23. Roberts AB, Sporn MB. The transforming growth factor-betas. In: Sporn MB, Roberts AB, eds. Peptide growth factors and their receptors: handbook of experimental pharmacology; vol 95. 1990:415–72.
24. Chen R-H, Ebner R, Derynck R. Inactivation of the type II receptor reveals two receptor pathways for the diverse TGF-β activities. Science 1993;26: 1335–8.
25. Lea RG, Flanders KC, Harley CB, Manuel J, Banwatt D, Clark DA. Release of a TGF-β2-related suppressor factor from postimplantation murine decidual tissue can be correlated with the detection of a subpopulation of cells containing RNA for TGF-β2. J Immunol 1992;148:778–87.
26. Slapsys RM, Richards CD, Clark DA. Active suppression of host-versus-graft reaction in pregnant mice, VIII. The uterine decidua-association suppressor cell is distinct from decidual NK cells. Cell Immunol 1985;99:140–9.
27. Hoskin DW, Brooks-Kaiser JC, Kaiser M, Murgita RA. Reactivity of monoclonal antibody IE5.B5 with a novel phenotypic marker expressed on a murine natural suppressor cell subset. Hybridoma 1992;11:203–15.
28. Slapsys RM, Younglai E, Clark DA. A novel suppressor cell is recruited to decidua by fetal trophoblast-type cells. Reg Immunol 1988;1:182–9.
29. Clark DA, Chaput A, Tutton D. Active suppression of host-versus-graft reaction in pregnant mice, VII. Spontaneous abortion of allogeneic DBA/2 × CBA/J fetuses in the uterus of CBA/J mice correlates with deficient non-T suppressor cell activity. J Immunol 1986;136:1668–75.
30. Clark DA, Slapsys RM, Croy BA, Rossant J. Suppressor cell activity in uterine decidua correlates with success or failure of murine pregnancies. J Immunol 1983;131:540–2.
31. Clark DA, Drake B, Head JR, Stedronska-Clark J, Banwatt D. Decidua associated suppressor activity and viability of individual implantation sites of allopregnant C3H mice. J Reprod Immunol 1990;17:253–64.
32. Clark DA, Banwatt D, Croy BA. Murine trophoblast failure and spontaneous abortion. Am J Reprod Immunol 1993;29:199–205.
33. Slapsys RM, Clark DA. Active suppression of host-versus-graft reaction in pregnant mice, IV. Local suppressor cells in decidua and uterine blood. J Reprod Immunol 1982;4:355–64.
34. Waites GT, Whyte A. Effect of pregnancy on collagen-induced arthritis in mice. Clin Exp Immunol 1987;67:467–76.
35. Young EJ, Gomez CI. Enhancement of herpes virus type 2 infection in pregnant mice (40461). Proc Soc Exp Biol Med 1979;160:416–20.
36. Wegmann TG, Lin H, Guilbert L, Mosmann TR. Bidirectional cytokine interactions in the maternal-fetal relationship: is successful pregnancy a $T_H2$ phenomenon? Immunol Today 1993;14:353–6.
37. Clark DA, Kiger N, Guenet J-L, Chaouat G. Local active suppression and successful vaccination against spontaneous abortion in CBA/J mice. J Reprod Immunol 1987;10:79–85.
38. Daya S, Clark DA. Identification of two species of suppressive factors of differing molecular weight released by in vitro fertilized human oocytes. Fertil Steril 1988;49:360–3.
39. Daya S, Rosenthal KL, Clark DA. Immunosuppressor factor(s) produced by decidua-associated suppressor cells—a proposed mechanism for fetal allograft survival. Am J Obstet Gynecol 1987;156:344–50.

40. Clark DA, Lea RG, Underwood J, et al. A subset of recurrent first trimester-aborting women show subnormal TGF-β2 suppressor activity at the implantation site associated with miscarriage [Abstract]. EOS 1992;12:83.
41. Bulmer JN, Longfellow M, Ritson A. Leukocytes and resident blood cells in endometrium. Ann NY Acad Sci 1992;622:57–68.
42. Vince G, Starkey P, Hirte H, Flanders K, Clark DA. TGF-β production by large granular lymphocytes from human decidua. BSI-MFIG meet, April 1993, Liverpool.
43. Daya S, Johnson PM, Clark DA. Trophoblast induction of suppressor type cell activity in human endometrial tissue. Am J Reprod Immunol 1989;19: 65–72.
44. Graham CH, Lala PK. Mechanism of control of trophoblast invasion in situ. J Cell Physiol 1991;148:228–34.
45. Christmas SE, Bulmer J, Meager A, Johnson PM. Phenotypic and functional analysis of human $CD3^-$ decidual leukocyte clones. Immunology 1990;71: 182–9.
46. Clark DA, Christmas S, Johnson PM, Banwatt D. Failure to detect TGF-β2 mRNA or TGF-β suppressive activity associated with human CD3 decidual leukocyte clones [Abstract]. EOS 1992;12:82.
47. Subiza JL, Viñuela-Rodriguez R, Figueredo JGMA, De La Concha E. Development of splenic natural suppressor (NS) cells in Ehrlich tumor-bearing mice. Int J Cancer 1989;44:307–14.
48. Kerbel RS, Theodorescu D. Tumor cell subpopulation interactions mediated by transforming growth factor β may contribute to clonal dominance of metastatically competent cells in primary tumours. In: Burger MM, Sordat B, Zinkernagel RM, eds. Cell to cell interaction. Basel: Karger, 1990:100–13.
49. Marquardt H, Lioubin MN, Ikeda T. Complete amino acid sequence of human transforming growth factor type β2. J Biol Chem 1987;262:12127–31.
50. Constam DB, Phillipp J, Malipiero UV, ten Dijke P, Schachner M, Fontana A. Differential expression of transforming growth factor-β1, -β2 and -β3 by glioblastoma cells, astrocytes, and microglia. J Immunol 1992;148:1404–10.
51. Bodmer S, Huber D, Heid I, Fontana A. Human glioblastoma cell derived transforming growth factor-β2: evidence for secretion of both high and low molecular weight biologically active forms. J Neuroimmunol 1991;34:33–42.
52. Hirte H, Clark DA. Generation of lymphokine-activated killer cells (LAK) in human ovarian carcinoma ascitic fluid: identification of transforming growth factor-beta (TGF-β) as a suppressive factor. Cancer Immunol Immunother 1990;32:296–302.
53. Hirte HW, Kaiser J, Rusthoven JJ, Mazurka J, O'Connell G. High molecular weight forms of transforming growth factor-α and transforming growth factor-β are present in ascites from human ovarian carcinoma [Abstract 1428]. Proc Am Assoc Cancer Res 1993;34:240.
54. McCune BK, Mullin BR, Flanders KC, Jaffurs WJ, Mullen LT, Sporn MB. Localization of transforming growth factor-β isotypes in lesions of the human breast. Hum Pathol 1992;1(suppl 23):13–20.
55. King RJB, Wang DY, Daly RJ, Darbre PD. Approaches to studying the role of growth factors in the progression of breast tumours from the steroid sensitive to insensitive state. J Steroid Biochem 1989;34:133–8.

56. Dublin EA, Barnes DM, Wang DY, King RJB, Levison DA. TGF alpha and TGF beta expression in mammary carcinoma. J Pathol 1993;170:15–22.
57. Xie J, Gallagher G. Transforming growth factor-β is not the major soluble immunosuppressor in the microenvironments of human breast tumours. Anticancer Res 1992;12:2117–22.

# 10

# Tumor Necrosis Factor α: Potential Relationships with Cancers of the Female Reproductive Tract

Joan S. Hunt, Hua-Lin Chen, Yaping Yang,
Katherine F. Roby, and Fernando U. Garcia

A wide range of polypeptide growth factors is synthesized in the female reproductive tract (reviewed in 1–5). Many of these were first identified as promoters of hematopoietic cell growth and differentiation in the bone marrow or as products of activated hematopoietic cells (6–8). Table 10.1 shows that genes encoding these factors are commonly, but not invariably, regulated by female sex steroid hormones. Pregnancy, where ovarian hormone levels remain elevated, is associated with increased expression of uterine cytokines. For example, immunological assays show that during mouse pregnancy uterine *colony stimulating factor I (CSF-I)* is increased 1000-fold over basal levels (9).

Intermittent hormonal stimulation of growth factor production during the menstrual cycle and constant stimulation during pregnancy may foster conditions favorable to the development of cancer. Tissues in both the cycling and pregnant reproductive tract are characterized by rapid cell proliferation that is undoubtedly stimulated by autocrine and/or paracrine growth factors and that markedly increases the potential for errors. Cancers are extremely common in breast, another hormonally targeted tissue with a high rate of cell turnover. Links between malignancy and autocrine growth factor production, as well as aberrant expression of growth factor receptors, have been clearly established in breast tumors (10, 11).

This chapter discusses the potential of a pleiotrophic, multifunctional cytokine, *tumor necrosis factor α* (TNFα), to contribute to tumor initiation or maintenance in the female reproductive tract. Data are presented showing that (i) cells in normal ovaries, oviducts, uteri and placentas of humans and experimental animals contain mRNA and protein derived from this gene; (ii) in some tissues, TNFα gene expression is regulated by female sex steroid hormones; (iii) neoplastic ovarian, endometrial, and

TABLE 10.1. Cellular expression and regulation of polypeptide growth factors in the cycling and pregnant uterus.

| Factor | Major producing cells | Regulation (species) | Ref. |
|---|---|---|---|
| CSF-I | Uterine epithelia | Hormonal (mouse, human) | 75, 76 |
| | Trophoblast | Unknown (human) | 76 |
| EGF | Uterine epithelia | Hormonal (mouse) | 77 |
| GM-CSF | Uterine epithelia | Seminal fluid (mouse) | 78 |
| IGF | Multiple cell types | Hormonal (mouse) | 79 |
| IL-1 | Trophoblast | Unknown (human) | 80 |
| | Uterus | Hormonal (mouse) | 47 |
| IL-1β | Placental macrophages | Unknown (human) | 81 |
| | Uterine leukocytes | Mating (mouse) | 82 |
| IL-6 | Uterine epithelia | Seminal fluid (mouse) | 77 |
| | Uterus | Hormonal (mouse) | 47 |
| IL-8 | Placental cells | Unknown (human) | 83 |
| LIF/DIF | Uterine glands | Mating (mouse) | 73 |
| TGFα | Multiple cell types | Hormonal (mouse) | 84 |
| TGFβ | Multiple cell types | Hormonal (mouse) | 85, 86 |
| TNFα | Multiple cell types | Hormonal (human, mouse) | 44, 47 |

placental cells contain TNFα gene products; and (iv) TNFα facilitates metastasis and in some instances serves as an autocrine growth factor for tumors arising in the female reproductive tract.

## Properties of TNF

TNF was first identified as a product of endotoxin-activated macrophages that lysed selected tumor cells (12–14). Further studies showed that many types of normal cells were also adversely affected by high concentrations of TNF and that this caused many of the symptoms of septic shock as well as cancer-associated tissue wasting (cachexia) (15–19). Thus, TNF acquired its name and reputation as a destructive molecule. A pair of genes coding for two species, TNFα and TNFβ, have been identified on human chromosome 6 centromeric to HLA-B (20). The TNF genes are highly conserved and are located in the same genomic regions of mice and rats (21, 22). Macrophage-derived TNFα and lymphocyte-derived TNFβ (lymphotoxin) bind to the same receptors with similar affinities and have essentially identical biological activities.

Production of biologically active 17.2-kd TNFα is closely regulated. In macrophages controls on the rate of gene transcription, transcript stability, message accumulation, translation, secretion, and bioactivity of the end product have been reported (23). Regulatory molecules include endotoxins, prostaglandins (24), and other cytokines, such as *interferons* (IFNs) and *transforming growth factor β* (TGFβ) (25). Controls in nonmacrophages have received little attention, although it has been shown that astrocyte synthesis of TNFα is, as in macrophages, stimulated by endotoxin (26). Soluble TNFα receptors have been identified in urine that interfere with ligand binding to cell surface receptors and are believed to provide a measure of defense against excessive TNFα (27).

## Distribution of TNFα mRNA and Protein in the Female Reproductive Tract

Tovey has proposed that in the absence of infection and cancer, TNFα serves as a mediator of normal-tissue homeostasis (28). This postulate has considerable experimental support. While best known for its toxic effects, TNFα stimulates the growth of some types of cells and has no effect on others (29, 30). Moreover, TNFα transcripts are present in normal tissues; many types of cells produce low levels of TNFα; cell differentiation is facilitated by autocrine TNFα; and expression of a variety of cellular genes is modulated by TNFα ligand/receptor interactions (31–36).

Consistent with this postulate, which would predict expression of the TNFα gene in tissues undergoing change, specific gene products have been identified in the cycling and pregnant female reproductive tract (reviewed in 37). Northern blot and in situ hybridization experiments have shown that TNFα message and protein are prominent in the ovary, oviduct, and cycling as well as pregnant uterus. Support for a role in pregnancy and embryonic development is found in studies showing that administration of exogenous TNFα overcomes fertility defects in a mouse model where resorption of embryos is common (38) and that administration of anti-TNFα inhibits development of mouse embryos (39).

In mouse ovaries, the granulosa and thecal cells that make up the follicle and the luteal and mesenchymal stromal cells express the TNFα gene (40). TNFα gene products are also present in oocytes, an unexpected finding that has been confirmed in the rat. Figure 10.1 shows that TNFα mRNA is present in various cell types in the rat ovary and appears to be polarized within the oocyte. Other studies have shown that TNFα message is absent in preimplantation mouse embryos tested by *reverse transcriptase-polymerase chain reaction* (RT-PCR) (41). This finding suggests that oocyte transcripts, which might originate in follicle cells or be synthesized within the oocyte itself, have been utilized or degraded. Interestingly, supernatant media obtained from in vitro-cultured human embryos contain readily detectable TNFα (42), suggesting that gene transcription and translation are reinstated at a later stage.

Epithelial and stromal cells in the mouse oviduct and rat, mouse, and human uterus are prominent sites of TNFα gene expression (41, 43–45). In cycling human endometrium transcripts are first identified in epithelial cells (early proliferative phase). At later stages both cell lineages contain TNFα gene products, with gene expression appearing to peak in the late

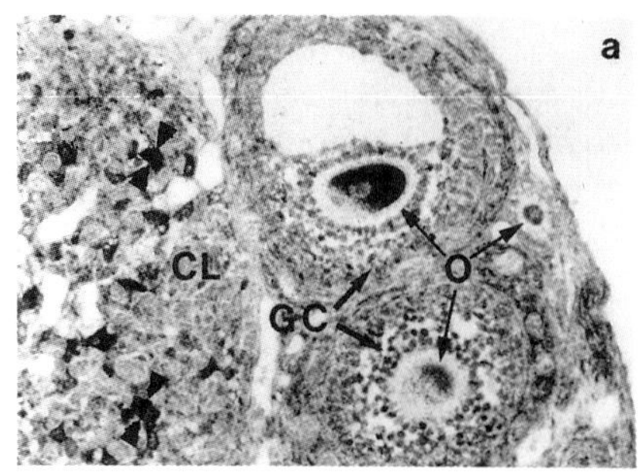


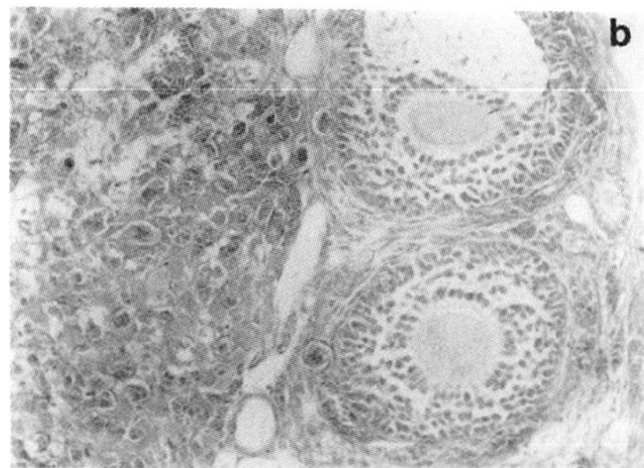


FIGURE 10.1. Identification of TNFα mRNA in rat ovary by in situ hybridization. *a:* Transcripts are prominent in granulosa cells (GC, large arrows), oocytes (O, small arrows), and corpus lutea cells (arrowheads). *b:* No hybridization signals are detectable in a semiserial section of the same tissue hybridized with the sense version of the TNFα riboprobe (×200).

secretory phase immediately prior to menstruation (44). Message and protein are low in postmenopausal endometrium and are confined to epithelial cells (Garcia, Hunt, unreported data).

During pregnancy TNFα mRNA and protein are present in the decidual cells and placentas of rats, mice, and humans (41, 43, 45). In first-trimester human placentas, TNFα gene products are prominent only in trophoblast cells, whereas both trophoblast and mesenchymal cells contain high levels of mRNA and protein near parturition (45). In mouse and rat placentas, transcripts are first identified in giant trophoblastic cells, then in certain spongy region trophoblast cells, and finally in labyrinthine trophoblast (41, 43). Thus, expression of the TNFα gene is gestation related and cell lineage restricted.

Embryos have not been thoroughly evaluated for expression of the TNFα gene. An early study identified transient gene transcription in mouse embryos by Northern blot hybridization (46). A more recent in situ hybridization and immunocytochemical study in our laboratory has shown that the TNFα gene is differentially expressed among mouse embryonic cells during the course of gestation (41).

## Regulation of TNFα Gene Expression in the Female Reproductive Tract

Regulation of the TNFα gene appears to be different in ovaries, uteri, and placentas. In mouse ovaries Northern blot hybridization failed to show any statistically significant relationship with synthesis of female sex steroid hormones (40), although steady state levels of TNFα mRNA were higher in diestrus, when both estrogens and progesterone are present.

Differential expression of the TNFα gene among ovarian cells in the mouse (40) and rat (shown above) seems to be related to stage of follicle development. Mechanisms regulating follicle development that might also control TNFα gene expression could include actions by ovarian hormones or hormones originating from the anterior pituitary, activins and inhibins, and other locally produced growth factors.

TNFα gene expression in the uterus is clearly regulated by female sex steroid hormones. Evidence for hormonal regulation of this gene was first obtained in studies on samples of cycling human endometrium (44). The results of in situ hybridization and immunocytochemical experiments indicated that message and protein fluctuate in concert with levels of ovarian hormones. Further experiments on mouse uteri showed that transcription of the TNFα gene is abrogated by ovariectomy and is restored by the administration of estrogen and progesterone (47 and Roby, Hunt, unreported data). There is as yet no evidence that the TNFα gene is regulated by other uterine cytokines. By using a mouse model where the CSF-I gene is nonfunctional—the osteopetrotic (*op/op*) mouse—expression of the TNFα gene in ovaries, oviduct, uteri, placentas, and embryos was shown to be unrelated to production of CSF-I by maternal cells (41).

Factors regulating placental expression of the TNFα gene are as yet unknown. Transcripts and protein in early gestation placentas are restricted to trophoblast cells, and expression is not uniform (45). Near parturition TNFα mRNA and protein are markedly increased in placental stromal cells that resemble macrophages (45). Macrophages in adjacent maternal tissues also synthesize vast quantities of TNFα as pregnancy nears completion (48). Collectively, these findings suggest the possibility of a causal relationship between induction of gene expression in macrophages and termination of pregnancy, a postulate that is supported by the observation that high levels of TNFα are associated with infection-mediated preterm labor (49).

## Expression and Regulation of TNFα Receptors

TNFα is bound by two species of high-affinity cell membrane receptors termed *p55/p60* and *p75/p80* to reflect their molecular masses (50, 51). The extracellular domains of these receptors are very similar, although the p75/p80 receptor has higher affinity ($1 \times 10^{-6}$ compared to $5 \times 10^{-6}$M). Their intracellular domains show little homology, suggesting that the two species may transmit distinctly different intracellular signals. This has often turned out to be the case (52–56). For example, the human p55/p60 receptor is associated with TNFα antiviral activity of hepatoma cells and adhesion to endothelial cells, while the p75/p80 receptor promotes TNFα cytotoxicity in HeLa cells.

TNFα receptors are expressed by essentially all cells (with the exception of erythrocytes), and most cells express both species, although in differing proportions. Little is known of the mechanisms that regulate expression of these genes. However, recent experiments in our laboratory have established that steady state levels of p55/p60 gene transcripts are markedly elevated in the uteri of ovariectomized mice by simultaneous administration of estrogen and progesterone (Roby, Hunt, unreported data), indicating that receptor as well as ligand expression is under strict control by ovarian hormones.

Austgulen et al. (57) have reported soluble p55/p60 as well as p75/p80 receptors in human amniotic fluid and the urine of pregnant women. While the cellular sources of these receptors have not yet been identified with certainty, it seems likely that they originate in placental cells (58). In situ hybridization and immunocytochemical experiments have shown that the p55/p60 receptor predominates in early gestation placentas, where it is synthesized by essentially all types of cells. Expression of this receptor species remains strong in term placentas. At term p75/p80 gene transcripts are prominent in mesenchymal cells that make up the core of placental villi. Taken together with the observation that TNFα is produced in this organ, the observations suggest that TNFα ligand/receptor interactions may be important to placental development and function.

Interactions between TNFα and its receptors might also facilitate gamete development. As noted above, TNFα is produced in the mouse oviduct. A recent study shows that mouse preimplantation blastocysts express p55/p60 TNFα receptors (59), which indicates that fertilized eggs will receive maternal TNFα signals. In the mouse testis differential expression of the TNFα and p55/p60 TNFα receptor genes has been observed in sperm precursors and their supporting cells (60).

## TNFα in Cancers of the Female Reproductive Tract

TNFα could be particularly useful to tumor cells because of its ability to stimulate cell growth, promote angiogenesis, influence the production of remodeling enzymes required for tumor expansion, direct the activities of immune cells, and facilitate metastasis (61–65). Therefore, it is not surprising that expression of the TNFα gene has been linked to cancers of the female reproductive tract.

### *TNFα in Ovarian Cancer*

The first identification of TNFα in tumors of the female reproductive tract was made in ovaries, where specific mRNA was shown to be present in neoplastic cells by using in situ hybridization (66, 67). Further experiments have shown that introduction of the TNFα gene into Chinese hamster ovarian cells confers a metastatic phenotype (64). This is also the case with chemically transformed mouse fibrosarcoma cells (65).

## *TNFα in Endometrial Cancers*

The TNFα gene is expressed by human epithelial cell-derived endometrial adenocarcinoma cells in situ (Garcia, Chen, Yang, Pace, Hu, Hunt, manuscript submitted). Transcripts are relatively scarce and are often confined to the nucleus in low-grade adenocarcinomas, whereas message is abundant in the cytoplasm of highly malignant growths, as illustrated in Figure 10.2. Specific protein is more readily detected in low-grade than in high-grade tumor cells. In a mixed mullerian tumor, the TNFα gene was expressed in both types of cancerous cells: epithelial and stromal cells. These observations show clearly that neoplastic endometrial cells express the TNFα gene and suggest that TNFα gene products may become more useful as the cells diverge from normal.

Several endometrial adenocarcinoma cell lines transcribe the TNFα gene. Testing of one of these lines has shown that the addition of estrogen plus progesterone stimulates low but detectable levels of TNFα protein, that low concentrations of recombinant TNFα stimulate cell growth, and that incorporation of $^{3}$H-thymidine at 24 h is inhibited by TNFα-specific oligonucleotide in a dose-dependent manner. Collectively, these data support the postulate that hormone receptor-positive adenocarcinoma cells have a modest requirement for TNFα at some point in progression through the cell cycle. In an adenocarcinoma cell line that does not transcribe the TNFα gene, estrogen stimulates enhanced expression of TNFα receptors (68). Thus, there are at least two routes by which utilization of TNFα might be enhanced in adenocarcinoma cells.

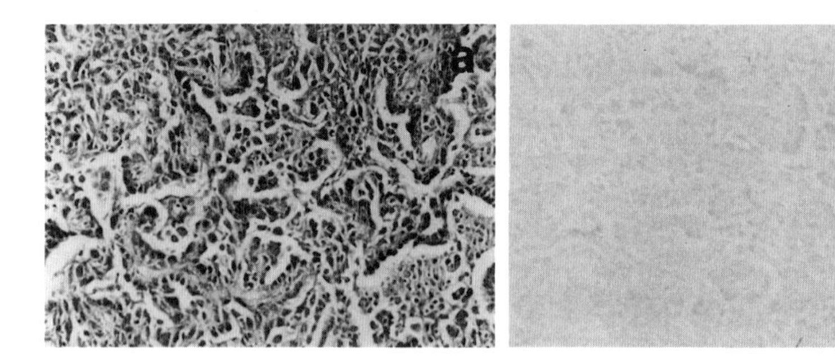


FIGURE 10.2. TNFα mRNA in high-grade (FIGO 3) adenocarcinoma cells. *a:* In situ hybridization with a 298-bp antisense TNFα RNA probe is shown. *b:* No hybridization signals are detectable in a semiserial section of the same tissue hybridized with the sense version of the TNFα riboprobe (×100).

## *TNFα in Trophoblastic Tumors*

Utilization of TNFα as an autocrine growth factor has been clearly described for choriocarcinoma cell lines (69), which are derived from trophoblastic tumors. The results obtained by testing two established cell lines, JAR and JEG-3, also support the postulate that TNFα gene expression is linked to rate of cell growth, as suggested by the studies on adenocarcinoma cells in situ. Although both JAR and JEG-3 cells transcribe and translate the TNFα gene (Fig. 10.3), rapidly proliferating JAR cells produce more immunoreactive TNFα than the slower-growing JEG-3 cells. JAR cells respond strongly to recombinant TNFα with increased proliferation, and their growth is markedly suppressed by 17-mer TNFα antisense oligonucleotide. JEG-3 cells respond less vigorously to recombinant TNFα, and their growth is less inhibited by TNFα-specific oligonucleotide.

In JAR cells, utilization of TNFα as a growth factor is mediated exclusively through p55/p60 TNFα receptors (69). While the cells also express p75/p80 TNFα receptors, antibody inhibition experiments failed to reveal any effects on cell proliferation. This does not rule out the possibility that binding of autocrine or paracrine TNFα via the p75/p80 receptor has other important consequences. It will therefore be of considerable interest to establish patterns of expression of the TNFα receptor genes in normal endometrial cells and to learn whether or not these patterns are altered by tumorigenic transformation.

Investigators have transfected various cytokine genes into tumor cells in order to determine if expression might enhance tumor destruction. Interestingly, but perhaps not unexpectedly in view of the findings re-

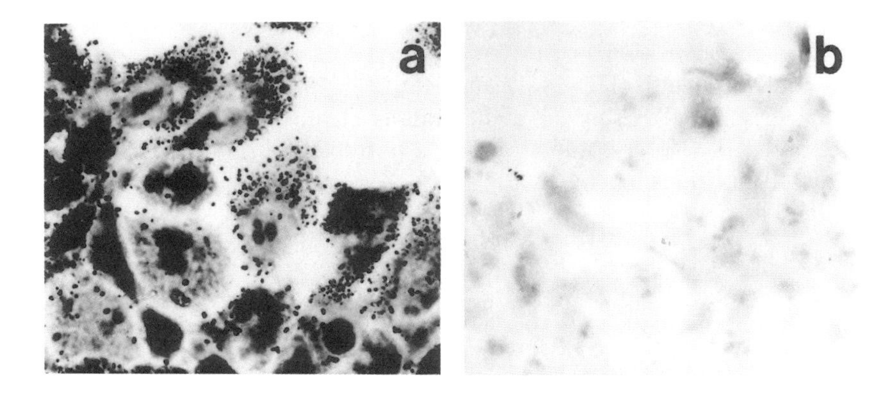


FIGURE 10.3. TNFα mRNA in the human choriocarcinoma cell line JAR identified by in situ hybridization. *a:* Hybridization with a 298-bp antisense TNFα riboprobe is shown. *b:* Hybridization with the sense orientation of the same sequence is shown (×400).

ported above, transfection of the TNFα gene failed entirely to facilitate tumor rejection (70).

## Perspectives

The specific functions of the various lymphohematopoietic factors produced in the female reproductive tract are as yet poorly understood. Presumably, these factors have important activities, possibly regulating cycle progression, facilitating implantation, and promoting pregnancy by autocrine and paracrine pathways. Potential roles for these factors have recently been examined in transgenic mice, where specific cytokine genes have been disrupted (71–73). With the exception of *leukemia inhibitory factor* (LIF/DIF), which appears to be absolutely essential to implantation (73), the results support the emerging concept that reproduction is protected by redundancy in local growth factor production.

The continual synthesis of growth factors in the reproductive tract, apparently required for successful pregnancy, may also pose a serious health risk to women. Loss of control over production and utilization of these potent factors could result in uncontrolled cell growth, and as mentioned above, many cytokines, including TNFα, are strictly regulated by ovarian hormones. It is therefore of major importance that the loss of hormone receptors accompanies tumor progression. New therapeutic strategies are under evaluation for bypassing this dangerous condition. For breast cancers these strategies include the targeting of antibodies and antisense RNA constructs to growth factors and their receptors (74). Possibly, similar approaches might be of value in treating TNFα-related tumors of the female reproductive tract.

*Acknowledgments.* This work is supported by grants from the National Institutes of Health (HD-24212 and HD-29156) to J.S.H. Y. Yang is the recipient of a fellowship from the Kansas Health Foundation, Wichita, KS, and K.L. Roby is supported by a postdoctoral fellowship from the National Institutes of Health.

## *References*

1. Brigstock DR, Heap RB, Brown KD. Polypeptide growth factors in uterine tissues and secretions. J Reprod Fertil 1989;85:747–58.
2. Hill DJ. Growth factors and their cellular actions. J Reprod Fertil 1989;85: 723–34.
3. Hunt JS. Cytokine networks in the uteroplacental unit: macrophages as pivotal regulatory cells. J Reprod Immunol 1989;16:1–17.
4. Pollard JW. Regulation of polypeptide growth factor synthesis and growth factor-related gene expression in the rat and mouse uterus before and after implantation. J Reprod Fertil 1990;88:721–31.

5. Simmen FA, Simmen RCM. Peptide growth factors and protooncogenes in mammalian conceptus development. Biol Reprod 1991;44:1–5.
6. Pollard JW. Lymphohematopoietic cytokines in the female reproductive tract. Curr Opin Immunol 1991;3:772–7.
7. Pampfer S, Arceci RJ, Pollard JW. Role of colony stimulating factor-1 (CSF-1) and other lympho-hematopoietic growth factors in mouse pre-implantation development. Bioessays 1991;13:535–40.
8. Hunt JS, Pollard JW. Macrophages in the uterus and placenta. In: Gordon S, Russell SW, eds. Current topics in microbiology and immunology. Berlin-Heidelberg: Springer-Verlag, 1992:39–63.
9. Bartocci A, Pollard JW, Stanley ER. Regulation of colony-stimulating factor 1 during pregnancy. J Exp Med 1986;164:956–61.
10. Read LD, Heith D Jr, Slamon DJ, Katzenellenbogen BS. Hormonal modulation of HER-2/neu protooncogene messenger ribonucleic acid and p185 protein expression in human breast cancer cell lines. Cancer Res 1990;50: 3947–51.
11. Lupu R, Colomer R, Zugmaier G, et al. Direct interaction of a ligand for the erbB2 oncogene product with the EGF receptor and p185erbB2. Science 1990;249:1552–5.
12. Beutler B, Mahoney J, Trang NL, Pekala P, Cerami A. Purification of cachectin, a lipoprotein lipase-suppressing hormone secreted by endotoxin-induced RAW 264.7 cells. J Exp Med 1985;161:984–95.
13. Sugarman BJ, Aggarwal BB, Hass PE, Figari IS, Palladino MA, Shepard HM. Recombinant human tumor necrosis factor-α: effects on proliferation of normal and transformed cells in vitro. Science 1985;230:943–5.
14. Ruggiero V, Latham K, Baglioni C. Cytostatic and cytotoxic activity of tumor necrosis factor on human cancer cells. J Immunol 1987;138:2711–7.
15. Beutler B, Cerami A. The biology of cachectin/TNF—a primary mediator of the host response. Annu Rev Immunol 1989;7:625–55.
16. Akira S, Hirano T, Taga T, Kishimoto T. Biology of multifunctional cytokines: IL 6 and related molecules (IL 1 and TNF). FASEB J 1990;4:2860–7.
17. Chouaib S, Branellec D, Buurman WA. More insights into the complex physiology of TNF. Immunol Today 1991;12:141–2.
18. Camussi G, Albano E, Tetta C, Bussolino F. The molecular action of tumor necrosis factor-alpha. Eur J Biochem 1991;202:3–14.
19. Tracey KJ. The acute and chronic pathophysiologic effects of TNF: mediation of septic shock and wasting (cachexia). In: Beutler B, ed. Tumor necrosis factors. New York: Raven Press, 1992:255–73.
20. Pennica D, Nedwin GE, Hayflick JS, et al. Human tumor necrosis factor: precursor structure, expression and homology to lymphotoxin. Nature 1984; 312:724–9.
21. Fransen L, Muller R, Marmenout A, et al. Molecular cloning of mouse tumor necrosis factor cDNA and its eukaryotic expression. Nucleic Acids Res 1985; 13:4417–29.
22. Shirai T, Shimizu N, Horiguchi S, Ito H. Cloning and expression in *Escherichia coli* of the gene for rat tumor necrosis factor. Agric Biol Chem 1989;53: 1733–6.
23. Beutler B, Han J, Kruys V, Giroir BP. Coordinate regulation of TNF biosynthesis at the levels of transcription and translation. In: Beutler B, ed. Tumor necrosis factors. New York: Raven Press, 1992:561–74.

24. Renz H, Gong J-H, Schmidt A, Nain M, Gemsa D. Release of tumor necrosis factor-α from macrophages. Enhancement and suppression are dose-dependently regulated by prostaglandin E2 and cyclic nucleotides. J Immunol 1988;141:2388–93.
25. Flynn RM, Palladino MA. TNF and TGF-β: the opposite sides of the avenue? In: Beutler B, ed. Tumor necrosis factors. New York: Raven Press, 1992: 131–44.
26. Chung IY, Benveniste EN. Tumor necrosis factor-α production by astrocytes: induction by lipopolysaccharide, IFN-γ and IL-1β. J Immunol 1990;144: 2999–3007.
27. Seckinger P, Zhang J-H, Hauptmann B, Dayer J-M. Characterization of a tumor necrosis factor α (TNF-α) inhibitor: evidence of immunological cross-reactivity with the TNF receptor. Proc Natl Acad Sci USA 1990;87:5188–92.
28. Tovey MG. Expression of the genes of interferons and other cytokines in normal and diseased tissues of man. Experientia 1989;45:526–35.
29. Vilcek J, Palombella V, Henriksen-De Stefano D, et al. Fibroblast growth-enhancing activity of tumor necrosis factor and its relationship to other polypeptide growth factors. J Exp Med 1986;163:632–43.
30. Hunt JS, Atherton RA, Pace JL. Differential responses of rat trophoblast cells and embryonic fibroblasts to cytokines that regulate proliferation and class I MHC antigen expression. J Immunol 1990;145:184–9.
31. Tovey MG, Content J, Gresser I, et al. Genes for IFNβ-2 (IL-6), tumor necrosis factor, and IL-1 are expressed at high levels in the organs of normal individuals. J Immunol 1988;141:3106–9.
32. Steffen M, Abboud M, Potter GK, Yung YP, Moore MAS. Presence of tumor necrosis factor or a related factor in human basophil/mast cells. Immunology 1989;66:445–50.
33. Keshav S, Lawson L, Chung LP, Stein M, Perry VH, Gordon S. Tumor necrosis factor mRNA localized to Paneth cells of normal murine intestinal epithelium by in situ hybridization. J Exp Med 1990;171:327–32.
34. Barath P, Fishbein MC, Cao J, Berenson J, Helfant RH, Forrester JH. Tumor necrosis factor gene expression in human vascular intimal smooth muscle cells detected by in situ hybridization. Am J Pathol 1990;137:503–9.
35. Hunt JS, Chen H-L, Hu X-L, Chen T-Y, Morrison DC. Tumor necrosis factor-α gene expression in the tissues of normal mice. Cytokine 1992;4: 340–6.
36. Myers MJ, Pullen JK, Ghildval N, Eustis-Turf E, Schook LB. Regulation of IL-1 and TNF-alpha expression during the differentiation of bone marrow derived macrophages. J Immunol 1989;142:153–60.
37. Hunt JS. Expression and regulation of the tumor necrosis factor-α gene in the female reproductive tract. Reprod Fertil Dev 1993;5:141–53.
38. Tartakovsky B, Ben-Yair E. Cytokines modulate preimplantation development and pregnancy. Dev Biol 1991;146:345–52.
39. De Kossodo S, Grau GE, Daneva T, et al. Tumor necrosis factor α is involved in mouse growth and lymphoid tissue development. J Exp Med 1992;176:1259–64.
40. Chen H-L, Marcinkiewicz J, Sancho-Tello M, Hunt JS, Terranova PF. Tumor necrosis factor-α gene expression in mouse oocytes and follicular cells. Biol Reprod 1993;48:707–14.

41. Hunt JS, Chen H-L, Hu X-L, Pollard JW. Normal distribution of tumor necrosis factor-α messenger ribonucleic acid and protein in virgin and pregnant osteopetrotic (*op*/*op*) mice. Biol Reprod 1993;49:441–52.
42. Witkin SS, Liu H-C, Davis OK, Rosenwaks Z. Tumor necrosis factor is present in maternal sera and embryo culture fluids during in vitro fertilization. J Reprod Immunol 1991;19:85–93.
43. Yelavarthi KK, Chen H-L, Yang Y, Fishback JL, Hunt JS. Tumor necrosis factor-alpha mRNA and protein in rat uterine and placental cells. J Immunol 1991;146:3840–8.
44. Hunt JS, Chen H-L, Hu X-L, Tabibzadeh S. Tumor necrosis factor-α messenger ribonucleic acid and protein in human endometrium. Biol Reprod 1992;47:141–7.
45. Chen H-L, Yang Y, Hu X-L, Yelavarthi KK, Fishback JL, Hunt JS. Tumor necrosis factor-alpha and protein are present in human placental and uterine cells at early and late stages of gestation. Am J Pathol 1991;139:327–35.
46. Ohsawa T, Natori S. Expression of tumor necrosis factor at a specific developmental stage of mouse embryos. Dev Biol 1989;135:459–61.
47. De M, Sanford TR, Wood GW. Interleukin-1, interleukin-6, and tumor necrosis factor α are produced in the mouse uterus during the estrous cycle and are induced by estrogen and progesterone. Dev Biol 1992;151:297–305.
48. Vince G, Shorter S, Starkey P, et al. Localization of tumor necrosis factor production in cells at the materno/fetal interface in human pregnancy. Clin Exp Immunol 1992;88:174–80.
49. Romero R, Manogue KR, Mitchell MD, et al. Infection and labor, IV. Cathectin-tumor necrosis factor in the amniotic fluid of women with intra-amniotic infection and preterm labor. Am J Obstet Gynecol 1989;161:336–41.
50. Pfizenmaier K, Himmler A, Schutze S, Scheurich P, Kronke M. TNF receptors and TNF signal transduction. In: Beutler B, ed. Tumor necrosis factors. New York: Raven Press, 1992:439–72.
51. Tartaglia LA, Goeddel DV. Two TNF receptors. Immunol Today 1992;13: 151–3.
52. Tartaglia LA, Weber RF, Figari IS, Reynolds C, Palladino MA Jr, Goeddel DV. The two different receptors for tumor necrosis factor mediate distinct cellular responses. Proc Natl Acad Sci USA 1991;88:9292–6.
53. Wong GHW, Tartaglia LA, Lee MS, Goeddel DV. Antiviral activity of tumor necrosis factor (TNF) is signaled through the 55-kDa receptor, Type I TNF. J Immunol 1992;149:3350–3.
54. Gehr G, Gentz R, Brockhaus M, Loetscher H, Lesslauer W. Both tumor necrosis factor receptor types mediate proliferative signals in human mononuclear cell activation. J Immunol 1992;149:911–7.
55. Heller RA, Song K, Fan N, Chang DJ. The p70 tumor necrosis factor receptor mediates cytotoxicity. Cell 1992;70:47–56.
56. Mackay F, Loetscher H, Stueber D, Gehr G, Lesslauer W. Tumor necrosis factor α (TNF-α)-induced cell adhesion to human endothelial cells is under dominant control of one TNF receptor type, TNF-R55. J Exp Med 1993;177: 1277–86.
57. Austgulen R, Liabakk N-B, Brockhaus M, Espevik T. Soluble TNF receptors in amniotic fluid and in urine from pregnant women. J Reprod Immunol 1992;22:105–16.

58. Yelavarthi KK, Hunt JS. Analysis of p60 and p80 tumor necrosis factor-α mRNA and protein in human placentas. Am J Pathol 1993;143:1131–41.
59. Pampfer S, Vanderheyden I, Wuu YD, Baufays L, De Hertogh R. Selective action of tumor necrosis factor-α on the inner cell mass of mouse blastocysts [Abstract]. Symposium on Immunobiology of Reproduction. Serono Symposia, USA, Boston, 1993.
60. De SK, Chen H-L, Pace JL, Hunt JS, Terranova PF, Enders GC. Expression of tumor necrosis factor alpha in mouse spermatogenic cells. Endocrinology 1993;133:389–96.
61. Frater-Schroder M, Risau W, Hallmann R, Gautschi P, Bohlen P. Tumor necrosis factor type α, a potent inhibitor of endothelial cell growth in vitro, is angiogenic in vivo. Proc Natl Acad Sci USA 1987;84:5277–81.
62. Dayer JM, Beutler B, Cerami A. Cachectin/tumor necrosis factor (TNF) stimulates collagenase and $PGE_2$ production by human synovial cells and dermal fibroblasts. J Exp Med 1985;162:2163–8.
63. Gordon C, Wofsy D. Effects of recombinant murine tumor necrosis factor-α on immune function. J Immunol 1990;144:1753–8.
64. Malik STA, Stuart Naylor M, East N, Oliff A, Balkwill FR. Cells secreting tumour necrosis factor show enhanced metastasis in nude mice. Eur J Cancer 1990;26:1031–4.
65. Orosz P, Echtenacher B, Falk W, Ruschoff J, Weber D, Mannel DN. Enhancement of experimental metastasis by tumor necrosis factor. J Exp Med 1993;177:1391–8.
66. Naylor MS, Malik STA, Jobling T, Stamp G, Balkwill F. In situ detection of tumour necrosis factor in human ovarian cancer specimens. Eur J Cancer 1990;26:1027–30.
67. Takeyama H, Wakamiya N, O'Hara C, et al. Tumor necrosis factor expression by human ovarian carcinoma in vivo. Cancer Res 1991;51:4476–80.
68. Ininns EK, Gatanaga M, Cappuccini F, et al. Growth of the endometrial adenocarcinoma cell line AN3 CA is modulated by tumor necrosis factor and its receptor is up-regulated by estrogen in vitro. Endocrinology 1992;130: 1852–6.
69. Yang Y, Yelavarthi KK, Chen H-L, Pace JL, Terranova PF, Hunt JS. Molecular, biochemical and functional characteristics of tumor necrosis factor-α produced by human placental cytotrophoblastic cells. J Immunol 1993;150: 5614–24.
70. Karp SE, Farber A, Salo JC, et al. Cytokine secretion by genetically modified nonimmunogenic murine fibrosarcoma. J Immunol 1993;150:896–908.
71. Shull MM, Ormsby I, Kier AB, et al. Targeted disruption of the mouse transforming growth factor-β1 gene results in multifocal inflammatory disease. Nature 1992;359:693–9.
72. Dalton DK, Pitts-Meek S, Keshav S, Figari IS, Bradley A, Stewart TA. Multiple defects of immune cell function in mice with disrupted interferon-γ genes. Science 1993;259:1739–42.
73. Stewart CL, Kaspar P, Brunet LJ, et al. Blastocyst implantation depends on maternal expression of leukaemia inhibitory factor. Nature 1992;359:76–9.
74. Lippman ME. The development of biological therapies for breast cancer. Science 1993;259:631–2.

75. Pollard JW, Bartocci A, Arceci R, Orlofsky A, Ladner MB, Stanley ER. Apparent role of the macrophage growth factor, CSF-1, in placental development. Nature 1987;330:484–6.
76. Daiter E, Pampfer S, Yeung YG, Barad D, Stanley ER, Pollard JW. Expression of colony stimulating factor-1 (CSF-1) in the human uterus and placenta. J Clin Endocrinol Metab 1992;74:850–8.
77. Huet-Hudson YM, Chakraborty C, De SK, Suzuki Y, Andrews GK, Dey SK. Estrogen regulates the synthesis of epidermal growth factor in mouse uterine epithelial cells. Mol Endocrinol 1990;4:510–23.
78. Robertson SA, Mayrhofer G, Seamark GF. Uterine epithelial cells synthesize granulocyte-macrophage colony-stimulating factor and interleukin-6 in pregnant and nonpregnant mice. Biol Reprod 1992;46:1069–79.
79. Kapur S, Tamada H, Dey SK, Andrews GA. Expression of insulin-like growth factor-I (IGF-I) and its receptor in the peri-implantation mouse uterus, and cell-specific regulation of IGF-I gene expression by estradiol and progesterone. Biol Reprod 1992;46:208–19.
80. Main EK, Strizki J, Schochet P. Placental production of immunoregulatory factors: trophoblast is a source of interleukin-1. Troph Res 1987;2:149–60.
81. Hu X-L, Yang Y, Hunt JS. Differential distribution of interleukin-1α and interleukin-1β proteins in human placenta. J Reprod Immunol 1992;22: 257–68.
82. McMaster MT, Newton RC, Dey SK, Andrews GK. Activation and distribution of inflammatory cells in the mouse uterus during the preimplantation period. J Immunol 1992;148:1699–1705.
83. Shimoya K, Matsuzaki N, Taniguchi T, et al. Human placenta constitutively produces interleukin-8 during pregnancy and enhances its production in intrauterine infection. Biol Reprod 1992;47:220–6.
84. Tamada H, Das SK, Andrews GK, Dey SK. Cell-type-specific expression of transforming growth factor-α in the mouse uterus during the peri-implantation period. Biol Reprod 1991;45:365–72.
85. Tamada H, McMaster MT, Flanders KC, Andrews GK, Dey SK. Cell type-specific expression of TGF-β1 in the mouse uterus during the periimplantation period. Mol Endocrinol 1990;4:965–72.
86. Das SK, Flanders KC, Andrews GK, Dey SK. Expression of transforming growth factor-β isoforms (β2 and β3) in the mouse uterus: analysis of the periimplantation period and effects of ovarian steroids. Endocrinology 1992; 130:3459–66.

# Part IV

# Placental Expression of Major Histocompatibility Complex and Associated Genes

# 11

# MHC Gene Expression in Placentas of Domestic Animals

D.F. Antczak, Juli K. Maher, Gabriele Grünig,
W.L. Donaldson, Julia Kydd, and W.R. Allen

The placenta affords protection for the developing fetus from attack and destruction by the maternal immune system through a variety of mechanisms. These include the physical separation of fetal and maternal blood supplies and local nonspecific immunosuppressive effects of high concentrations of hormones in the uterus (1, 2). Additionally, there are components of these protective mechanisms that pertain directly to the mother's immune system and the ability of the fetoplacental unit to be recognized as foreign.

One critical aspect concerns the type of local immune response that can occur in the uterus. The uterus is part of the common mucosal immune system, but it has developed a unique repertoire of immune cells and mechanisms. In humans the predominant leukocyte found in the uterus is the *large granular lymphocyte*, an unusual cell with many attributes of *natural killer* (NK) cells (3, 4). Classical $CD4^+$ and $CD8^+$ *major histocompatibility complex* (MHC)-restricted lymphocytes are relatively uncommon. The function of these uterine granular lymphocytes is not known, but it has been hypothesized that they are important in the establishment of placentation (4). The granulated metrial gland cells of the mouse appear to share some of the characteristics and functions of the human uterine large granular lymphocytes (5, 6). Much less research has been conducted on uterine leukocytes in other species; those results are summarized below.

A second consideration is the question of trophoblast-specific molecules. Efforts to generate monoclonal antibodies directed against molecules expressed only by trophoblast—and, therefore, potential antigenic targets for antifertility vaccines—have resulted in only limited success (7). This may be because such molecules are very rare in any species (8). If a trophoblast cell can carry out its tissue-specific functions using few tissue-specific molecules, it reduces its chance of being recognized as foreign by the maternal immune system. *Monoclonal antibodies* (mAbs) that distinguish trophoblast subpopulations have been easier to produce, and some of these recognize epitopes shared by humans and horses (9).

A third and, perhaps, most important immunological aspect of the fetomaternal relationship is the expression of MHC antigens by the trophoblast, the tissue layer that forms the outer surface of the placenta and comprises the interface between mother and fetus during pregnancy. This subject has been a continuing topic of investigation in reproductive immunology, and during the past five years, fascinating results have begun to emerge, many of which are based on molecular biological approaches (10, 11). Most research has been carried out in humans and rodents. In no species have MHC class II antigens been detected on trophoblast tissues. However, for the MHC class I antigens, there is no consensus, as different results have been obtained in the various species studied.

In humans, there is now strong evidence that certain trophoblast populations express a novel, nonpolymorphic MHC class I molecule, HLA-G (12). This molecule is expressed only by the extravillous trophoblast cells, but mRNA has also been detected in the villous cytotrophoblast (13). Polymorphic MHC class I antigens have not been reported on any trophoblast cells of humans.

In the rat evidence has been presented for the expression by trophoblast of a unique class I antigen encoded by the *Pa* locus (14). Kanbour and colleagues have reported that the classical MHC class I antigens of the rat are not expressed on the cell surface of the trophoblast in semiallogeneic pregnancies and that this modulation is induced and not internally programmed (14). Furthermore, the expression of MHC class I genes in rat trophoblast may be governed by imprinted genes (15). These observations have very important implications and demand further investigation.

In mice in vivo studies have provided clear evidence for expression of polymorphic MHC class I antigens by the spongiotrophoblast (16). The reasons for the apparently different results obtained in these two closely related species are unknown.

## Immunological Implications and Consequences

The failure of most trophoblast tissues to express the classical polymorphic MHC antigens has two very important immunological consequences. The first and obvious one is that transplantation responses to MHC antigens will seldom be generated, and if generated, they will be ineffective. This has been borne out by observation (reviewed in 17). Pregnancy-induced maternal antibody responses to paternal MHC antigens occur infrequently in most species, including humans (reviewed in 18, 19). Most antibodies to paternal antigens generated in females as a result of pregnancy are directed against MHC class I antigens. Antibodies to MHC class II antigens and non-MHC-linked alloantigens are rarely detected.

The second and less obvious consequence of the lack of expression of MHC antigens by trophoblast cells is that immune responses to antigens that must be recognized in the context of MHC antigens will also be eliminated. MHC-restricted T cell responses to the so-called minor histocompatibility antigens fall into this category. Thus, by selective suppression of expression of the polymorphic MHC molecules, the trophoblast renders itself virtually invulnerable to recognition and destruction by the T cell arm of the mother's immune system.

## Summary of Domestic Species

The common domestic animals (dogs and cats, horses, pigs, cattle, and sheep) have evolved reproductive strategies considerably different from those of humans and rodents. Of greatest relevance to the immunobiology of reproduction are differences in the anatomy and physiology of the placentas and in the uterine mucosal systems of these species. The placentas of mammals range from very invasive types, typified by the hemochorial placentas of humans and rodents, to the noninvasive epitheliochorial types of pigs, horses, and ruminants. The carnivores fall between these two extremes (20).

### *Ruminants*

The placentas of ruminants (cattle, sheep, and goats) have two features of particular relevance to reproductive immunology. First, the mature placenta consists of focal areas of intimate contact between placenta and uterus called *placentomes*. Placentomes comprise endometrial caruncles that accommodate the fetal cotyledons (20). It is in the placentomes that nutrients are transferred to the fetal circulation and fetal waste is removed to the mother's circulation. In the interplacentomal region there is less intimate contact between maternal endometrial epithelium and fetal trophoblast. Comparisons of MHC antigen expression and maternal leukocyte accumulations between these two sites are thus possible.

Second, in the placentomes there is fusion of the binucleate cells of the trophoblast with endometrial epithelial cells to form trinucleate cells containing two copies of the diploid fetal genome and one copy of the diploid maternal genome. The morphology of these cells has been described (21), but no studies of gene regulation in these unusual cells have been carried out. This is clearly an area of considerable interest for future research.

Maternal antibody to paternally inherited fetal MHC antigens is detectable in sera from about 30% of primiparous cows (22). Antibodies are first detected just after calving or late in pregnancy, but seldom prior to 10 days before parturition (23). In multiparous cows the frequency of maternal sensitization increases to about 65%, but antibody titers for these animals are not higher than for primiparous cows. Antibodies can be detected earlier in gestation in multiparous cows, however, suggesting that an anamnestic immune response to the conceptus occurs (24). Antibodies to paternal MHC antigens are also generated as a result of pregnancy in sheep (25, 26); about 50% of pregnant ewes become sensitized.

The investigations cited above suggest that paternally inherited MHC antigens expressed on the ruminant trophoblast might be responsible for maternal sensitization. Localization studies using mAbs failed to detect MHC class II antigens on trophoblast cells of cattle (27) or sheep (28) at any stage of gestation. For MHC class I antigens, there is an apparent discrepancy between the two species. In the sheep no MHC class I antigens were detected on any trophoblast cells from conceptuses recovered between days 9 and 125 of gestation (28), while in cows with mid- and late gestational fetuses, MHC class I antigens were detected in the interplacentomal region, but not on trophoblast cells of the cotyledons (27). The mAb used in this latter study recognizes framework determinants on MHC class I molecules and, therefore, could not determine if the class I antigens on bovine trophoblast were classical polymorphic antigens or nonpolymorphic class I molecules similar to the HLA-G antigen (12). Weak expression of MHC class I antigens was also detected on day 6–9 blastocysts using alloantisera (29) and mAbs (27). The alloantiserum results suggest that the MHC class I antigens detected later in gestation using the mAb might also be polymorphic.

If the MHC class I antigens expressed on bovine trophoblast are polymorphic and paternally inherited, it is necessary to seek an explanation for the low frequency of maternal sensitization that has been observed. Two possibilities are local immunosuppressive effects of hormones (1) or immunoregulatory effects of the leukocytes inhabiting the ruminant uterus. In the sheep few $CD4^+$ or $CD8^+$ lymphocytes were found in the uterus. The predominant leukocyte was a $CD45R^+$ large granular lymphocyte, a cell that may be similar to the large granular lymphocytes of the human uterus. These cells were found more frequently in the interplacentomal regions than in the placentomes or in the nonpregnant uterus (28). In cows $CD45^+$ cells were found in the intercaruncular region, similar to the finding in the sheep (27).

In an intriguing study Joosten and colleagues (30) have provided evidence suggesting that the pathological condition known as *retained placenta* occurs more frequently in pregnancies in which the calf carries no MHC class I antigens that can be recognized as foreign by the dam

(MHC-compatible pregnancies). MHC class II antigen compatibility did not influence the development of retained placenta. Cows carrying MHC class I incompatible pregnancies in which the paternally inherited allele was the same as the allele that the mother did not inherit from its dam also appeared to be at increased risk. This latter type of MHC-incompatible pregnancy could behave like an MHC-compatible pregnancy through the development of tolerance in the mother to noninherited maternal histocompatibility antigens, as has been described in humans (31). The hypothesis forwarded to explain these findings is that maternal allorecognition of foreign class I MHC antigens aids in the normal expulsion of the placenta. Placentas retained for longer than normal after parturition frequently result in bacterial infections and decreases in subsequent fertility for cows having this condition. The economic importance of retained placenta in cattle production and the ready possibility of testing this interesting hypothesis should spur further research to confirm this immunogenetic correlate.

## *Swine*

Relatively little research has been conducted on MHC expression in the trophoblast of the pig, which has a noninvasive epitheliochorial placenta. *Beta-2 microglobulin* (β2 microglobulin), the nonpolymorphic light chain of the MHC class I molecular complex, was detected on preimplantation pig embryos between the early blastocyst stage and the beginning of elongation of the conceptus (day 12), but not on younger embryos (32). Antibodies to the MHC class I heavy chain were not used in that study. NK cell activity increases during early pregnancy in the pig and then declines by day 30 (33), perhaps in association with the disappearance of uterine intraepithelial lymphocytes (34).

Population studies of breeding swine have revealed effects on litter size and other reproductive parameters that could be attributed to MHC genes (35). The effects were small, however, and could not be easily explained through the expression of MHC antigens on pig trophoblast or the lack thereof.

## *Carnivores*

Little information is available on placental expression of MHC antigens in dogs, cats, or other carnivores. Pregnant bitches can become sensitized to paternal MHC antigens as a result of pregnancy, but the frequency and timing of sensitization and the specificity of the antibodies have not been thoroughly investigated (36). Monoclonal antibodies to dog and cat MHC class I and class II antigens and some gene probes have been produced, however, and these reagents should permit studies of placental expression of MHC genes in the placentas of carnivores to be conducted.

## Contributions of the Horse to Reproductive Immunology

Studies in equids have helped to provide answers to three important questions in reproductive immunology. Those studies are summarized below, but it is appropriate to consider at the outset why the horse has been the subject of investigations in this field. The answer lies in the type of placentation that has evolved in this genus, which contains horses, donkeys, and zebras. First, the developing horse conceptus makes firm attachment with the uterus relatively late in pregnancy, at about day 50 after ovulation (37). This permits conceptuses of varying stages of development to be collected by nonsurgical methods. Second, the equine placenta provides populations of trophoblast cells with very different biological behavior that can be easily identified, isolated, and therefore studied both in situ and in vitro (38). Additionally, a panel of mAbs has been produced that can be used to identify the major trophoblast subpopulations of the horse (8, 9, 39, 40).

Equine trophoblast consists of three distinct forms: (i) the noninvasive trophoblast of the allantochorion, (ii) the rapidly dividing invasive cells of the chorionic girdle, and (iii) the mature *equine chorionic gonadotropin*

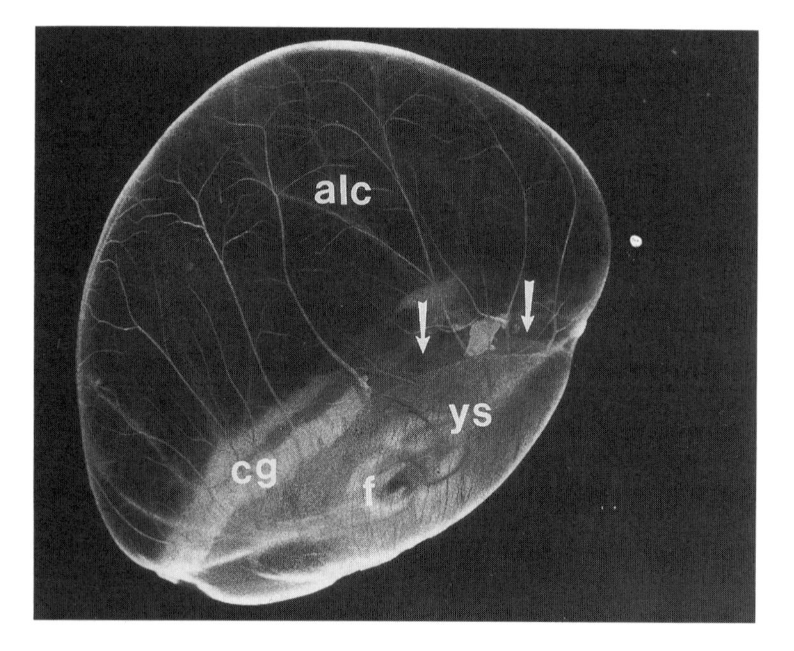


FIGURE 11.1. Intact horse conceptus. The specimen was recovered on day 36 after ovulation. (f = fetus; ys = yolk sac; alc = allantochorion; cg = chorionic girdle.) Note the patches where the chorionic girdle, having already invaded the endometrium, is missing (arrows).

(eCG)-secreting endometrial cup cells that are derived from the chorionic girdle cells (41). The chorionic girdle appears as a raised band of rapidly dividing cells that encircles the conceptus at about day 25 of gestation (Fig. 11.1). The girdle cells form a stratified columnar epithelium that increases in height and cell number until days 36–38, when the girdle cells invade the endometrium to form the endometrial cups (42). The cells on the surface of the girdle differentiate into binucleate cells that once in the endometrium cease their migration, become greatly enlarged, and begin to secrete large quantities of eCG into the maternal circulation. Typically, 10–20 endometrial cups of 0.5–1.0 cm in diameter form in a ring at the base of the pregnant horn of the uterus (38).

The endometrial cup trophoblast cells are completely detached from the rest of the placenta, similar to the extravillous trophoblast cells of human pregnancy. Unlike the extravillous trophoblast cells, however, the endometrial cup trophoblasts remain aggregated after the invasion of the endometrium and associate in closely packed clusters that can be visualized as pale structures protruding slightly above the surface of the endometrium. The natural history of the endometrial cups and the mare's response to their invasion of the uterus make a compelling argument for maternal immunological recognition of the developing conceptus.

## *Is the Uterus Immunologically Inert?*

In distinct contrast to the situation in most species, virtually all mares make strong, reproducible maternal antibody responses to paternal MHC antigens early in pregnancy (43). These antibody responses are directed almost exclusively toward paternal MHC class I antigens (44). Antibodies to MHC class II antigens have not been detected in horse pregnancy sera, but antibodies to non-MHC alloantigens can be generated if the appropriate immunogenetic difference between mother and fetus occurs (45).

The antipaternal MHC class I antigen responses of equine pregnancy arise early in gestation, usually by day 60 (Fig. 11.2). This timing follows shortly after the invasion of the endometrium by the chorionic girdle cells to form the endometrial cups. Immunohistochemical studies using mAbs and alloantisera have demonstrated that the invasive trophoblast cells of the chorionic girdle express high levels of MHC class I antigens, while the mature endometrial cup cells and the noninvasive trophoblast of the allantochorion do not (46, 47). MHC class II molecules cannot be detected on trophoblast cells from any stage of horse pregnancy. Classical polymorphic MHC class I molecules can be precipitated from radiolabeled chorionic girdle cells using either mAbs or alloantisera (47), and injection of chorionic girdle cells into *third-party* horses leads to the production of alloantibodies to both paternal and maternal polymorphic class I antigens (48). Thus, the localization of paternal polymorphic MHC class I antigens

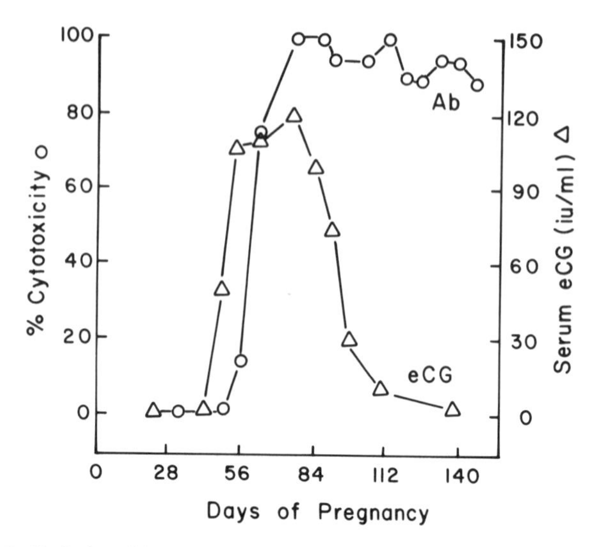


FIGURE 11.2. Relationship between the appearance of maternal antibody to paternal MHC antigens and the secretion of eCG by the endometrial cup cells. The horizontal axis indicates days of pregnancy. The left vertical axis indicates antibody reactivity to paternal MHC antigens in maternal serum measured by the lymphocyte microcytotoxicity assay, as described in reference 43; the right vertical axis indicates levels of eCG, measured in the same sample of maternal serum. Reprinted with permission from Antczak (63).

on the invasive trophoblast cells of the chorionic girdle and the timing of the maternal antibody responses to those antigens are consistent with a local immune response to MHC antigens in the pregnant horse uterus.

The maternal antibody response to paternal MHC antigens in horse pregnancy has an anamnestic component. Strong secondary responses can be detected in a second pregnancy of the same antigenic character as the first (43) and in first pregnancy after priming by allografting with skin from an MHC homozygous mating stallion (44). The secondary antibody responses have much higher titers than the primary responses, and they occur earlier in gestation, but never before the time of formation of the chorionic girdle. This is further indirect evidence that the first expression of polymorphic MHC class I antigens on horse trophoblast occurs on the developing chorionic girdle.

## *Do Uterine Lymphocytes Recognize Fetal Antigens?*

Equine pregnancy provides what is perhaps the most striking example yet described of a maternal cellular response to the placenta in the endometrial cup reaction (Fig. 11.3). Endometrial cups have a life span of about 70 days, roughly from day 40 to day 120 of gestation. The cups are

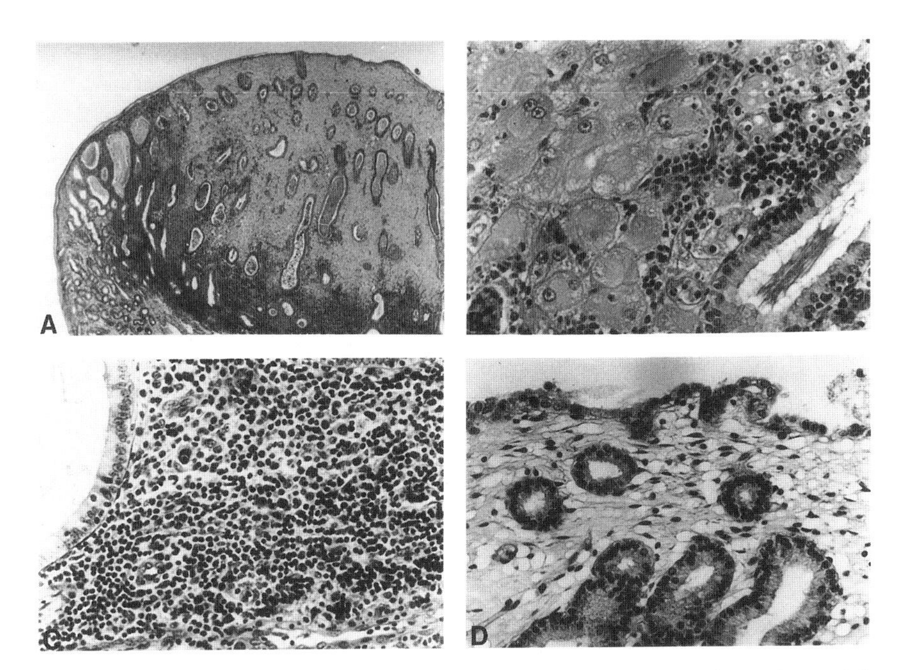


FIGURE 11.3. Maternal cellular response to the equine conceptus: the endometrial cup reaction. All sections are from tissues recovered surgically from a 60-day pregnancy in a primiparous pony mare mated to a stallion that carried no MHC antigens shared by the mare (MHC-incompatible mating). Tissues were fixed in Bouin's solution, sectioned, and stained with hematoxylin-eosin. *A:* Shown is one-half of a large endometrial cup. Note the dilated endometrial glands and lymphatics throughout the endometrial cup and the thick band of maternal leukocytes surrounding the cup (×10). *B:* Shown is a high-power view of the fetal-derived endometrial cup cells, infiltrating maternal lymphocytes, and a dilated gland. Note the binucleate endometrial cup cell at the upper left (×100). *C:* Shown is a high-power view of lymphocytes accumulated at the base of the cup (×100). *D:* Shown is a high-power view of a section of endometrium distal to the endometrial cup. The overlying allantochorion was removed before fixation of the tissue. Note the sparse distribution of lymphocytes in comparison to the area around the endometrial cup (×100). Reprinted with permission from Antczak (63).

the site of focal accumulations of maternal leukocytes that appear to wall off and destroy the cup cells (38). No such accumulations are observed in the endometrium along the border with the allantochorion. Early in the life of the endometrial cups, the leukocytes comprise mostly horse $CD4^+$ and $CD8^+$ lymphocytes, in a ratio of about 4:1 (Grünig, Antczak, unpublished observations). Granulated and nongranulated intraepithelial lymphocytes can be found at the time of girdle invasion, and the horse uterus is also rich in lymphatic vessels (37). In the late stages of the cup reaction, large numbers of neutrophils and accumulations of eosinophils are also present in and around the dying, necrotic cups. The maternal leukocyte response to the endometrial cups shares many histological features with organ grafts undergoing immunological rejection.

Member species of the genus *Equus* can interbreed to produce viable but usually infertile hybrids, such as the mule. The endometrial cup reaction in mares carrying mule pregnancies appears more vigorous than that observed in intraspecies horse pregnancy, possibly as a result of the greater antigenic disparity between the horse mother and her semixenogeneic fetus (49). When a mare is mated to a jack donkey in consecutive pregnancies, the level and duration of detection of eCG in maternal blood are reduced in the second pregnancy (18). This is consistent with an earlier demise of the mule endometrial cups caused by a secondary cellular immune response.

It has not been possible, unfortunately, to alter the endometrial cup reaction in intraspecies horse pregnancy, either by prior immunization of the mare to increase the cellular response or by the establishment of MHC-compatible matings to decrease the cellular response. The histological appearance of maternal lymphocyte accumulations around the endometrial cups at day 60 in MHC-compatible pregnancies is indistinguishable from that in MHC-incompatible pregnancies (50). This may not be surprising when one considers that the mature endometrial cups do not express MHC class I antigens and, therefore, would probably be a poor antigenic target for MHC-restricted cytotoxic lymphocytes. In order to detect differences between MHC-compatible and MHC-incompatible pregnancies, it may be necessary to look earlier in gestation, shortly after the girdle cells migrate into the uterus, for example. The early endometrial cup cells still express MHC class I antigens, although the expression of these molecules is declining rapidly at this stage (51).

Thus, the question of whether the lymphocytes that accumulate around the endometrial cups have specificity for paternal histocompatibility or placental-specific antigens has yet to be fully resolved. However, the close proximity of maternal lymphocytes to the invading MHC class I antigen-positive chorionic girdle cells on days 36–38 appears to provide ideal conditions for maternal immunological sensitization to paternal histocompatibility antigens (42).

## *How Are MHC Genes Regulated in the Trophoblast?*

Several pieces of evidence support the hypothesis that the reduction and eventual loss of expression of MHC class I antigens in the maturing endometrial cup trophoblast cells are developmentally controlled and transcriptionally regulated. First, the loss of MHC class I antigens from endometrial cup cells occurs in both MHC-compatible and MHC-incompatible pregnancies, indicating that the down-regulation is not caused by and is not dependent on maternal antipaternal MHC antibodies (51). Second, the differentiation process leading to loss of expression of MHC class I molecules by the binucleate endometrial cup cells occurs when chorionic girdle cells are cultured in vitro, indicating that maternal factors in the endometrium are not required for this event (51). Third, the down-regulation of MHC class I antigens in the mature cup cells occurs in an area of the endometrium in which there is a dramatic focal accumulation of maternal $CD4^+$ and $CD8^+$ lymphocytes. The maternal endometrium epithelium and glands show an increase in the expression of MHC class I and class II antigens around and within the endometrial cups at this time (51). The endometrial cup trophoblast cells appear to be refractory to the cytokine signaling that up-regulates MHC gene expression in other tissues (53–55).

The expression of polymorphic MHC class I genes in the trophoblast of the horse is restricted spatially to the invasive cells of the chorionic girdle and early endometrial cups. This pattern of expression suggests that the display of MHC antigens by trophoblast is carefully controlled to ensure the survival of the fetal allograft and that aberrant MHC gene expression may be involved in cases of immunologically mediated pregnancy loss.

Evidence for aberrant expression of MHC antigens by trophoblast populations that are normally MHC negative has been difficult to document in any species. In the horse, however, the MHC class I antigen-negative trophoblast of the noninvasive allantochorion membrane does express class I antigens in two situations. First, during normal equine pregnancy MHC antigens can be detected in small patches immunohistochemically using mAbs (47, 52). The class I expression is weak compared to that of nearby MHC class I-positive maternal endometrium epithelium, and it appears to occur most often in areas where the normal interdigitation of trophoblast and endometrial epithelium has failed to take place. No functional significance has been attributed to this expression; in fact, because the MHC antigens were detected using mAbs to framework determinants of horse MHC class I molecules, it is not certain that these class I antigens are polymorphic and paternally inherited.

The second case involves failing donkey-in-horse extraspecies pregnancies established by embryo transfer (56). These pregnancies are created by transfer of normal donkey embryos collected at days 6–7 to the uteri of normal horse mares. These fully xenogeneic pregnancies result in a nongenetic developmental defect in placental development. The donkey chorionic girdle develops normally in these pregnancies, but it fails to invade the endometrium of the surrogate horse mother. This results in the absence of endometrial cups and their principal hormonal product, eCG (57). Most, but not all, donkey-in-horse pregnancies end in abortion between days 70 and 95. MHC class I antigens are expressed over large areas of the poorly interdigitated, noninvasive allantochorion trophoblast in these pregnancies, where they appear to induce a vigorous maternal lymphocytic reaction. In the rare successful donkey-in-horse pregnancies, the interdigitation between chorionic villi and endometrial crypts required for successful implantation occurs, and MHC class I antigens are not detectable on the donkey trophoblast in such cases (52). The accelerated pregnancy loss in mares carrying a second unsuccessful donkey pregnancy can be explained as a manifestation of an anamnestic immune response (58). This suggests that the maternal immune system is participating in the abortion process, perhaps through recognition of paternal MHC class I antigens expressed abnormally on the noninvasive allantochorion trophoblast.

The two cases described above suggest that the control of class I gene expression may be less stringent in allantochorion trophoblast than in endometrial cup cells. In the endometrial cups MHC class I antigen loss is uniform on all cells of an endometrial cup, and class I antigens are seldom detected after terminal differentiation occurs. Therefore, MHC antigen expression may be controlled by two different mechanisms in the two forms of MHC class I-negative trophoblast. In situ hybridization studies using recently cloned horse MHC class I genes have failed to detect MHC class I mRNA in mature endometrial cups, providing further evidence that the cup cells regulate MHC class I genes at the transcriptional level (Maher, Barbis, Antczak, unpublished observations).

## Conclusions

The data on the control of expression of MHC antigens in trophoblast cell populations from the species considered above suggest that there are diverse ways of dealing with this issue. While down-regulation of MHC class I expression in the majority of trophoblast cells seems to be the common theme, this can be accomplished in many ways, including direct regulation of MHC class I genes via the action of transcription factors, posttranscriptional regulation by a variety of mechanisms, or negative regulation of the other genes required for MHC class I expression, such as β2 microglobulin (59–61).

One untoward effect of the suppression of MHC antigen expression by the trophoblast is that this tissue might become a safe harbor for virus pathogens. MHC-restricted cytotoxic T cells would not be able to recognize MHC-negative virus-infected trophoblast cells as foreign. Thus, the down-regulation of MHC antigens on trophoblast becomes an immunological gamble for the developing conceptus. The evasion of maternal immune responses must be balanced against the failure to provide the correct molecular environment for T cell recognition and consequent increased vulnerability to infection. If $T_{H2}$ responses are preferentially induced by pregnant females (62), this could further compromise the ability of the maternal immune system to provide adequate host defense during this critical time.

The signals and mechanisms controlling MHC gene expression in the placenta are of considerable interest to immunobiologists of many persuasions. The trophoblast is a simple, early differentiating tissue that may serve as a model for studies of gene regulation that can be applied to clinical organ transplantation and to tumor immunology. For the reasons described above, the horse represents an excellent species in which to study the control of MHC antigen expression in the trophoblast and, in addition, to study an animal whose maternal antifetal immune responses may have important consequences for fetal survival.

*Acknowledgments.* The research described in this chapter was supported in part by the National Institutes of Health (HD-15799), the Zweig Memorial Fund for Equine Research in New York State, and the Dorothy Russell Havemeyer Foundation, Inc. The authors thank Ms. Dorothy Scorelle for assistance with manuscript preparation.

## References

1. Low BG, Hansen PJ. Actions of steroids and prostaglandins secreted by the placenta and uterus of the cow and ewe on lymphocyte proliferation in vitro. Am J Reprod Immunol Microbiol 1988;18:71–5.
2. Hansen PJ, Stephenson DC, Low BG, Newton GR. Modification of immune function during pregnancy by products of the sheep uterus and conceptus. J Reprod Fertil Suppl 1989;37:55–61.
3. King A, Balendran N, Wooding P, Carter NP, Loke YW. CD3-leukocytes present in the human uterus during early placentation: phenotypic and morphologic characterization of the CD56+ population. Dev Immunol 1991;1: 169–90.
4. King A, Loke YW. On the nature and function of human uterine granular lymphocytes. Immunol Today 1991;12:432–5.
5. Redline RW, Lu CY. Localization of fetal major histocompatibility complex antigens and maternal leukocytes in murine placenta: implications for maternal-fetal immunological relationship. Lab Invest 1989;61:27–36.
6. Croy BA, Guilbert LJ, Browne MA, et al. Characterization of cytokine production by the metrial gland and granulated metrial gland cells. J Reprod Immunol 1991;19:149–66.
7. Anderson DJ, Johnson PM, Alexander NJ, Jones WR, Griffin PD. Monoclonal antibodies to human trophoblast and sperm antigens: report of two WHO-sponsored workshops, June 30, 1986-Toronto, Canada. J Reprod Immunol 1987;10:231–57.
8. Oriol JG, Poleman JC, Antczak DF, Allen WR. A monoclonal antibody specific for equine trophoblast. Eq Vet J Suppl 1989;8:14–8.
9. Oriol JG, Donaldson WL, Dougherty DA, Antczak DF. Molecules of the early equine trophoblast. J Reprod Fertil Suppl 1991;44:455–62.
10. Hunt JS, Hsi BL. Evasive strategies of trophoblast cells: selective expression of membrane antigens. Am J Reprod Immunol 1990;23:57–63.

11. Hunt JS, Orr HT. HLA and maternal-fetal recognition. FASEB J 1992;6:2344–8.
12. Kovats S, Main EK, Librach C, Stubblebine M, Fisher SJ, DeMars R. A class I antigen, HLA-G, expressed in human trophoblast. Science 1990;248:220–3.
13. Hunt JS, Fishback JL, Chumbley G, Loke YW. Identification of class I MHC mRNA in human first trimester trophoblast cells by in situ hybridization. J Immunol 1990;144:4420–5.
14. Kanbour A, Ho H-N, Misra DN, MacPherson TA, Kunz HW, Gill TJ. Differential expression of MHC class I antigens on the placenta of the rat. J Exp Med 1987;166:1861–82.
15. Kanbour-Shakir A, Zhang X, Rouleau A, et al. Gene imprinting and major histocompatibility complex class I antigen expression in the rat placenta. Proc Natl Acad Sci USA 1990;87:444–8.
16. Raghupathy R, Singh B, Leigh JB, Wegmann TG. The ontogeny and turnover kinetics of paternal H-2K antigenic determinants on the allogeneic murine placenta. J Immunol 1981;127:2074–9.
17. Clark DA, Chaput A, Slapsys R, et al. Local intrauterine suppressor cells and suppressor factors in survival of the fetal allograft. Colloque INSERM 1987; 154:77–88.
18. Antczak DF, Allen WR. Invasive trophoblast in the genus *Equus*. Ann Immunol (Paris) 1984;135D:325–31; 341–2.
19. Antczak DF. Maternal antibody responses in pregnancy. Curr Opin Immunol 1989;1:1135–40.
20. Mossman HW. Vertebrate fetal membranes. New Brunswick, NJ: Rutgers University Press, 1987.
21. Wooding FBP. The role of the binucleate cell in ruminant placental structure. J Reprod Fertil Suppl 1982;31:31–9.
22. Newman MJ, Hines HC. Production of foetally stimulated lymphocytotoxic antibodies by primiparous cows. Anim Genet 1979;10:87–92.
23. Newman MJ, Hines HC. Stimulation of maternal anti-lymphocyte antibodies by first gestation bovine fetuses. J Reprod Fertil 1980;60:237–41.
24. Hines HC, Newman MJ. Production of foetally stimulated lymphocytotoxic antibodies by multiparous cows. Anim Genet 1981;12:201–6.
25. Ford CH, Elves MW. The production of cytotoxic antileucocyte antibodies by parous sheep. J Immunogenet 1974;1:259–64.
26. Stear MJ, Spooner RL. Occurrence of cytotoxic antilymphocyte antibodies in sheep. Res Vet Sci 1983;34:218–23.
27. Low BG, Hansen PJ, Drost M, Gogolin-Ewens KJ. Expression of major histocompatibility complex antigens on the bovine placenta. J Reprod Fertil 1990;90:235–43.
28. Gogolin-Ewens KJ, Lee CS, Mercer WR, Brandon MR. Site-directed differences in the immune response to the fetus. Immunology 1989;66:312–7.
29. Templeton JW, Tipton RC, Garber T, Bondioli K, Kraemer DC. Expression and genetic segregation of parental BoLA serotypes in bovine embryos. Anim Genet 1987;18:317–22.
30. Joosten I, Sanders MF, Hensen EJ. Involvement of major histocompatibility complex class I compatibility between dam and calf in the aetiology of bovine retained placenta. Anim Genet 1991;22:455–63.

31. Claas FHJ, Gijbels Y, van der Velden-de Munck J, van Rood JJ. Induction of B cell unresponsiveness to noninherited maternal HLA antigens during fetal life. Science 1988;241:1815–7.
32. Meziou W, Chardon R, Flechon J-E, Kalil J, Vaiman M. Expression of β2-microglobulin on preimplantation pig embryos. J Reprod Immunol 1983;5: 73–80.
33. Yu Z, Croy A, Chapeau C, King GJ. Elevated endometrial natural killer cell activity during early porcine pregnancy is conceptus-mediated. J Reprod Fertil 1993;24:153–64.
34. King GJ. Reduction in uterine intraepithelial lymphocytes during early gestation in pigs. J Reprod Immunol 1988;14:41–6.
35. Vaiman M, Renard C, Bourgeaux N. SLA, the major histocompatibility complex in swine: its influence on physiological and pathological traits. In: Warner CM, Rothschild MF, Lamont SJ, eds. The molecular biology of the major histocompatibility complex of domestic animal species. Ames: Iowa State University Press, 1987:23–38.
36. Vriesendorp HM, Grosse-Wilde H, Dorf ME. The major histocompatibility system of the dog. In: Götze D, ed. The major histocompatibility system in man and animals. Berlin: Springer-Verlag, 1977:129–63.
37. Enders AC, Liu IKM. Lodgement of the equine blastocyst in the uterus from fixation through endometrial cup formation. J Reprod Fertil 1991;44: 427–38.
38. Allen WR. Immunological aspects of the equine endometrial cup reaction. In: Edwards RG, Howe C, Johnson MH, eds. Immunobiology of the trophoblast. Cambridge: Cambridge University Press, 1975:217–53.
39. Antczak DF, Poleman JC, Stenzler LM, Volsen SG, Allen WR. Monoclonal antibodies to equine trophoblast. Troph Res 1987;2:199–214.
40. Antczak DF, Oriol JG, Donaldson WL, et al. Differentiation molecules of the equine trophoblast. J Reprod Fertil Suppl 1987;35:371–8.
41. Allen WR, Moor RM. The origin of the equine endometrial cups, I. Production of PMSG by fetal trophoblast cells. J Reprod Fertil 1972;29:313–6.
42. Enders AC, Liu IKM. Trophoblast-uterine interactions during equine chorionic girdle cell maturation, migration, and transformation. Am J Anat 1991;192:366–81.
43. Antczak DF, Miller JM, Remick LH. Lymphocyte alloantigens of the horse, II. Antibodies to ELA antigens produced during equine pregnancy. J Reprod Immunol 1984;6:283–97.
44. Donaldson WL, Crump AL, Zhang C, et al. At least two loci encode polymorphic class I MHC antigens in the horse. Anim Genet 1988;19: 379–90.
45. Antczak DF. Lymphocyte alloantigens of the horse, III. ELY-2.1: a lymphocyte alloantigen not coded by the MHC. Anim Genet 1984;15:103–15.
46. Crump A, Donaldson WL, Miller J, Kydd J, Allen WR, Antczak DF. Expression of major histocompatibility complex (MHC) antigens on equine trophoblast. J Reprod Fertil Suppl 1987;35:379–88.
47. Donaldson WL, Zhang CH, Oriol JG, Antczak DF. Invasive equine trophoblast expresses conventional class I major histocompatibility complex antigens. Development 1990;110:63–71.

48. Donaldson WL, Oriol JG, Pelkaus CL, Antczak DF. Paternal and maternal major histocompatibility complex class I antigens are expressed codominantly by equine trophoblast. Placenta 1993.
49. Allen WR. Maternal recognition of pregnancy and immunological implications of trophoblast-endometrium interactions in equids. In: Maternal recognition of pregnancy. Ciba Foundation Series 64. Excerpta Medica, May 1979: 323–52.
50. Allen WR, Kydd J, Miller J, Antczak DF. Immunological studies on fetomaternal relationships in equine pregnancy. In: Crighton DB, ed. Proc 38th Easter School, University of Nottingham, Immunological Aspects of Reproduction in Mammals. London: Butterworths, 1984:183–93.
51. Donaldson WL, Oriol JG, Plavin A, Antczak DF. Developmental regulation of class I major histocompatibility complex antigen expression by equine trophoblastic cells. Differentiation 1992;52:69–78.
52. Kydd JH, Butcher GW, Antczak DF, Allen WR. Expression of major histocompatibility complex (MHC) class I molecules on early trophoblast. J Reprod Fertil Suppl 1991;44:463–77.
53. Zuckerman FA, Head JR. Expression of MHC antigens on murine trophoblast and their modulation by interferon. J Immunol 1986;137:846–53.
54. Hunt JS, Atherton RA, Pace JL. Differential responses of rat trophoblast cells and embryonic fibroblasts to cytokines that regulate proliferation and class I MHC antigen expression. J Immunol 1990;145:184–9.
55. Mattsson R, Holmdahl R, Scheynius A, Bernadotte F, Mattsson A, Van der Meide PH. Placental MHC class I antigen expression is induced in mice following in vivo treatment with recombinant interferon-gamma. J Reprod Immunol 1991;19:115–29.
56. Allen WR. Immunological aspects of the equine endometrial cup reaction and the effect of xenogeneic pregnancy in horses and donkeys. J Reprod Fertil 1982;31:57–94.
57. Antczak DF, Allen WR. A non-genetic developmental defect in trophoblast formation in the horse: immunological aspects of a model of early abortion. In: Beard RW, Sharp F, eds. Early pregnancy loss: mechanisms and treatment. Proc 18th Study Group of the Royal College of Obstetricians and Gynaecologists, October 1987. Ashton-under-Lyme, UK: Peacock Press, 1988:123–40.
58. Allen WR, Kydd J, Antczak DF. Maternal immunological response to the trophoblast in xenogeneic equine pregnancy. In: Gill TJ, Wegmann T, eds. Immunoregulation and fetal survival. New York: Oxford University Press, 1987:263–85.
59. Jaffe L, Robertson EJ, Bikoff EK. Distinct patterns of expression of MHC class I and β-2 microglobulin transcripts at early stages of mouse development. J Immunol 1991;147:2740–50.
60. Boucraut J, Hawley S, Robertson K, Bernard D, Loke YW, Le Bouteiller P. Differential nuclear expression of enhancer A DNA-binding proteins in human first trimester trophoblast cells. J Immunol 1993;150:3882–94.
61. Drezen JM, Babinet C, Morello D. Transcriptional control of MHC class I and β2-microglobulin genes in vivo. J Immunol 1993;150:2805–13.

62. Wegmann TG, Lin H, Guilbert L, Mosmann TR. Bidirectional cytokine interactions in the maternal-fetal relationship: is successful pregnancy a $T_H2$ phenomenon? Immunol Today 1993;14:353–6.
63. Antczak DF. The major histocompatibility complex of the horse. In: Plowright W, Rossdale PD, Wade JF, eds. Equine infectious diseases VI, proc 6th int conf. Newmarket, UK: R & W Publications, 1992:99–112.

# 12

# MHC-Linked Genes and Their Role in Growth and Reproduction

THOMAS J. GILL III, HONG-NERNG HO, AMAL KANBOUR-SHAKIR, AND HEINZ W. KUNZ

The region of the genome that contains the *major histocompatibility complex* (MHC), which has been highly conserved throughout evolution, and its linked genes is critical for self-recognition, growth, and development in a variety of species (1–3). We have developed a hypothesis to relate these various functions, and it is shown in Figure 12.1. The MHC genes are primarily involved in the differentiation of self and nonself, and the regulation of their expression plays a major role in the unique immunogenic profile of the placenta. They are also important in the resistance to infectious diseases. The MHC-linked genes influence growth, development, and resistance to cancer. In the rat, which is the experimental animal that we have used, the MHC-linked region is designated the *growth and reproduction complex* (*grc*). In the mouse the *t*-haplotypes display many similar properties, but they are fundamentally different because they cause segregation distortion and suppress recombination, which the *grc* does not (4). Similar genes are also found in other species, and they are discussed below. In the human there is evidence that the genes of the MHC behave in groups that have been variously called *complotypes* (5) or *extended haplotypes* (6, 7). Recombination appears to be more frequent between blocks of MHC genes than within blocks, and extended haplotypes have been extensively associated with susceptibility to a variety of diseases.

The structure of the MHC in the human, rat, and mouse is very similar (1). The arrangement of the loci is different, however, between the rat and the mouse on the one hand and the human and all other species on the other. This appears to have been due to a translocation, such that the class II and class III genes in the rat and the mouse are between the two major class I genes, whereas in the human they occupy a region of the chromosome distinct from the region carrying the class I genes. Considering this translocation, the genes in the class II and class III regions of the

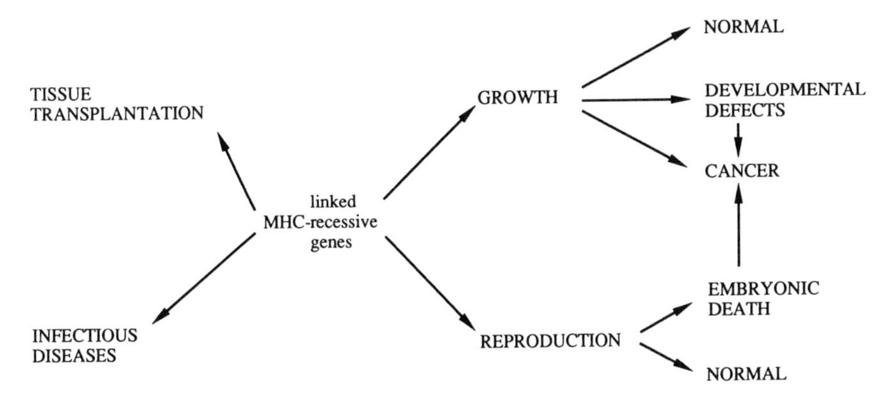


FIGURE 12.1. Hypothesis showing the relationship between MHC-linked recessive lethal genes or genetic defects and reproduction, developmental defects, and susceptibility to cancer. Reprinted with permission from Gill III (2).

TABLE 12.1. Mechanisms controlling MHC antigen expressions in the rat placenta.

| |
|---|
| Constitutive suppression |
| All class II antigens |
| Allele-specific class I antigens, Pa antigen, and RT1.E antigen in the labyrinthine trophoblast |
| Inducible suppression |
| Allele-specific class I antigens on the membrane of the basal trophoblast cells in allogeneic pregnancies |
| Genomic imprinting |
| Paternal (not maternal) allele-specific class I antigens and the monomorphic Pa antigen are expressed in the basal trophoblast cells |
| Monomorphic RT1.E antigen is not imprinted |

human (HLA-B to HLA-DQ) fall into a homologous position in both the rat and the mouse. This is a critical point to understand in examining the comparative genetics of growth, reproduction, and resistance to cancer.

## Placental Antigens

The mechanisms controlling the expression of MHC antigens in the rat placenta have been identified in experiments employing all four crosses between the DA and WF strains, and they are outlined in Table 12.1.

### *Constitutive Suppression*

All of the class II antigens and the labyrinthine class I antigens in the rat are constitutively suppressed (Table 12.2). The expression of the class II

TABLE 12.2. Constitutive suppression of MHC class II antigens and labyrinthine class I antigens in the placenta of the rat.

| Mating combination | Location on basal trophoblast | Class II (basal and labyrinth) | Class I on labyrinth (A, Pa, E) |
|---|---|---|---|
| WF × DA | M | 0 | 0 |
| | C | 0 | 0 |
| DA × WF | M | 0 | 0 |
| | C | 0 | 0 |
| DA × DA | M | 0 | 0 |
| | C | 0 | 0 |
| WF × WF | M | 0 | 0 |
| | C | 0 | 0 |

*Source:* Data from Kanbour-Shakir, Kunz, and Gill III (10); Kanbour, Ho, Misra, Macpherson, Kunz, and Gill III (11); and unpublished observations.

antigens in the human and the mouse are also constitutively suppressed (1).

## *Inducible Suppression: RT1.A Locus*

The experimental systems used to examine the expression of class I antigens in the placenta were natural matings and embryo transfers (9, 10). In the embryo transfer studies, mechanical stimulation was used to induce pseudopregnancy, and the zygote was transferred at the 2-cell stage (24 h). Specific monoclonal antibodies, light microscopy, and electron microscopy were used to detect the presence and location of the various antigens.

The studies on inducible suppression are summarized in Table 12.3. The $A^a$ and $A^u$ antigens in the F1 hybrids WF × DA and DA × WF, respectively, are not expressed on the surface of the basal trophoblast, but they are present in the cytoplasm. By contrast, the Pa antigen in the WF × DA mating is expressed on the surface of the basal trophoblast, as well as being present in the cytoplasm. This suppression occurs only in natural matings, and there is no suppression of surface antigen expression in the embryo transfer experiments. The DA × DA → WF transfer is equivalent to the WF × DA mating in which the paternal antigens come from the DA strain. In the WF × DA → WF transfer, the paternal antigens are contributed by the DA strain of the transferred embryo, and in the DA × WF → WF mating, the paternal antigens are contributed by the WF strain of the transferred embryo.

The results of the embryo transfer experiments indicate the importance of the normal uterine environment for the induction of RT1.A antigen

TABLE 12.3. Inducible suppression and genomic imprinting of MHC class I antigens on the basal trophoblast of the rat.

| Mating combination[a] | Location on basal trophoblast[b] | Antigen expression | | | |
|---|---|---|---|---|---|
| | | Polymorphic | | Monomorphic | |
| | | $A^a$ | $A^u$ | Pa | $E^u$ |
| Natural mating | | | | | |
| WF × DA | M | 0 | 0 | + | + |
| | C | + | 0 | + | + |
| DA × WF | M | 0 | 0 | 0 | + |
| | C | 0 | + | 0 | + |
| DA × DA | M | + | 0 | + | 0 |
| | C | + | 0 | + | 0 |
| WF × WF | M | 0 | + | 0 | + |
| | C | 0 | + | 0 | + |
| Embryo transfer | | | | | |
| WF × DA → WF | M | + | 0 | + | |
| | C | + | 0 | + | |
| DA × WF → WF | M | 0 | + | 0 | |
| | C | 0 | + | 0 | |
| DA × DA → WF | M | + | 0 | + | |
| | C | + | 0 | + | |
| DA × DA → DA | M | + | 0 | + | |
| | C | + | 0 | + | |

*Note:* Gene order: *Glo1*, *A*, *Pa*, *E-grc*, and *G/C*.
[a] Female strain written first.
[b] M = cell membrane; C = cytoplasm.
*Source:* Data from Kanbour-Shakir, Zhang, Rouleau, Armstrong, Kunz, and Gill III (9); Kanbour-Shakir, Kunz, and Gill III (10); and Kanbour, Ho, Misra, Macpherson, Kunz, and Gill III (11).

suppression: The presence of the conceptus in the oviduct during the first 24 h or the presence of seminal fluid may be crucial. Additional evidence for the importance of the uterine environment comes from studies on class I antigen expression when the embryo is fertilized in vivo compared to when it is fertilized in vitro prior to transfer (12). In syngeneic pregnancies there is no difference in antigen expression, but in allogeneic pregnancies antigen expression is greater after in vivo fertilization than after in vitro fertilization.

The RT.1E$^u$ antigen does not show any inducible suppression (10). It is expressed both on the basal trophoblast membrane and in the cytoplasm in allogeneic pregnancies (Table 12.3).

## Genomic Imprinting

Genomic imprinting was first demonstrated by experiments in mice in which androgenetic and gynecogenetic zygotes were generated by nuclear transplantation and then transferred into pseudopregnant females (13, 14). The androgenetic zygotes developed a normal placenta but a vestigial fetus, whereas the gynecogenetic zygotes developed a relatively normal fetus but a very small placenta. The conclusion from these studies was that the paternal genome controlled primarily placental development, and the maternal genome controlled primarily the development of the fetus.

The same experimental system that was used to study inducible suppression was used to study genomic imprinting (Table 12.3). The *A* and *Pa* loci are imprinted: In the allogeneic natural matings (WF × DA and DA × WF) and in the allogeneic embryo transfers (WF × DA → WF, DA × WF → WF, and DA × DA → WF), only the paternal class I antigens are expressed. By contrast, studies using natural matings showed that the *E* locus is not imprinted; that is, both the paternal and maternal antigens are expressed. These results suggest that genomic imprinting is programmed within the cell; that is, no external factor is required for imprinting.

The expression of only the paternal antigens in the rat trophoblast is consistent with the results from the nuclear transplantation studies in mice and with the observation that hydatidiform moles, which may lead to choriocarcinoma, are of paternal origin. The demonstration that the *A* and *Pa* loci are imprinted and the *E* locus is not means that there is differential imprinting for homologous loci in a multigene family. This is not due to the monomorphic nature of the *E* locus since the *Pa* locus is also monomorphic, and it is imprinted. The best working hypothesis at the moment for the mechanism of imprinting is that the structure of the DNA proximate to the imprinted locus is critical for determining imprinting.

## MHC-Linked Genes

### *Growth and Reproduction*

A variety of studies in the rat using chromosome walking and pulse field gel electrophoresis (15) have shown that there is a deletion in the MHC-linked region in those strains that have defects in growth, reproduction, and resistance to cancer. A probe for the MHC-linked region was isolated from a $grc^+$ library on the basis of comparative mapping with $grc^-$ strains and then was tested on 49 inbred, congenic, or recombinant strains to determine its correlation with the $grc^+$ phenotype (Table 12.4). There was an excellent correlation between the reactivity of the probe and the normal phenotype and between lack of reactivity and the abnormal phenotype. The fact that the probe did not react with 10 $grc^+$ strains suggests that there is some expansion and contraction of the MHC-linked region; a similar phenomenon has been described in the mouse (16).

TABLE 12.4. Correlation between *grc* phenotype and reactivity with probe pGRC1.7 in 49 inbred, congenic, or recombinant strains of rats.

| | *grc* phenotype | |
|---|---|---|
| Reactivity with pGRC1.7 | Normal (+) | Abnormal (−) |
| Reactive (+) | 35 | 0 |
| Not reactive (−) | 10 | 4 |

TABLE 12.5. Test for segregation distortion in the R16 ($grc^{-}$) × R16 (a/n) ($grc^{-/+}$) backcross.

| | | | | Haplotypes of offspring | | | | | |
|---|---|---|---|---|---|---|---|---|---|
| | | | | Male | | Female | | Total | |
| Number of mating pairs | Number of litters | Number of born | Offspring weaned | a/a | a/n | a/a | a/n | a/a | a/n |
| 30 | 56 | 281 | 272 (96.8%) | 68 | 58 | 81 | 64 | 149 | 122 |
| | | | | (Ratio = 1.17) | | (Ratio = 1.26) | | (Ratio = 1.22) | |

These MHC-linked deletions in the rat do not act through the same mechanisms as do the *t*-haplotypes in the mouse since there is no segregation distortion (Table 12.5) or suppression of recombination (4, 17) associated with the deletions. Nonetheless, a region similar to the *grc* is present in the mouse since the $grc^{+}$ probe reacts with DNA from a panel of mouse strains, and the use of congenic and MHC recombinant strains places the $grc^{+}$ DNA within the *TL* region with class I loci proximal and distal to it (18).

Animals homozygous for the $grc^{-}$ deletion have defects in reproduction and in development (1). There is a 30% mortality of $grc^{-}$ homozygotes in the immediate postnatal period, and the surviving males and females are small throughout their lives (Fig. 12.2). The males are infertile due to a complete block of spermatogenesis at the pachytene stage of meiotic prophase I: The spermatocytes do not develop beyond the stage of the primary spermatocyte. The females have a block at a similar place in the development of the ovum, but the block is not complete; hence, the females have reduced fertility.

There is evidence for similar MHC-linked traits in various species (Table 12.6), including the human (Table 12.7). This evidence comes from correlating the various manifestations that have been described for the $grc^{-}$ deletion in the rat with the sharing of MHC antigens. An extensive study in an ethnically homogeneous population of Chinese in Taiwan showed that the major correlation was between sharing 2 or more of the HLA-B, -DR, and -DQ antigens and *recurrent spontaneous abortions* (RSAs) and failure of in vitro fertilization and tubal embryo transfer. The lack of correlation with the *HLA-A*, *-B*, and *-DR* loci suggests

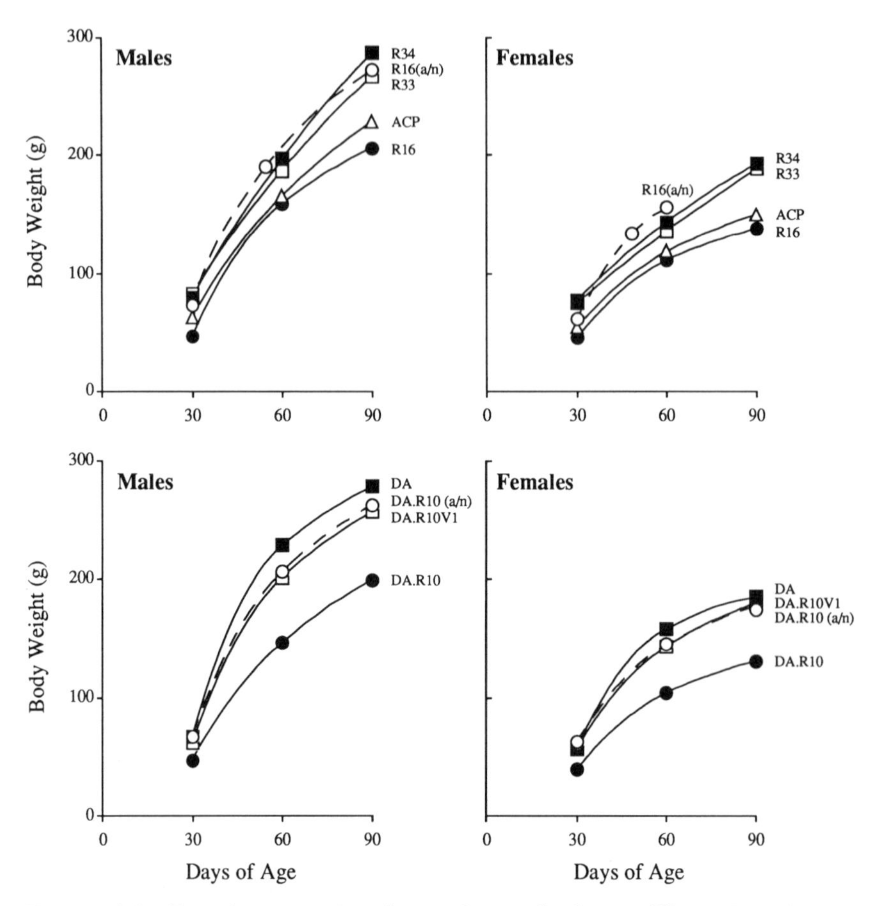


FIGURE 12.2. Growth curves of strains used to study the *grc*. The $grc^-$ strains are R16 and DA.R10; all other strains are $grc^+$. The standard error of the mean for these measurements falls within the size of the symbol used to designate the various weights.

that the *HLA-B*, *-DR*, and *-DQ* region contains the critical genes or genetic defects affecting reproduction.

Other evidence for the importance of MHC-linked genes affecting growth and development in the human comes from an examination of population isolates: They either have deficiencies of HLA homozygotes or increased HLA sharing in the couples with fertility problems (Table 12.8). Studies of HLA sharing between husband and wife in preeclampsia or eclampsia also indicate that there is an excess of HLA sharing in these diseases (Table 12.9).

TABLE 12.6. MHC-linked traits in various animal species that affect growth, development, and susceptibility to cancer.

| Species | Gene | Phenotype |
|---|---|---|
| Rat[a] | *rcc* | Increased susceptibility to chemical carcinogens |
| | *ft* | Decreased fertility |
| | *dw-3* | Small body size |
| Mouse | *T* | Skeletal defects |
| | *t* | Lethal or semilethal |
| | *grc* homologue | |
| | *Ped* | Differential egg cleavage |
| | | Aging (survival) |
| | | Mating preference |
| Horse | | Segregation distortion |
| Swine | | Deficiency of MHC homozygotes |
| | | Smaller litters |
| | | Different ovulation rates |
| Chicken | | Different ovulation rates |
| | | Different egg sizes |
| Rabbit | | Early embryonic losses |
| | | Different ovulation rates |
| Cattle | | Deficiency of MHC homozygotes |
| | | Decreased fertility |
| Cheetah | | Deficiency of MHC homozygotes |

[a] The *grc* comprises the *rcc*, *ft*, and *dw-3* genes.
*Source:* Reprinted from Gill III (1).

TABLE 12.7. Summary of data in the literature on the sharing of HLA-A, -B, and -DR antigens in couples with RSAs and in normally fertile couples.

| | Recurrent aborters | | | Normally fertile | | |
|---|---|---|---|---|---|---|
| | HLA-A | HLA-B | HLA-DR | HLA-A | HLA-B | HLA-DR |
| Review of the literature | 227/528 | 164/528 | 147/346 | 297/623 | 181/623 | 120/359 |
| Single study | 90/123 | 39/123 | 58/123 | 28/51 | 14/51 | 18/51 |
| Total | | | | | | |
| Number | 317/651 | 203/651 | 205/469 | 325/674 | 195/674 | 138/410 |
| Percentage | 48.7 | 31.2 | 43.7 | 48.2 | 28.9 | 33.6 |
| Difference from normal (*P*-value) | 0.9 | 0.4 | <0.003 | | | |

*Source:* Reprinted from Gill III (1).

# Cancer

A series of studies in the rat has shown that resistance to chemical carcinogens is linked to the MHC (17). Animals carrying the *grc*$^-$ deletion were highly susceptible to the induction of cancer by *aminoacetylfluorene* (AAF) in males, *diethylnitrosamine* (DEN) in males, and *dimeth-*

TABLE 12.8. MHC-linked genes in humans that affect growth and development.

| Population | Phenotype |
|---|---|
| Pooled populations (predominantly Caucasian) | HLA sharing in RSAs |
| Caucasian | HLA sharing in RSAs |
| Taiwan Chinese | HLA sharing in RSAs and in gestational trophoblastic tumors |
| Hutterites | HLA sharing in longer time to conception, decreased fertility, and deficiency in HLA homozygotes |
| Kes Kummer Tauregs | Deficiency of HLA homozygotes |
| Newfoundland Caucasian population isolates | Deficiency of HLA homozygotes and a familial aggregate of lymphomas and immunodeficiency syndromes |
| Brazilian Amerindian population isolates | Deficiency of HLA homozygotes |
| Havasupai Amerindian (Arizona) population isolates | Deficiency of HLA homozygotes |

*Source:* Reprinted from Gill III (1).

TABLE 12.9. HLA sharing between husband and wife in preeclampsia or eclampsia in Caucasian populations.

| HLA antigen shared | Status | Couples with shared Antigen/total (%) |
|---|---|---|
| B | Affected | 9/26 (35) |
| | Normal | 3/40 (8) |
| DR4 | Affected | 44/94 (47) |
| | Normal | 26/94 (28) |
| ABC | Affected | 16/26 (62) |
| | Normal | 10/40 (25) |
| ABCDRDQ | Affected | 23/26 (88) |
| | Normal | 19/40 (48) |

*Source:* Reprinted from Gill III (1).

*ylbenzathracine* (DMBA) in females. Since all three studies indicated that the same MHC-linked genetic defect was responsible for the loss of resistance to the chemical carcinogens, the induction of liver cancer by DEN in male rats was used as a prototype to map the gene(s) influencing *resistance to chemical carcinogens* (*rcc*). The strains chosen for this study, their genotypes, and their susceptibilities to DEN are summarized in Table 12.10. These rats also showed the phenotypic defects associated with the $grc^-$ deletion: reduced body weight (Fig. 12.2), male sterility, and reduced female fertility. The segregation of resistance to DEN mapped the *rcc* gene between *RT1.E* and *ft*.

TABLE 12.10. Genotypes of recombinant strains and their susceptibility to DEN carcinogenesis.

| | Genotype | | | | | | | |
|---|---|---|---|---|---|---|---|---|
| | MHC (RT1) | | | | *grc* | | | |
| Strain | A | B | D | E | *rcc* | *ft* | *dw-3* | Susceptibility to DEN (%) |
| R16 | a | a | a \| | − | − | − | − | 68 |
| R33 | a | a | a \| | u | + | + | + | 20 |
| R34 | a | a | a | − | − | + | + | 70 |
| DA | a | a | a | − | + | + | + | 8 |
| DA.R10 | n \| | l | l | − | − | − | − | 60 |
| DA.R10V1 | n | l | l | − \| | + | + | + | 25 |
| ACP | a | a | a \| | − | + | + | + | 6 |
| R21 | l | l | l | u | + | + | + | 0 |

Vertical lines indicate position of recombination.
*Source:* Data from Melhem, Kunz, and Gill III (17).

TABLE 12.11. HLA sharing in Taiwan Chinese couples with RSAs and in couples in whom the woman developed a gestational trophoblastic tumor.

| | ABDRDQ ≥ 3 | | BDRDQ ≥ 2 | | ABDR ≥ 2 | |
|---|---|---|---|---|---|---|
| Population | No./total | *P*-value[a] | No./total | *P*-value[a] | No./total | *P*-value[a] |
| Recurrent aborters | 54/123 | 0.004 | 66/123 | 0.006 | 64/123 | 0.053 |
| IVF-TET failures | 16/36 | 0.021 | 22/36 | 0.006 | 17/36 | 0.239 |
| Gestational trophoblastic tumors | 13/26 | 0.012 | 18/26 | 0.002 | 14/26 | 0.126 |
| Normally fertile couples | 11/51 | | 16/51 | | 19/51 | |

[a] Compared to normally fertile couples (Fisher exact test, 1-tail).
*Source:* Data from Gill III (1) and Ho, Yang, and Hsieh, et al. (8).

In humans there is an association between HLA sharing and gestational trophoblastic tumors (Table 12.11). The strongest correlation was, again, with the sharing of 2 or more of the HLA-B, -DR, and -DQ antigens.

In order to test the association between HLA sharing and defects in growth, reproduction, and resistance to cancer, a population-based study was undertaken to determine whether there was a higher prevalence of RSAs, cancer, and congenital anomalies in the first-, second-, and third-degree relatives of couples having RSAs or in whom the woman developed a gestational trophoblastic tumor (19). There was a higher prevalence of all three diseases in the relatives of the probands compared to those of normally fertile couples (Fig. 12.3). This association indicates that the gene(s) or the genetic defects influencing RSA, cancer, and congenital anomalies are segregating in the relatives of the index couples

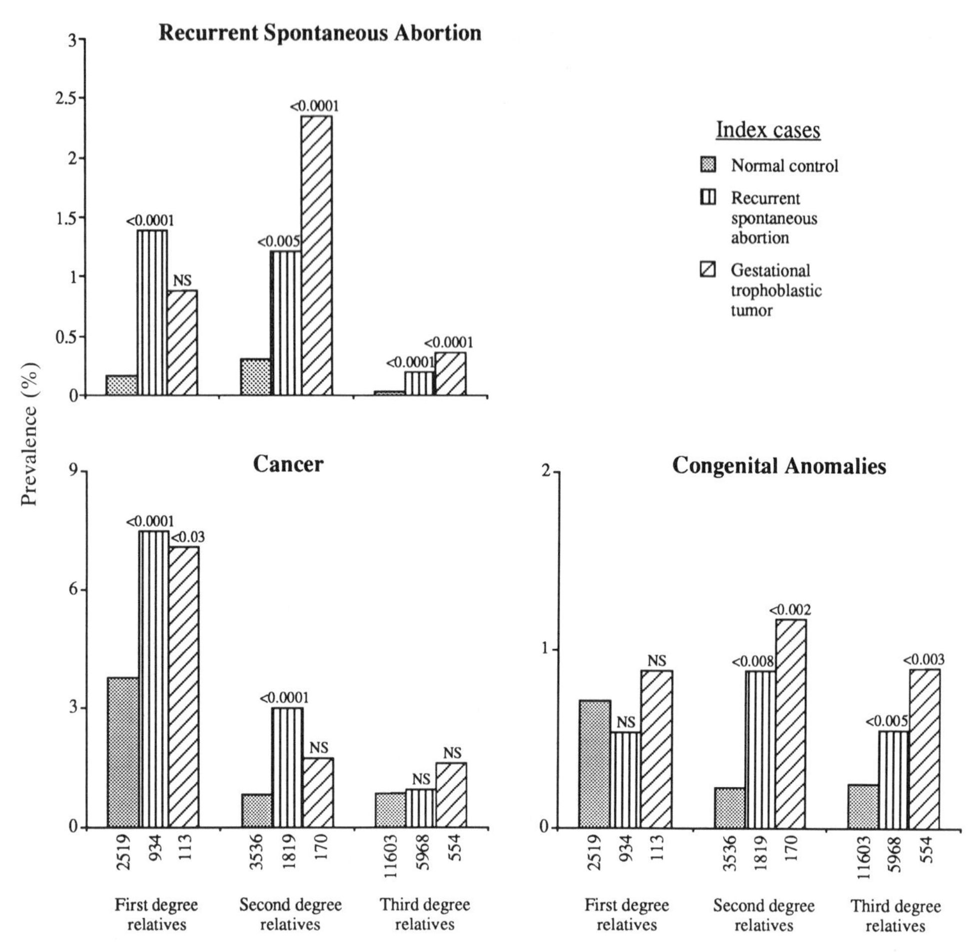


FIGURE 12.3. Prevalence of RSA, cancer, and congenital anomalies in the first-, second-, and third-degree relatives of couples who were normally fertile, who experienced RSAs, or in whom a gestational trophoblastic tumor developed in the woman. Reprinted with permission from Ho, Gill III, Hsieh, Yang, and Lee (19).

at a higher frequency than they are in the relatives of normally fertile couples.

## *Comparative Genetics*

The comparative genetics of the MHC of the rat, mouse, and human shows that the critical region influencing growth, development, and resistance to chemical carcinogens is in the same relative place in all three species, considering the MHC translocation that occurred in the rat and the mouse (1, 8). In the human the class II and class III loci are in a region distinct from the region in which all of the class I loci are located. In the rat and the mouse, the class II and class III regions are translocated such that they are between the two major class I antigens (RT1.A and RT1.E in the rat; H-2K and H-2D in the mouse). Thus, the *grc* region in the rat and the *Ped* locus in the mouse would fall into the same part of the MHC as the class II and III regions in the human if this translocation had not occurred. These observations should provide some insight into the design of experiments to study further genes influencing growth, reproduction, and susceptibility to cancer in humans.

## Conclusions

The evidence in experimental animals and in humans substantiates the hypothesis presented in Figure 12.1: There are genes linked to the MHC that influence resistance to cancer, fertility, and growth. The genes in this region may perform a gatekeeper function in which they act as negative growth control genes in growth and reproduction and as tumor-suppressor genes in resistance to cancer.

The control of MHC antigen expression in the placenta involves all of the known mechanisms that regulate gene expression (Table 12.1). This extraordinary diversity of regulation of the same class of antigens is particularly interesting in view of the cellular differentiation of the placenta. Initially, one cell—the germinal cytotrophoblast—differentiates into two major morphologically and functionally distinct cell lines: the basal trophoblast and the labyrinthine trophoblast. Each of these lines then differentiates into sublines that have different properties. Thus, the diversity of mechanisms affecting MHC antigen expression provides a useful system for studying differentiation and organogenesis, and the evolution of placental growth and development may provide a unique system for examining the aging process.

Finally, there is a high level of evolutionary conservation of the MHC and, most likely, of the MHC-linked region, also. One portion (MHC) encodes polymorphic antigens that determine biological individuality and resistance to infection; the other portion (MHC-linked) controls growth—both normal and abnormal—and development. This pattern appears to be critical for all mammalian species.

*Acknowledgments.* The work in the authors' laboratory was supported by grants from the National Institutes of Health (HD-08662, HD-09880, and CA-18659), the Tim Caracio Memorial Cancer Fund, the Pathology Education and Research Foundation, and the National Science Council of the Republic of China (NSC79-0412-BOO2-122 and NSC80-0412-BOO2-038).

## *References*

1. Gill TJ III. Reproductive immunology and immunogenetics. In: Knobil E, Neill JD, eds. The physiology of reproduction. 2nd ed. New York: Raven Press, 1994 (in press).
2. Gill TJ III. MHC-linked genes affecting reproduction, development, and susceptibility to cancer. In: Coulam C, Faulk WP, McIntyre JA, eds. Immunological obstetrics. New York: Norton, 1992:103–12.
3. Gill TJ III. Invited editorial: influence of MHC and MHC-linked genes on reproduction. Am J Hum Genet 1992;50:1–5.
4. Gill TJ III, Macpherson TA, Ho HN, et al. Immunological and genetic factors affecting implantation and development in the rat and human: a unique trophoblast antigen and genes regulating development. In: Gill TJ III, Wegmann TG, eds. Immunoregulation and fetal survival. New York: Oxford University Press, 1987:137–55.
5. Fraser PA, Awdeh ZL, Ronco P, et al. C4B gene polymorphism among Africans and African-Americans: HLA-Bw42-DRw18 haplotypes. Immunogenetics 1991;34:52–6.
6. Tokunaga K, Saueracker G, Kay PH, Christiansen FT, Anand R, Dawkins RL. Extensive deletions and insertions in different MHC supratypes detected by pulsed field gel electrophoresis. J Exp Med 1988;188:933–40.
7. Degli-Esposti MA, Leelayuwat C, Dawkins RL. Ancestral haplotypes carry haplotypic and haplospecific polymorphisms of BAT1: possible relevance to autoimmune disease. Eur J Immunogenet 1992;19:121–7.
8. Ho HN, Yang YS, Hsieh RP, et al. Sharing of human leukocyte antigens (HLA) in couples with unexplained infertility affects the success of in vitro fertilization and tubal embryo transfer. Am J Obstet Gynecol (in press).
9. Kanbour-Shakir A, Zhang X, Rouleau A, Armstrong DT, Kunz HW, Gill TJ III. Gene imprinting and MHC class I antigen expression in the rat placenta. Proc Natl Acad Sci USA 1990;87:444–8.
10. Kanbour-Shakir A, Kunz HW, Gill TJ III. Differential genomic imprinting of MHC class I antigens in the placenta of the rat. Biol Reprod 1993;48:977–86.
11. Kanbour A, Ho HN, Misra DN, Macpherson TA, Kunz HW, Gill TJ III. Differential expression of MHC class I antigens on the placenta of the rat: a mechanism for the survival of the fetal allograft. J Exp Med 1987;166: 1861–82.
12. Gill TJ III, Kanbour-Shakir A, Zhang X, Armstrong DT, Kunz HW. Gene imprinting of class I antigens in the placenta of the rat. In: Chauoat G, Mowbray J, eds. Cellular and molecular biology of the materno-fetal relationship. Colloque INSERM 1991;212:31–8.
13. Surani MA, Kothary R, Allen ND, et al. Genome imprinting and development in the mouse. Dev Suppl 1990:89–98.
14. Solter D. Differential imprinting and expression of maternal and paternal genomes. Annu Rev Genet 1988;22:127–46.
15. Vardimon D, Locker J, Kunz HW, Gill TJ III. Physical mapping of the MHC and *grc* by pulse field electrophoresis. Immunogenetics 1992;35:166–75.
16. Flaherty L, Elliot E, Tine J, Walsh A, Waters J. Immunogenetics of the Q and Tla regions of the mouse. Crit Rev Immunol 1990;20:131–75.

17. Melhem MF, Kunz HW, Gill TJ III. An MHC-linked gene critically influences resistance to DEN carcinogenesis. Proc Natl Acad Sci USA 1993;90:1967–71.
18. Vincek V, Figueroa F, Gill TJ III, Cortese-Hassett AL, Klein J. Identification and mapping of the growth and reproduction complex (*grc*) homolog in the house mouse. Immunogenetics 1990;32:293–5.
19. Ho HN, Gill TJ III, Hsieh CY, Yang YS, Lee TY. The prevalence of recurrent spontaneous abortions, cancer, and congenital anomalies in the families of couples with recurrent spontaneous abortions or gestational trophoblastic tumors. Am J Obstet Gynecol 1991;165:461–6.

# 13

# Molecular Regulatory Mechanisms That Repress Classical HLA Class I Gene Expression in Human Placenta

PHILIPPE LE BOUTEILLER, THIERRY GUILLAUDEUX, MARYSE GIRR, CÉCILE DEMEUR, AND ANNE-MARIE RODRIGUEZ

The human *major histocompatibility complex* (MHC) class I genomic region, located on the short arm of chromosome 6 (6p21.3 position), is composed of about 2000 kb of DNA. Besides pseudogenes and recently evidenced non-HLA class I genes, such as P5, R1, B30 (1), OTF3 (2), and S (3), this region contains both classical and nonclassical HLA class I genes (4). Polymorphic classical HLA-A, -B, -C class I genes are constitutively expressed (or inducible) on most somatic tissues, although at different levels, and play a critical role in cancer and viral immunity. Expression of these genes is developmentally regulated: They are not expressed on the cell surface of either male or female germinal cells (5, 6) or trophoblast cells (7). Current evidence favors repression of their transcription in all subpopulations of trophoblast cells in situ: Neither cytotrophoblast (from both term and first-trimester placentas) nor syncytiotrophoblast expresses these molecules (8). This suggests that such a repression plays a role in the maintenance of pregnancy. In contrast, the nonclassical HLA-E, -F, -G genes have a more restricted somatic tissue distribution, exhibit a low polymorphism, and are not yet associated with a known immunological or other biological function (7). Although nothing is known yet about expression of these genes in germinal cells, it is now clear that cytotrophoblast expresses HLA-G (8), and HLA-E and -F transcripts have been reported in the human placenta (9).

Studies on the molecular mechanisms that prevent expression of polymorphic HLA class I genes and up-regulate expression of nonpolymorphic HLA-E, -F, -G genes in human trophoblast are of particular interest. These may elucidate how the semiallogenic mammalian fetus is successfully tolerated by the mother and the functional role, if any, of nonclassical HLA class I genes.

This chapter examines some molecular regulatory mechanisms that may explain the repression of classical class I genes in different human trophoblast subpopulations, drawing comparisons with the choriocarcinoma cell line JAR that is also devoid of classical MHC class I molecules. Both *cis*- and *trans*-acting regulatory mechanisms may play a role in this repression.

## *Cis*-Acting Regulatory Mechanisms

Among the various *cis*-acting regulatory mechanisms that may repress the transcription of classical HLA class I genes in the human trophoblast are (i) methylation of CpG islands, (ii) genomic rearrangements, and (iii) changes in chromatin structure. This chapter focuses on the methylation at CpG sites.

*CpG islands* are parts of the eukaryote genome containing a high density of unmethylated CpG dinucleotides that surround the transcription start site of many genes (10), including MHC classical and nonclassical class I genes (4). These regions are rich in restriction endonuclease sites that are rare and methylation sensitive (4). Many studies have established an inverse relationship between the level of CpG island methylation and the transcriptional activity of associated coding sequences (11).

Site-specific methylation studies involving both viral and eukaryotic gene promoters have shown that methylation of only one or a few cytosines at CpG sites can alter transcriptional activity (12). Differences in methylation of the nonclassical HLA-F and -G class I genes have recently been evidenced between human sperm and somatic cell DNA (13).

In order to test whether methylated CpGs were responsible for the absence of classical HLA class I gene expression in the human trophoblast, we performed Southern blotting analysis of genomic DNA from different trophoblast subpopulations, double-digested by *Hind*III and different methylation-sensitive rare cutter enzymes, followed by agarose gel electrophoresis, transfer, and hybridization with HLA locus-specific probes (4). We first tested the following control HLA-expressing and HLA-nonexpressing human cell lines: HHK, which is an HLA homozygous lymphoblastoid cell line (4), and the HLA-negative trophoblast-derived choriocarcinoma cell line JAR (4). The fact that exogenous transfected HLA-A or -B heavy chain genes can be stably expressed at the cell surface of JAR in association with endogenous β2 microglobulin and can be up-regulated by *tumor necrosis factor* $\alpha$ (TNF$\alpha$) and/or *interferon* $\gamma$ (IFN$\gamma$) strongly suggests that both constitutive and inducible class I transcription factors are present and functional (14). Thus, the negative regulatory mechanism(s) that prevents HLA class I expression in JAR is likely to involve *cis* modifications.

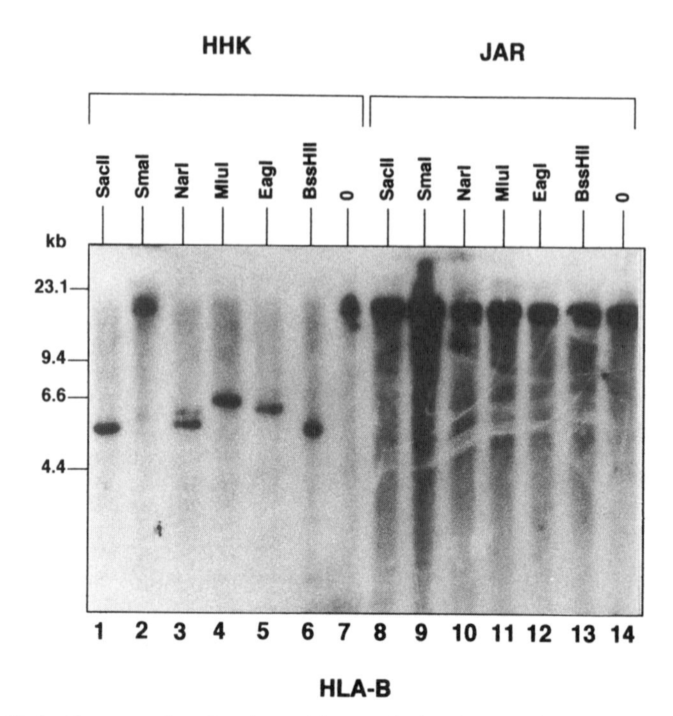


FIGURE 13.1. Comparative Southern blot analysis of HHK (lanes 1–7) and JAR (lanes 8–14) genomic DNA digested by *Hind*III alone (0) or double-digested by *Hind*III and different methylation-sensitive rare cutter enzymes (*Sac*II, *Sma*I, *Nar*I, *Mlu*I, *Eag*I, and *BssH*II) present in the 5′ part of the HLA-B gene, followed by hybridization with an HLA-B locus-specific probe.

The methylation status of the HLA-B gene in HHK and JAR genomic DNA is shown in Figure 13.1. The expected 21-kb *Hind*III fragment containing the HLA-B gene was detected in both HHK and JAR DNA (lanes 7 and 14). Following the second digestion the *Hind*III fragment completely disappeared in HHK (lanes 1–6), just as it did in human PBL (13), indicating the presence of cleavable sites, which was as expected from the sequence data. In contrast, the second digestion never affected the size of this fragment in JAR DNA (lanes 8–13), showing that CpG dinucleotides within these sites were all methylated. Hybridization of the same blot with an HLA-A, -C, -F, -G, or -H locus-specific probe gave the same results; that is, methylation at CpG sites in JAR and the absence of methylation in HHK (4). When the HLA-E locus-specific probe was used, the patterns obtained with JAR and HHK were similar; that is, unmethylated (4). Northern blot analysis of JAR mRNA indeed revealed the presence of 2 HLA-E transcripts (4). These results, together with those showing that treatment with the demethylating agent 5′-

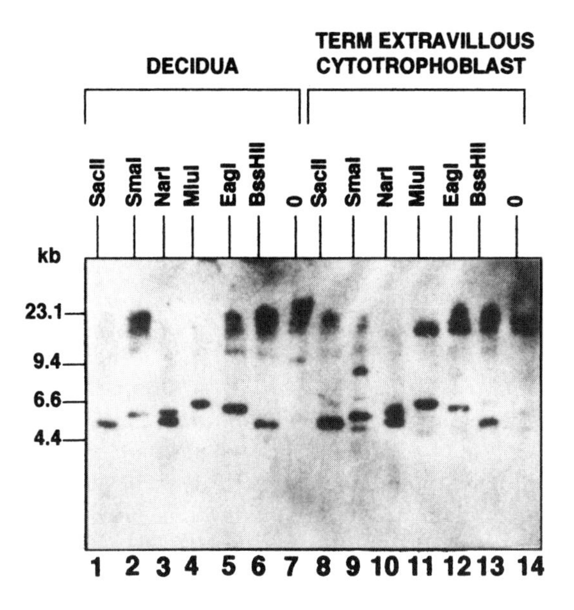


**HLA-B**

FIGURE 13.2. Comparative Southern blot analysis of term decidual tissue (lanes 1–7) and chorion membrane extravillous cytotrophoblast cell (lanes 8–14) genomic DNA from the same placenta digested by *Hind*III alone (0) or double-digested by *Hind*III and different methylation-sensitive rare cutter enzymes (*Sac*II, *Sma*I, *Nar*I, *Mlu*I, *Eag*I, and *BssH*II), followed by hybridization with an HLA-B locus-specific probe.

azacytidine induced reexpression of HLA class I molecules at the cell surface of JAR (4), clearly showed that CpG island methylation was the major cause of HLA class I repression (except for HLA-E) in this trophoblast-derived human cell line.

The same experiments were performed on different trophoblast subpopulations from term human placentas. Term extravillous cytotrophoblast—isolated from the amniochorion—and maternal decidual cells (15) were first used. Genomic DNA was double-digested with *Hind*III and a methylation-sensitive restriction enzyme, as described above, and hybridized with an HLA-B locus-specific probe (Fig. 13.2). The three following sites appeared partially methylated in the cytotrophoblast: *Mlu*I (lane 11), *Eag*I (lane 12), and *BssH*II (lane 13). This was not the case for the *Mlu*I site in decidual cells: Following digestion with this enzyme, the *Hind*III fragment completely disappeared (lane 4). Other sites tested (*Sac*II, *Sma*I, and *Nar*I) were similarly cleavable in both cytotrophoblast and decidual cells (lanes 1–3 and 8–10). When the same blot was hybridized with HLA-A or -E locus-specific probes, none of the sites appeared methylated (data not shown). Following hybridization with HLA-G locus-specific probe, only the *BssH*II site appeared partially methylated, just as it did in PBL (data not shown).

We also looked for the methylation patterns of the HLA class I genes in term cytotrophoblast isolated from the villi of cesarian placentas (after 12h of culture), as well as in term villous syncytiotrophoblast differentiated from the same cytotrophoblast (after 4 days of culture) (16). Flow cytometry analysis performed on this cytotrophoblast population using various antitrophoblast *monoclonal antibodies* (mAbs) (Fig. 13.3) revealed a reproducible degree of purity: Less than 3% of the cells were unstained with GB25, GB20, JEG13, and JAR02. A comparative flow cytometry analysis performed on the trophoblast-derived choriocarcinoma cell line JEG-3 (Fig. 13.4) and JAR (data not shown) revealed similar profiles. The cytotrophoblast population was also CD45 (leukocyte common antigen) negative (data not shown). Moreover, very few of these cells were stained by anti-HLA class I, anti-HLA class II, or antihuman β2 microglobulin mAbs as compared to the lymphoblastoid control cell line HHK (Fig. 13.5), suggesting that they expressed very low or undetectable levels of HLA-G surface molecules (17). (HLA-G heavy chain is recognized by the W6/32 mAb and is β2 microglobulin associated.)

These cytotrophoblast cells differentiate in culture and fuse to form functional syncytiotrophoblast (16). This was assessed by the secretion of *human chorionic gonadotropin* (hCG) (Fauvel, collaboration).

We studied the methylation status of different HLA class I genes in the genomic DNA of these two trophoblast subpopulations. Following hybridization with HLA-B locus-specific probe, the same results were obtained in term villous cytotrophoblast (Fig. 13.6) and syncytiotrophoblast (Fig. 13.7) as in term extravillous cytotrophoblast (Fig. 13.2): *BssH*II (lane 3) and *Mlu*I (lane 4) sites were partially methylated, whereas *Nar*I, *Sma*I, and *Sac*II (lanes 5–7) were not. Following hybridization with HLA-A, -E, or -G locus-specific probes, the same results were obtained as those observed in term extravillous cytotrophoblast (discussed above).

In conclusion, whereas methylation at CpG islands correlates with the repression of transcription of both classical HLA-A, -B, -C and nonclassical HLA-F and -G (but not HLA-E) genes in the JAR cell line, it remains to be proven that the partial methylations observed in term villous and extravillous cytotrophoblast, as well as in term syncytiotrophoblast, are related to the absence of expression of classical HLA class I genes in these tissues. It is possible that these partial methylations reflect a genomic imprinting (the preferential expression of the maternally or paternally inherited alleles) similar to the one described in the rat trophoblast, where the maternal class I antigens are suppressed (18). Such partial methylations may also interact with *trans*-acting factors or affect chromatin structure (11). Use of *Hpa*II and *Msp*I isoschizomer restriction

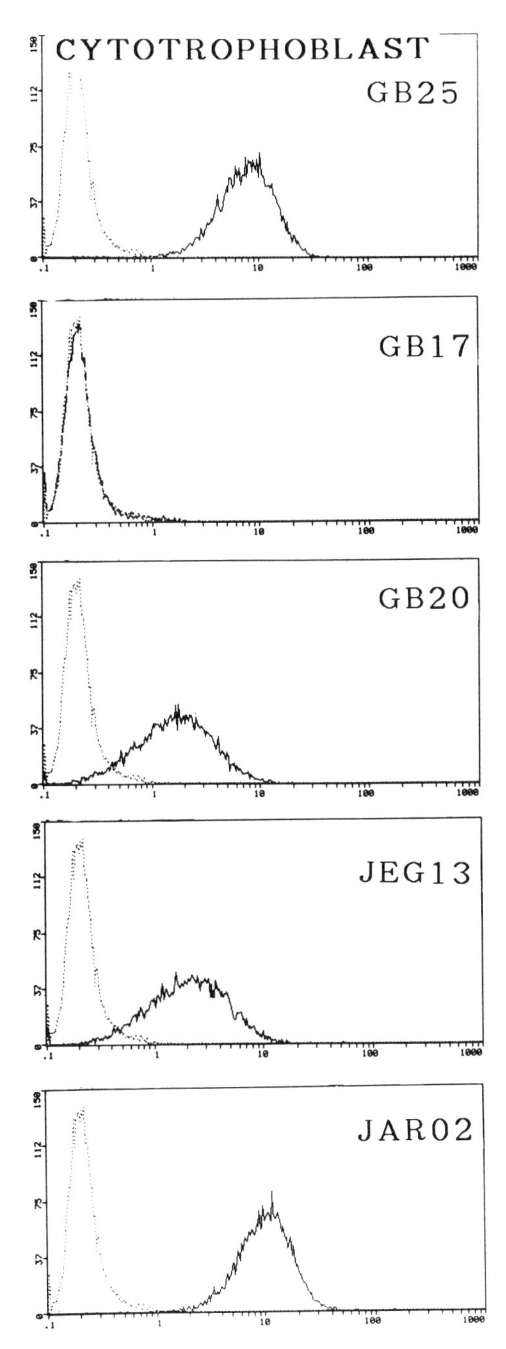


FIGURE 13.3. Flow cytometry analysis of purified term villous cytotrophoblast cells (after 12 h of culture) stained with the following mAbs: GB25 (specific for villous cytotrophoblast and syncytiotrophoblast), GB17 (specific for syncytiotrophoblast), GB20 (placental alkaline phosphatase), JEG13 (specific for JEG-3 cell line), JAR02 (specific for JAR cell line), and 10.3.6 (anti-Ia$^k$) used as negative control (dotted lines).

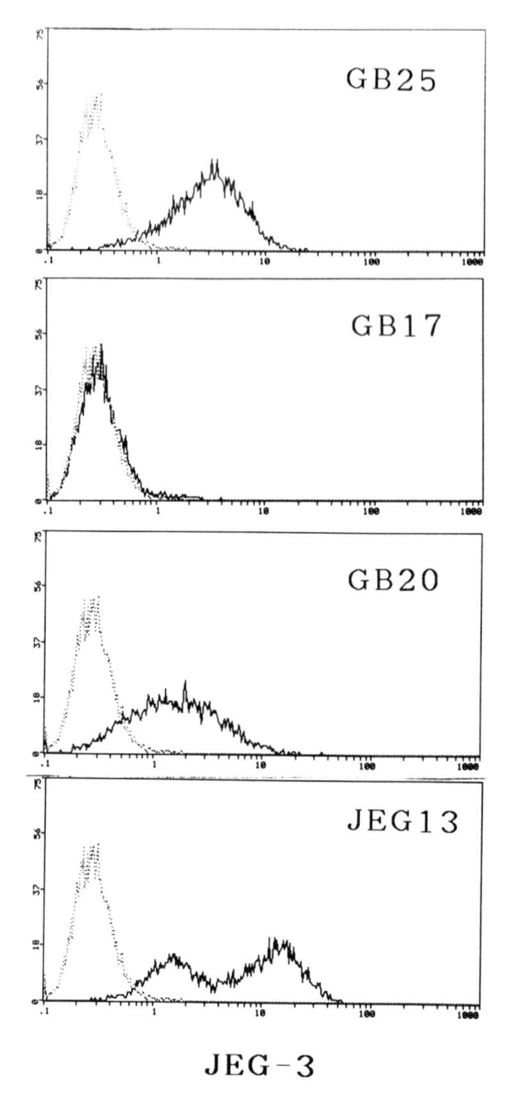


FIGURE 13.4. Flow cytometry analysis of JEG-3 cells using the same mAbs as in Fig. 13.3.

enzymes (11) and genomic sequencing technique on each strand of DNA (methylated dC in a nucleotide sequence renders this site refractory to hydrazine cleavage) may help to elucidate this hypothesis. The CpG methylation in first-trimester trophoblasts should also be investigated.

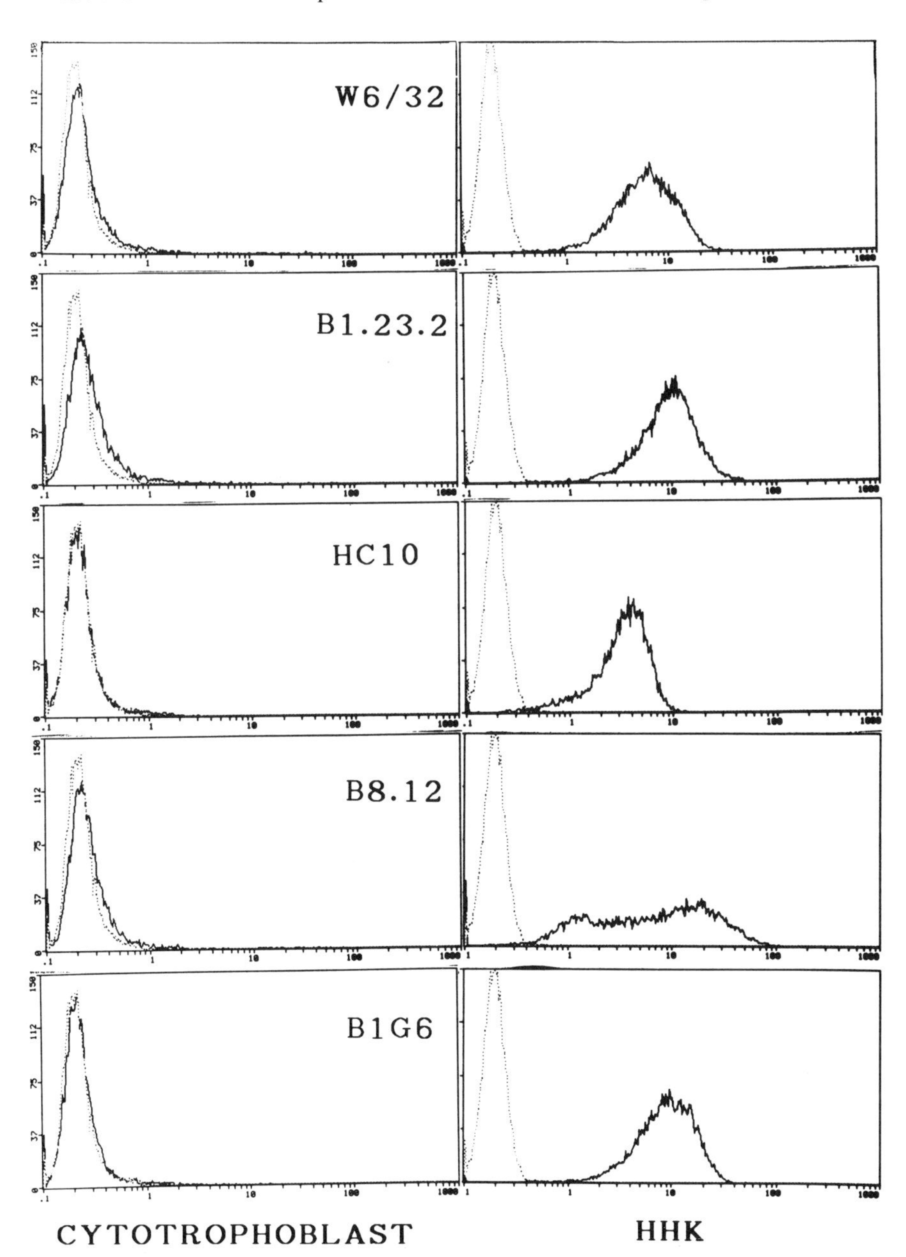


FIGURE 13.5. Flow cytometry analysis of purified term villous cytotrophoblast cells (after 12 h of culture) as compared to human HHK lymphoblastoid cell line stained with the following mAbs: W6/32 (monomorphic HLA-A, -B, -C class I); B1.23.2 (monomorphic HLA-B, and -C class I); B8.12 (monomorphic HLA class II); B1G6 (human β2 microglobulin), as described in reference 14; and HC10 (monomorphic HLA class I denaturated heavy chain).

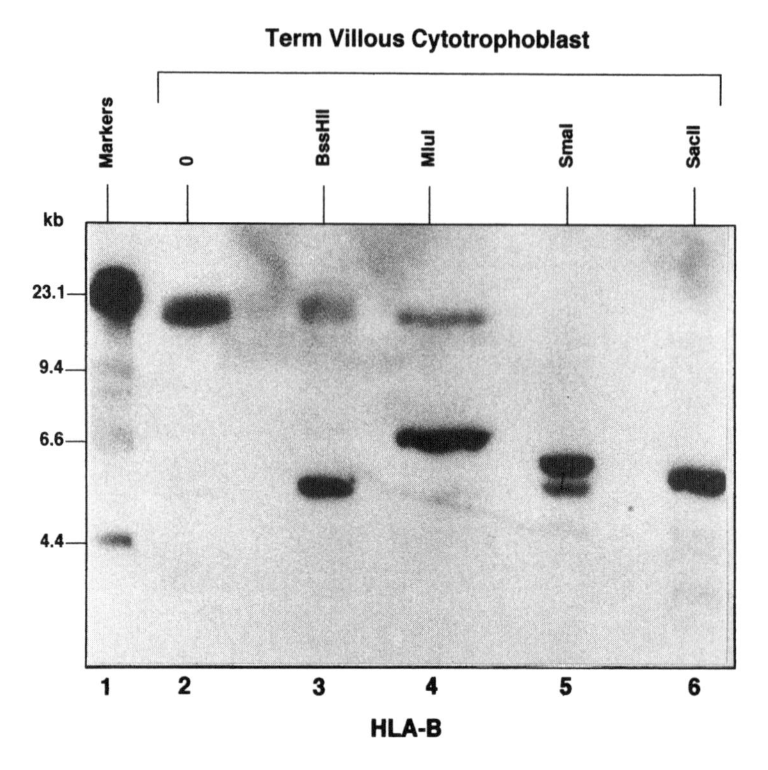


FIGURE 13.6. Southern blot analysis of purified term villous cytotrophoblast genomic DNA digested by *Hind*III alone (0) or double-digested by *Hind*III and different methylation-sensitive rare cutter enzymes (*BssH*II, *Mlu*I, *Sma*I, and *Sac*II), followed by hybridization with HLA-B locus-specific probe.

## *Trans*-Acting Suppressive Regulatory Elements and Factors

Transcriptional regulation of class I genes, like other eukaryotic genes, operates through a complex array of nuclear DNA-binding proteins that specifically recognize several conserved upstream regulatory elements present in their promoter region (19). *Trans*-acting suppression could be due to (i) recessive mutations in transcription factors controlling the expression of HLA class I genes, (ii) the presence of *trans*-acting repressor factor(s) bound to *trans*-dominant extinguisher region(s), and (iii) the prevention of DNA binding of functional (constitutive and/or inducible) stimulatory *trans*-acting factor(s) due to deletion or mutation of *trans*-stimulating region(s).

One of the best-studied class I DNA-binding factors is the KB-F1 protein that binds to a palindromic element located within the enhancer

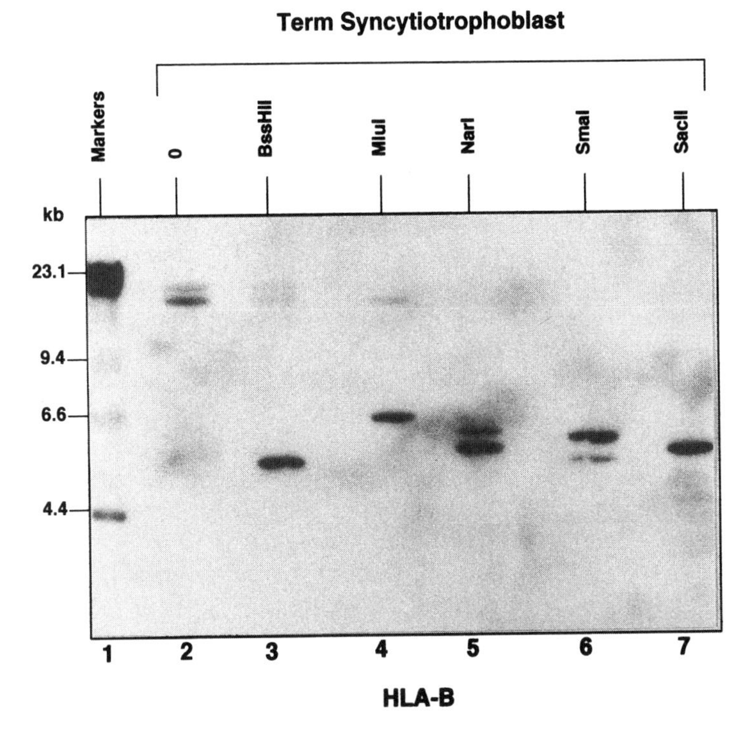


FIGURE 13.7. Southern blot analysis of term in vitro-differentiated syncytiotrophoblast genomic DNA digested by *Hind*III alone (0) or double-digested by *Hind*III and different methylation-sensitive rare cutter enzymes (*BssH*II, *Mlu*I, *Nar*I, *Sma*I, and *Sac*II), followed by hybridization with HLA-B locus-specific probe.

A promoter region and is very well conserved among mouse and human classical MHC class I genes (20). KB-F1 is a homodimer protein made of two p50 DNA-binding subunits that are also part of the NF-KB transcription factor (20). These DNA-binding subunits belong to the same family of proteins as the c-*rel* factor and the *Drosophila* maternal morphogen *dorsal* (20). Subunit p50 activity has been detected in the nuclei of most human, mouse, and rat differentiated cells tested. By contrast, p50 was not detected in undifferentiated embryonal carcinoma cells in which MHC class I and β2 microglobulin are absent (21).

In order to investigate the possible involvement of such DNA-binding factors in the repression of classical HLA class I genes in the human trophoblast, we analyzed the nuclear expression of the p50-related proteins in purified extravillous cytotrophoblast and villous syncytiotrophoblast subpopulations from first-trimester human placenta, as compared with MHC class I expressing maternal decidual cell and embryonic fibroblasts or MHC class I-negative choriocarcinoma human cell line JAR (19).

First, using the double-stranded enhancer A nucleotidic sequence that contains the KB site, we demonstrated by band-shift assay that binding activity, inhibited by the addition of anti-p50 polyclonal serum, was present in cytotrophoblast, as well as in control maternal decidual cells, embryonic fibroblasts, and the JAR cell line, but undetectable in syncytiotrophoblast nuclear extracts. The specificity of this DNA-protein complex was demonstrated by its disappearance upon competition with an excess of cold enhancer A competitor oligonucleotide. Other nucleoprotein complexes were also detected in all nuclear extracts, including syncytiotrophoblast, that were competed out by an excess of cold enhancer A competitor oligonucleotide, but were not affected by the addition of anti-p50 or anti-NF-KB antisera, suggesting the presence of additional enhancer A-binding factors different from the KB-F1/NF-KB/c-*rel* family (19). Second, immunofluorescence cell staining using the same anti-p50 serum showed a positive staining in both the cytoplasm and nucleus of cytotrophoblast and its absence in syncytiotrophoblast (Fig. 13.8). Finally, using a Western immunoblot analysis, a doublet of ~85 kd was specifically stained by the same anti-p50 serum in cytotrophoblast, maternal decidual cells, embryonic fibroblasts, and the JAR nuclear extracts, whereas no signal was obtained in syncytiotrophoblast (Fig. 13.9). We hypothesized that this enhancer A DNA-binding factor might represent the c-*rel trans*-acting factor (19).

The p50-related proteins are present in first-trimester extravillous cytotrophoblast nuclei and bind to the enhancer A promoter element. Nevertheless, these cells do not express classical HLA class I genes. It is thus likely that this repression is due to negative regulatory mechanisms other than a defect of p50-related proteins. Alternative negative *trans*-regulatory mechanisms could be caused by the presence of a negative regulatory factor (discussed above).

The p50-related proteins were undetectable in the nuclei of first-trimester villous syncytiotrophoblast. This absence means either that the genes encoding these proteins were silent in this tissue due to mutations, deletions, rearrangements, or changes in chromatin structure or that the proteins were present in an inactive form in the cytosol, which is associated with an inhibitor (19). The absence of these *trans*-acting factors in the nucleus might explain the repression of class I genes in this tissue. However, it does not exclude other negative regulatory mechanisms, such as those mentioned above.

We observed a differential p50-related expression in first-trimester trophoblast subpopulations. Whether such differential expression will persist during the course of gestation also remains to be determined.

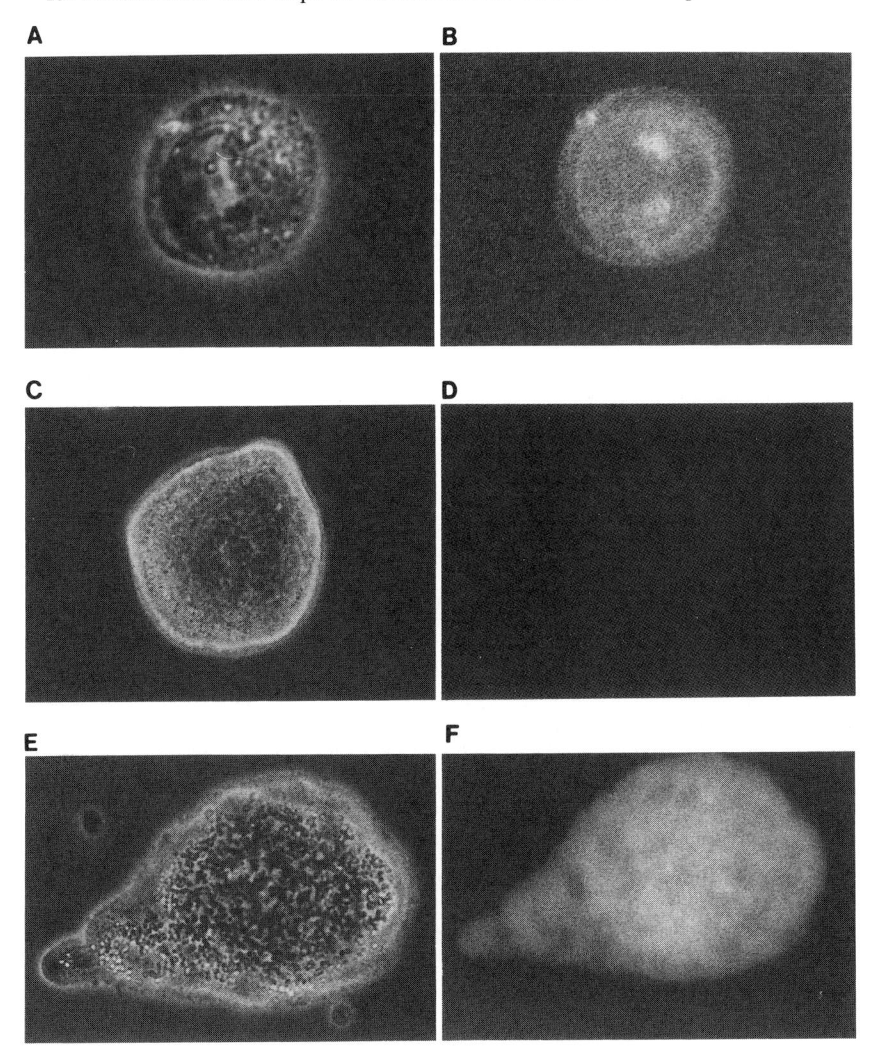


FIGURE 13.8. Immunofluorescent staining of first-trimester extravillous cytotrophoblast: phase contrast (×600) (*A*) and stained with the polyclonal anti-p50 serum 3 (×600) (*B*). Immunofluorescent staining of first-trimester villous syncytiotrophoblast is as follows: phase contrast (×370) (*C*); stained with the anti-p50 serum 3 (×370) (*D*); phase contrast (×260) (*E*); and stained with antihuman pregnancy-specific β1 glycoprotein SP1 (×260) (*F*). Reprinted with permission from Boucraut, Hawley, Robertson, Bernard, Loke, and Le Bouteiller (19), Copyright 1993, The Journal of Immunology.

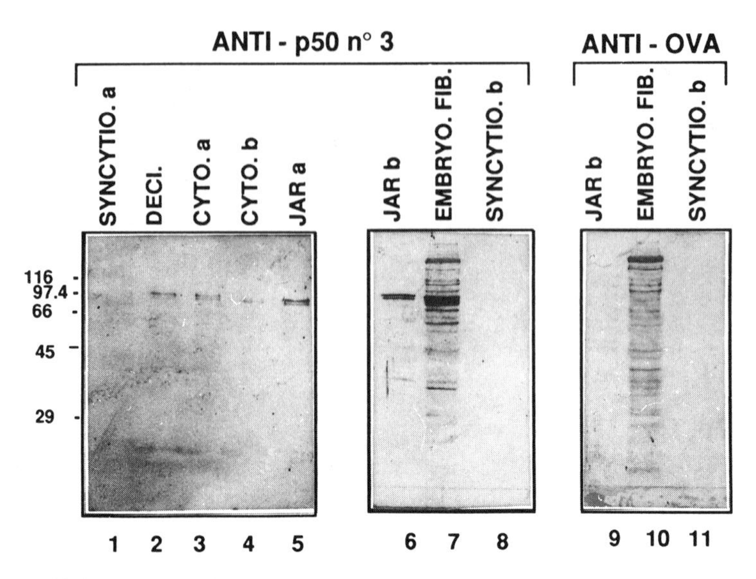


FIGURE 13.9. Western immunoblot analysis. Nuclear extracts from purified first-trimester syncytiotrophoblast (lanes 1, 8, and 11), maternal decidual (lane 2), first-trimester extravillous cytotrophoblast cells (lanes 3 and 4), JAR cell line (lanes 5, 6, and 9), and embryonic fibroblasts (lanes 7 and 10) were electrophoresed and immunostained with the anti-p50 serum 3 (lanes 1–8) or the control anti-ovalbumin serum (lanes 9–11). The letters (a and b) indicate nuclear extracts from different placentas or cell line preparations. Reprinted with permission from Boucraut, Hawley, Robertson, Bernard, Loke, and Le Bouteiller (19), Copyright 1993, The Journal of Immunology.

## Conclusions

Taken together, our observations confirm that the negative regulatory mechanisms that prevent transcriptional expression of classical HLA class I genes in human trophoblast are multiple and nonexclusive, involving both *cis-* and *trans-*acting regulatory elements and depending on the time of gestation and on the origin of the trophoblast subpopulation (villous or extravillous cytotrophoblast or syncytiotrophoblast). Further studies are therefore required in order to clarify these various possibilities.

*Acknowledgments.* In this work we have collaborated with Drs. J. Boucraut, Faculté de Médecine de la Timone, Marseille, France; Pr. R. Fauchet, Laboratoire d'Hématologie-Immunologie, Rennes, France; S.A. Ellis, Institute For Animal Health, Compton, UK; and I.L. Sargent, John Radcliffe Hospital, Oxford, UK; and S. Hawley, K. Robertson, and

Y.W. Loke, Department of Pathology, University of Cambridge, UK. We thank Drs. D. Evain-Brion and E. Alsat, Ecole Normale Supérieure, Paris, for their help in the isolation and culture of term cytotrophoblast and H. Brun and G. Cassar for the flow cytometry analysis. We would also like to thank members of the Departments of Obstetrics of Hôpital Lagrave and Clinique Ambroise Paré, Toulouse; in particular, Drs. H. and B. Grandjean for their kind cooperation in supplying us with cesarian placentas. For their respective gifts we also thank Dr. B.L. Hsi (GB25, GB17, and GB20), Pr. D. Bellet (JAR02), Pr. H. Ploegh (HC10), and Dr. A. Israël (serum 3). This work was supported by INSERM and grants to P.L.B. from ARC, FEGEFLUC, and Conseil Régional de la Région Midi Pyrénées. T.G. was supported by a fellowship from The Ligue Nationale Contre Le Cancer and A.M.R. by a fellowship from Association Pour la Recherche Contre le Cancer.

## *References*

1. Vernet C, Boretto J, Mattéi MG, et al. Evolutionary study of multigenic families mapping close to the human MHC class I region. J Mol Evol (in press).
2. Guillaudeux T, Mattéi MG, Depetris D, Le Bouteiller P, Pontarotti P. In situ hybridization localizes the human OTF3 to chromosome 6p21.3/p22 and OTF3L to 12p13. Cytogenet Cell Genet 1993;63:212–4.
3. Zhou Y, Chaplin DD. Identification in the HLA class I region of a gene expressed late in keratinocyte differentiation. Proc Natl Acad Sci USA (in press).
4. Boucraut J, Guillaudeux T, Alizadeh M, et al. HLA-E is the only class I gene that escapes CpG methylation and is transcriptionally active in the trophoblast-derived human cell line JAR. Immunogenetics 1993;38:117–30.
5. Ohashi K, Saji F, Kato M, Wakimoto A, Tanizawa O. HLA expression on human ejaculated sperm. Am J Reprod Immunol 1990;23:29–32.
6. Roberts JM, Taylor CT, Melling GC, Kingsland CR, Johnson PM. Expression of the CD46 antigen and absence of class I MHC antigen on the human oocyte and preimplantation blastocyst. Immunology 1992;75:202–5.
7. Hunt JS. Immunobiology of pregnancy. Curr Opin Immunol 1992;4:591–6.
8. Hunt JS, Yelavarthi KK, Chen HL, Yang Y, Fishback JL. Analysis of HLA-G mRNA in placentas by in situ hybridization. In: Tsuji K, Aizawa M, Sasazuki T, eds. HLA 1991; vol 1. New York: Oxford Science Publication, 1992:1034–9.
9. Wei X, Orr HT. Differential expression of HLA-E, HLA-F, and HLA-G transcripts in human tissue. Hum Immunol 1990;29:131–42.
10. Bird A. CpG islands and the function of DNA methylation. Trends Genet 1987;3:342–7.
11. Razin A, Cedar H. DNA methylation and gene expression. Microbiol Rev 1991;55:451–8.
12. Davies K. Imprinting makes its mark. Nature 1993;363:94.
13. Guillaudeux T, d'Almeida M, Girr M, Pontarotti P, Fauchet R, Le Bouteiller P. Differences between human sperm and somatic cell DNA in CpG methyla-

tion within the HLA class I chromosomal region. Am J Reprod Immunol (in press).
14. Boucraut J, Hakem R, Gauthier A, Fauchet R, Le Bouteiller P. Transfected trophoblast-derived human cells can express a single HLA class I allelic product. Tissue Antigens 1991;37:84–9.
15. Ellis SA, Sargent IL, Redman CWG, McMichael AJ. Evidence for a novel HLA antigen found on human extravillous trophoblast and a choriocarcinoma cell line. Immunology 1986;59:595–601.
16. Kliman HJ, Nestler JE, Sermasi E, Sanger JM, Strauss III JF. Purification, characterization, and in vitro differentiation of cytotrophoblasts from human term placentae. Endocrinology 1986;118:1567–82.
17. Kovats S, Librach C, Fisch P, et al. Expression and possible function of the HLA-G alpha chain in human cytotrophoblasts. In: Chaouat G, Mowbray J, eds. Cellular and molecular biology of the materno-fetal relationship. Colloque INSERM 1991;212:21–9.
18. Kanbour-Shakir A, Zhang X, Rouleau A, et al. Gene imprinting and major histocompatibility complex class I antigen expression in the rat placenta. Proc Natl Acad Sci USA 1990;87:444–8.
19. Boucraut J, Hawley S, Robertson K, Bernard D, Loke YW, Le Bouteiller P. Differential nuclear expression of enhancer A DNA-binding proteins in human first trimester trophoblast cells. J Immunol 1993;150:3882–94.
20. Kieran M, Blank Y, Logeat F, et al. The DNA binding subunit of NF-KB is identical to factor KBF1 and homologous to the rel oncogene product. Cell 1990;62:1007–18.
21. David-Watine B, Israël A, Kourilsky P. The regulation and expression of MHC class I genes. Immunol Today 1990;11:286–92.

# Part V

# Experimental Models of MHC Gene Expression

# 14

# Developmental Regulation of MHC Class I Gene Expression in Mice

ELIZABETH K. BIKOFF AND ELIZABETH J. ROBERTSON

The *major histocompatibility complex* (MHC) is a large multigene complex that encodes several cell surface molecules involved in immune recognition. Originally discovered by their ability to provoke tissue graft rejection, it is now understood that these highly polymorphic membrane glycoproteins play an essential role in the presentation of antigenic peptides to T lymphocytes. In adults class I molecules are expressed on the surface of virtually all somatic cells. The absence of expression of class I transplantation antigens at early stages of development—particularly in the extraembryonic cell lineages, such as those that contribute to the placenta—is thought to play a critical role in maternal tolerance of the fetal allograft. For this reason, considerable effort has been directed toward describing expression of paternal alloantigens present at the fetomaternal interface. This chapter describes our recent work in analyzing the developmental regulation of MHC class I gene expression in mice.

Class I molecules are heterodimers consisting of a 40- to 45-kd glycosylated heavy chain noncovalently associated with a 12-kd light chain, β2 microglobulin. The H-chains associate with β2 microglobulin shortly after synthesis in the rough *endoplasmic reticulum* (ER). This interaction is thought to be essential for conformational stability and transport to the cell surface. Highly polymorphic residues in the α1 and α2 domains are present in the peptide-binding groove located on the top surface of the class I molecule (1). This is the site that binds peptide and is recognized by T cells. Recent x-ray crystal studies have clearly demonstrated that peptide ligand is an essential structural component of the class I molecule (1–3).

The recent discovery of MHC-linked peptide transporters has provided new insight into the class I biosynthetic pathway. These homologous genes mapping within the class II regions of the rat, human, and mouse

MHC encode highly conserved transmembrane proteins related to the ABC superfamily of transporters (4–7). The MHC-linked peptide transporters closely resemble the mammalian *multidrug resistance* (*Mdr*) genes, and each of the two MHC-encoded transporter molecules corresponds to exactly half of the P-glycoprotein. Consistent with this, recent experiments have demonstrated that the peptide transporter is a heterodimer comprised of two closely related gene products that associate to form a channel across the ER membrane (8, 9). Considerable progress has been made recently concerning the biochemical analysis of ATP-dependent translocation of specific peptides (10, 11). Additionally, two closely linked genes encoding polymorphic components of a proteolytic complex similar to the 19s cylinder-type particle, multicatalytic proteinase (proteasome), are thought to be responsible for generating peptides (12–14). It is envisioned that intact, probably ubiquitin-conjugated cytoplasmic proteins degraded by the 19s multicatalytic proteinase (proteasome) and the resultant peptides are subsequently transported into the lumen of the ER, where they can promote assembly of newly synthesized MHC class I H-chains and β2 microglobulin.

Studies analyzing developmentally regulated expression of MHC class I molecules have been hampered in part because of the difficulty of obtaining sufficient quantities of fetal tissues from genetically defined strains of mice. Another problem is the difficulty in assessing expression of individual members of a complex multigene family. Developmental regulation of β2 microglobulin and MHC class I gene expression has been extensively analyzed using teratocarcinoma-derived *embryonal carcinoma* (EC) cell lines, such as F9, that exhibit various properties characteristic of early embryo cells. It has been shown that F9 cells do not express β2 microglobulin or conventional MHC class I transcripts; however, MHC class I and β2 microglobulin mRNA are expressed after induction of differentiation by retinoic acid (15, 16). Coordinate regulation of β2 microglobulin and MHC class I gene expression in this system is known to be controlled at the level of mRNA transcription by common *trans*-acting factors that recognize homologous *cis*-acting sequences located in the 5′ flanking regions upstream of β2 microglobulin and conventional MHC class I genes (reviewed in 17). Although these studies have provided significant new insights into the molecular mechanisms responsible for controlling MHC class I gene expression in vitro, by comparison, relatively little is known with respect to transcriptional regulation of MHC class I expression in the various cell types present during embryonic development in vivo.

## Results and Discussion

In situ hybridization using RNA probes offers the advantage that it provides a sensitive, reliable method for assessing the presence of RNA transcripts in individual cells in the intact embryo. We recently used in situ hybridization techniques to describe the temporal and spatial pattern of expression of β2 microglobulin mRNA in the developing mouse embryo (18). We found that the β2 microglobulin transcripts initially detected in the extraembryonic derivatives of the early primitive streak stage embryo were specifically restricted to the tissues of the ectoplacental cone and chorion. Both the chorion and the ectoplacental cone are known to originate from the extraembryonic ectoderm, suggesting that the β2 microglobulin gene may be selectively transcribed in this cell lineage. In addition to the component contributed by the extraembryonic ectoderm, the chorion also contains cells derived from the extraembryonic mesoderm. It is certainly possible that β2 microglobulin transcripts are expressed in both cell types within the chorion. A potentially important point is that none of the other derivatives of the extraembryonic mesoderm present at this stage, such as the amnion and the allantois, were found to express β2 microglobulin. Expression of the β2 microglobulin gene continues in the derivative of the chorion and ectoplacental cone during formation of the definitive placenta.

The second major site of β2 microglobulin expression is in the visceral yolk sac. The visceral yolk sac first begins to form during gastrulation, when the extraembryonic mesoderm migrates around the exocoelom and comes to underlie the visceral endoderm. Although this distinctive tissue layer, including structures recognizable as blood islands, can be identified in day 8.5 early somite stage embryos, β2 microglobulin expression was not detectable prior to day 10.5 of development, by which time, the fetal circulation is well established. The β2 microglobulin mRNA was present at high levels in the tissues of the visceral yolk sac starting at day 10.5 in both the visceral yolk sac endoderm and the extraembryonic mesoderm. Thus, the induction of β2 microglobulin mRNA expression is not coincident with the onset of hematopoiesis per se.

Significant levels of β2 microglobulin mRNA were not detected until day 9.5 in the embryo proper. At this stage, there was a strong signal over the liver rudiment, and the remaining somatic tissues expressed low levels of transcripts.

To determine whether MHC class I gene expression and β2 microglobulin gene expression are coordinately regulated during early postimplantation development, it was important to examine the temporal and spatial pattern of expression of MHC class I transcripts. We analyzed the onset and pattern of expression of MHC class I transcripts using a probe derived from the highly conserved α3 coding region in order to determine the overall pattern of expression of the MHC class I gene family (19). Surprisingly, we found that MHC class I and β2 microglobulin show remarkably distinct patterns of expression in early mouse embryos. Thus, MHC class I transcripts were initially detected at day 9.5 p.c. in the primary and secondary trophoblast giant cell populations. At this stage none of the remaining fetal components of the developing embryo expressed MHC class I mRNA. This finding was especially striking since the outer zone of trophoblast giant cells is the only embryo-derived tissue in the developing placenta that does not express detectable levels of β2 microglobulin mRNA.

It seems likely that the absence of expression of β2 microglobulin provides a *fail-safe* mechanism to protect the fetus from being recognized by the maternal immune system since in this case paternal alloantigens cannot be expressed at the cell surface. This is consistent with the idea that the outer zone of trophoblast giant cells functions as an immunologic barrier. The onset of MHC class I expression is not coincident with terminal differentiation to the nonproliferative giant cell type because these cells are present from implantation onward. Rather, MHC class I expression seems to be temporally regulated, possibly by external stimuli.

The question as to which MHC class I gene products are expressed in the trophectodermal derivatives of the early mammalian embryo has proven to be particularly controversial. Although our in situ hybridization analysis described above provided clear evidence that secondary trophoblast giant cells surrounding the embryo express MHC class I mRNA, these experiments used a probe made up of the highly conserved α3 coding region and, thus, could not discriminate among individual members of the MHC class I gene family. To further characterize class I transcripts expressed at the maternal fetal interface, we used well-characterized RNA probes in RNAse protection assays that specifically detect conventional H-2D$^{d}$, H-2K$^{b}$, or H-2K$^{k}$ transcripts. Expression of paternally derived class I products was assessed to avoid the inevitable contribution from contaminating maternal tissues. Total cellular RNA was prepared from populations of secondary trophoblast giant cells carefully isolated from dissected embryos at 10.5 days p.c. recovered from matings of CBA/J (H-2$^{k}$) (Fig. 14.1) or C.B-20 (H-2$^{d}$) (Fig. 14.2) males crossed to MF1 (H-2$^{b}$) females. In both cases the results clearly demonstrate that paternally derived conventional class I gene products are expressed by the secondary trophoblast giant cells.

Extremely low levels of MHC class I transcripts were detectable by in situ hybridization analysis in the visceral yolk sac. Thus, MHC class I mRNA expression was not visible by gross inspection of visceral yolk sac tissues. However, determination of the silver grain density demonstrated a 2-fold increase in the hybridization signal over the background. There was no significant increase in the level of MHC class I mRNA expression in the visceral yolk sac from days 10.5 through 13.5 of development. These findings were surprising in light of several reports documenting expression of MHC class I transcripts in the visceral yolk sac at midgestation. Consistent with results of other investigators, as shown in Figure 14.3, we also found that MHC class I mRNA expression was readily detectable in visceral yolk sac tissues using an RNAse protection assay that is far more sensitive in comparison with in situ hybridization analysis.

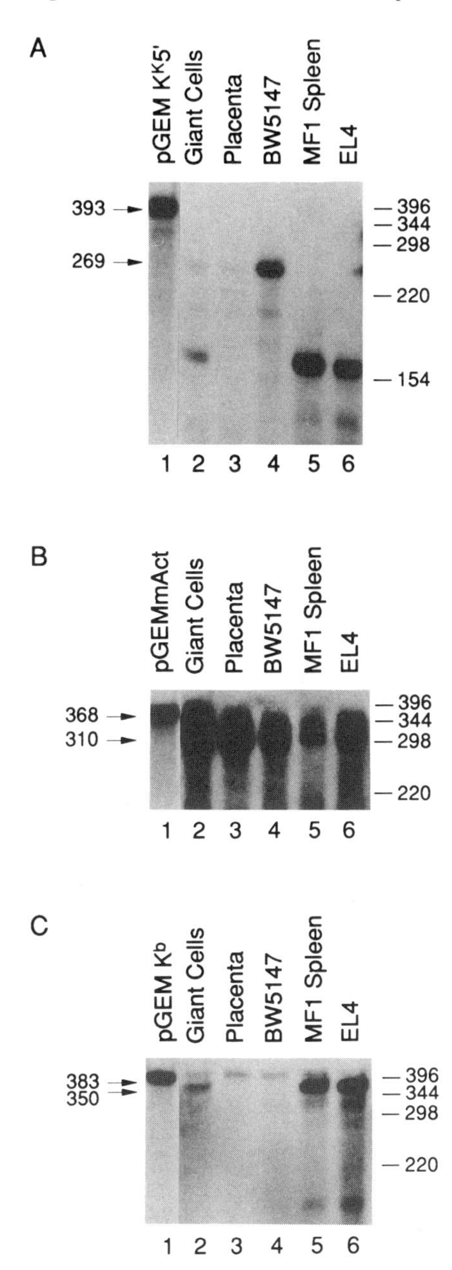


FIGURE 14.1. RNAse protection analysis of MHC class I transcripts in secondary trophoblast giant cells. Probes specific for H-2K$^{k}$ (*A*), cytoplasmic actin (*B*), or H-2K$^{b}$ (*C*) were hybridized with total RNA (10 μg) prepared from adult MF1 (H-2$^{b}$) spleen, EL4 (H-2$^{b}$) lymphoma cells, BW5147 (H-2$^{k}$) thymoma cells, placenta, or secondary trophoblast giant cells isolated from matings of CBA/J (H-2$^{k}$) males crossed to MF1 (H-2$^{b}$) females at day 10.5 p.c. The arrows indicate probe sizes and predicted full-length protected fragments.

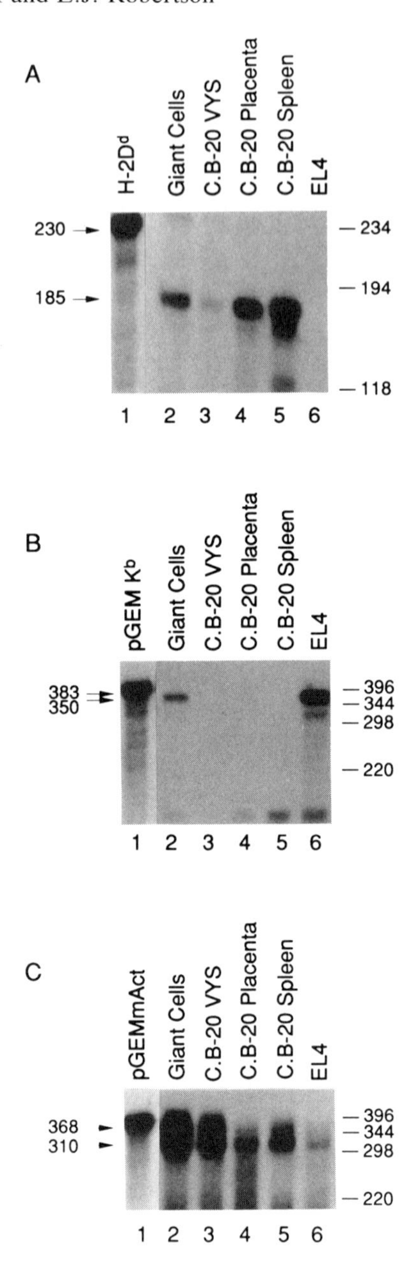


FIGURE 14.2. RNAse protection analysis of MHC class I transcripts in secondary trophoblast giant cells. Probes specific for H-2D$^{d}$ (*A*), H-2K$^{b}$ (*B*), or cytoplasmic actin (*C*) were hybridized with total RNA (10 µg) prepared from adult C.B-20 (H-2$^{d}$) spleen, EL4 (H-2$^{b}$) lymphoma cells, visceral yolk sacs, placenta, or secondary trophoblast giant cells isolated from matings of C.B-20 (H-2$^{d}$) males crossed to MF1 (H-2$^{b}$) females at day 10.5 p.c. The arrows indicate probe sizes and predicted full-length protected fragments.

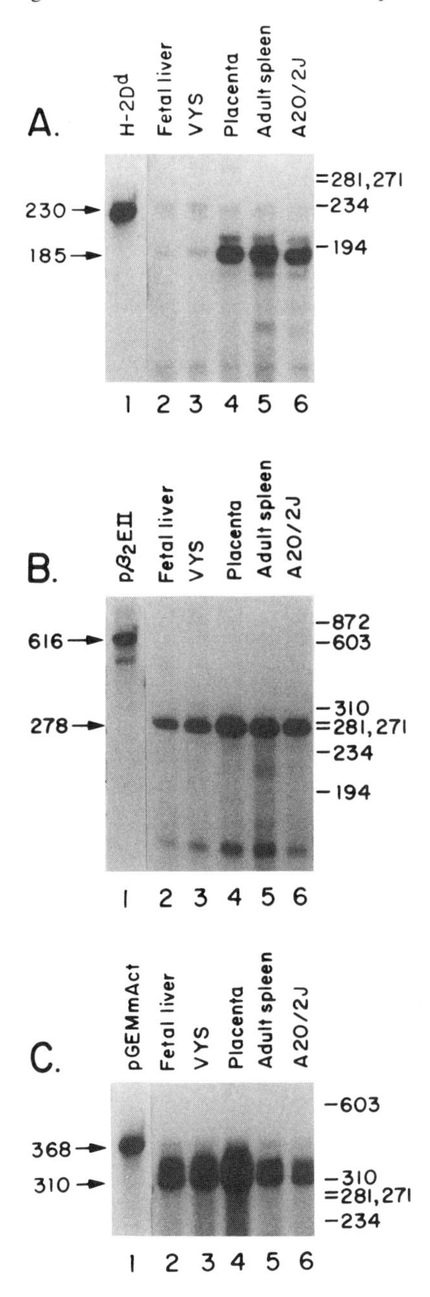


FIGURE 14.3. RNAse protection analysis of MHC class I transcripts in adult and fetal tissues. Probes specific for H-2D$^d$ (*A*), β2 microglobulin (*B*), or cytoplasmic actin (*C*) were hybridized with total RNA (5 μg) prepared from adult C.B-20 (H-2$^d$) spleen, isolated fetal tissues at day 13.5 p.c., or A20/2J (H-2$^d$) lymphoma cells. The arrows indicate probe sizes and predicted full-length protected fragments. The smaller protected fragments present in (*B*) are due to RNAse cleavage at an AT-rich sequence within the β2 microglobulin exon II.

These findings indicate that the onset and pattern of MHC class I and β2 microglobulin gene expression in early postimplantation stage embryos are not coordinately regulated.

Interestingly, coincident expression of high levels of β2 microglobulin and MHC class I mRNA was only observed in the mature placenta and in the developing lymphoid organs, such as the thymus and spleen. Thus, these in situ hybridization experiments identified a limited number of tissues in the conceptus having the potential for expression of MHC class I surface antigens. However, as discussed above, coordinately regulated expression of β2 microglobulin and MHC class I H-chains in the absence of peptide is insufficient to promote assembly and surface expression.

To evaluate whether MHC class I transcripts documented in these studies encode functional H-chains that are assembled with β2 microglobulin and transported to the cell surface, it is essential to know whether the fetal tissues previously shown to express MHC class I and β2 microglobulin transcripts also express all the components of the MHC class I peptide loading machinery. As discussed above, the MHC-encoded peptide transporters are members of the ABC superfamily of transporters. Similar ATP-driven transporters are present in diverse organisms throughout evolution. Current models envision that the MHC-linked peptide pump is a heterodimer composed of two closely related gene products (in the mouse, HAM1 and HAM2). Closely linked genes located in the class II region encode proteasome-like proteins. Sequence data suggest that these genes may have arisen by duplication of an initial transporter and proteasome gene pair. This is similar to the situation for bacterial transport systems, where genes coding for a soluble substrate-binding protein and the membrane-bound ATP-driven permease components frequently constitute a single operon.

The exact relationship between the two MHC-encoded proteasome components thought to be involved in antigen processing and the ubiquitous multicatalytic high-molecular weight complexes responsible for intracellular protein degradation is unknown. In light of the evidence that MHC-encoded peptide transporters are necessary for class I assembly and stability at the cell surface, we decided to evaluate developmentally regulated HAM1 mRNA expression.

We subcloned a 255-bp *Pvu*II/*Pst*I fragment encompassing the 3′ portion of HAM1 coding sequence (site position 1417–1672) into pGEM4. The resultant plasmid (designated pGEM-HIPP) was linearized with *Eco*RI and transcribed using T7 polymerase to yield the 296-nt antisense RNA probe. As expected, RNA prepared from A20/2J and EL4 lymphoma cells strongly expressing class I surface antigens protect the predicted full-length 255-nt fragment (Fig. 14.4). In a panel of embryo-derived cell lines, including CCE, a pluripotent *embryonic stem* (ES) cell line derived from a 129/Sv (H-$2^b$) blastocyst stage embryo (20); F9, an EC cell line derived from a 129/Sv (H-$2^b$) teratocarcinoma (21); and

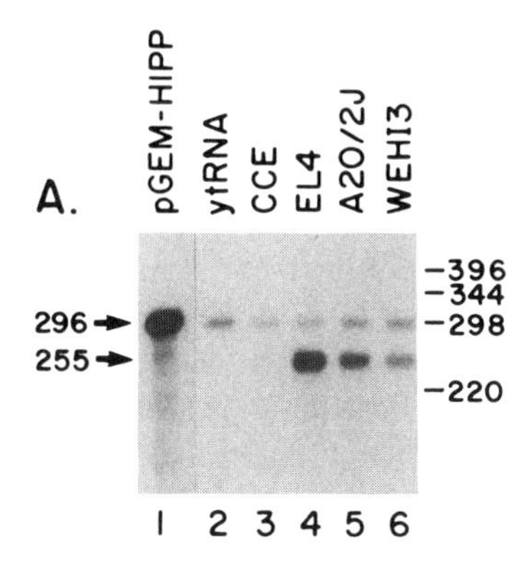


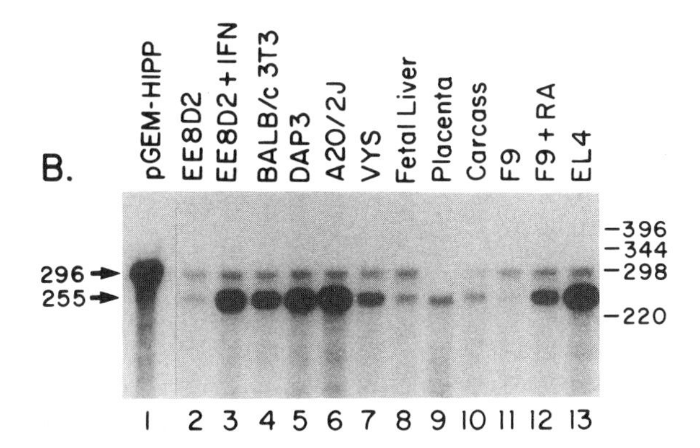


FIGURE 14.4. RNAse protection analysis of HAM1 mRNA expressed by various embryo-derived cell lines and isolated fetal tissues. The HAM1-specific probe (designated pGEM-HIPP) was hybridized with total RNA (10 μg) from the indicated cell lines or embryonic tissues prepared at day 13.5 p.c. Where indicated, mouse interferon (500 units/mL) (cat. #20071 $\alpha+\beta$; Lee Biomolecular, San Diego, CA) was added 48 h before RNA extraction. The F9 teratocarcinoma cells were treated with retinoic acid as described. The arrows indicate probe size and the predicted full-length protected fragment.

EE8D2, a HaSv-transformed fibroblastlike cell line isolated from C3H/HeJ (H-$2^k$) day 8 embryos (22), all expressed barely detectable levels of HAM1 transcripts. HAM1 mRNA expression was inducible by treating the cells with interferon or during differentiation in vitro. We have previously reported that in vitro differentiation and interferon treatment of these embryonic cell lines result in a large induction of MHC class I surface expression (23, 24). Thus, overall, there is a good correlation between the level of HAM1 transcripts and expression of MHC class I surface antigens as analyzed using a panel of embryo-derived cell lines.

We also used the RNAse protection assay to examine HAM1 mRNA expression in various embryonic tissues. We found HAM1 transcripts were present at relatively low levels in restricted tissues, including the visceral yolk sac, fetal liver, and placenta at 13.5 days p.c. The finding of such a low level of mRNA in these heterogenous cell populations might reflect uniformly low levels of expression in all cell types or, alternatively, a significant level of expression restricted to a selected subpopulation.

To distinguish these possibilities and to describe further the onset and pattern of HAM1 expression in early postimplantation stage embryos, in situ hybridization experiments were carried out. The pattern of hybridization was examined starting at day 6.5 of development. Prior to day 16.5 p.c., HAM1 mRNA was not present at detectable levels in the embryo or extraembryonic tissues as assessed by grain counts. Since we had previously shown that the only tissue showing high levels of expression of both MHC class I and β2 microglobulin is the placenta at midgestation, it was particularly important to determine the onset and pattern of HAM1 mRNA expression in the developing placenta. Significant levels of HAM1 transcripts were present from day 16.5 onward. In contrast to MHC class I and β2 microglobulin mRNA, which are most strongly expressed in the outer spongiotrophoblast layer, HAM1 transcripts were restricted to the inner labyrinth region that is infiltrated by fetal blood vessels and surrounded by maternal blood sinuses. As expected, HAM1 mRNA was expressed in the developing thymus and fetal liver at later stages of gestation.

As shown in Figure 14.5, we similarly found very low levels of HAM1 mRNA expressed in adult nonlymphoid tissues, such as heart, liver, and uterus. A strong argument can be made that this amount of HAM1 mRNA is sufficient for MHC class I assembly and transport since these

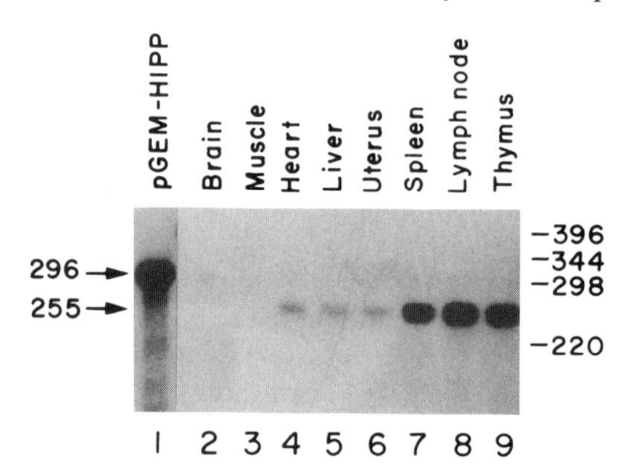


FIGURE 14.5. RNAse protection analysis of HAM1 mRNA expressed by adult tissues. The HAM1-specific probe (designated pGEM-HIPP) was hybridized with total RNA (10 µg) from various adult somatic tissues. The arrows indicate probe size and the predicted full-length protected fragment.

tissues have been shown to express detectable levels of MHC class I surface antigens.

To further document the onset and pattern of expression of the various components required for peptide-dependent MHC class I assembly, we developed probes specific for the HAM2 transporter and LMP2 proteasome component by subcloning appropriate fragments from the cDNA clones into riboprobe vectors. RNAse protection experiments demonstrate selective strong expression of HAM2 and LMP2 mRNA in the developing thymus at day 17.5 p.c. By contrast, other fetal tissues, including kidney, heart, liver, and placenta, were found to express relatively low levels of HAM2 and LMP2 transcripts.

In conclusion, a coordinate up-regulation of β2 microglobulin and MHC class I mRNA expression has been demonstrated during in vitro differentiation and following interferon treatment of embryo-derived cell lines. This also results in the induction of HAM1 mRNA expression. By contrast, we observe distinct patterns of expression of MHC class I, β2 microglobulin, and HAM1 transcripts in early mouse embryos. As expected, HAM1 mRNA is most strongly expressed in the developing thymus and fetal liver. Interestingly, in similar RNAse protection experiments using specific probes, we found that the onset and pattern of HAM1, HAM2, and LMP2 mRNA expression at the early stages of development appear to be coordinately regulated. Overall, these findings support the idea that tolerance of the fetal allograft is, in part, controlled posttranslationally at the level of peptide-dependent MHC class I assembly.

*Acknowledgments.* We thank Enuma Okoye for excellent technical assistance. This work was supported by a grant from the National Institutes of Health (HD-25926 to E.K.B. and E.J.R.). E.J.R. is the recipient of a Leukemia Society Fellowship.

## References

1. Bjorkman PJ, Saper MA, Samraoui B, Bennett WS, Strominger JL, Wiley DC. Structure of the human class I histocompatibility antigen, HLA-A2. Nature 1987;329:506–12.
2. Garrett TPJ, Saper MA, Bjorkman PJ, Strominger JL, Wiley DC. Specificity pockets for the side chains of peptide antigens in HLA-Aw68. Nature 1989; 342:692–6.
3. Madden DR, Gorga JC, Strominger JL, Wiley DC. The structure of HLA-B27 reveals nonamer self-peptides bound in an extended conformation. Nature 1991:353:321–5.
4. Deverson EV, Gow IR, Coadwell WJ, Monaco JJ, Butcher GW, Howard JC. MHC class II region encoding proteins related to the multidrug resistance family of transmembrane transporters. Nature 1990;348:738–41.

5. Trowsdale J, Hanson I, Mockridge I, Beck S, Townsend A, Kelly A. Sequences encoded in the class II region of the MHC related to the ABC superfamily of transporters. Nature 1990;348:741–4.
6. Spies T, Bresnahan M, Bahram S, et al. A gene in the human major histocompatibility complex class II region controlling the class I antigen presentation pathway. Nature 199;348:744–7.
7. Monaco JJ, Cho S, Attaya M. Transport protein genes in the murine MHC: possible implications for antigen processing. Science 1990;250:1723–6.
8. Kelly A, Powis SH, Kerr LA, et al. Assembly and function of the two ABC transporter proteins encoded in the human major histocompatibility complex. Nature 1992;355:641–4.
9. Spies T, Cerundolo V, Colonna M, Cresswell P, Townsend A, DeMars R. Presentation of viral antigens by MHC class I molecules is dependent on a putative peptide transporter heterodimer. Nature 1992;355:644–6.
10. Neefjes JJ, Momburg F, Hammerling GJ. Selective and ATP-dependent translocation of peptides by the MHC-encoded transporter. Science 1993; 261:769–71.
11. Shepherd JC, Schumacher TNM, Ashton-Rickardt PG, et al. TAP-1 dependent peptide translocation in vitro is ATP dependent and peptide selective. Cell 1993;74:577–84.
12. Brown MG, Driscoll J, Monaco JJ. Structural and serological similarity of the MHC-linked LMP and proteasome (multicatalytic proteinase) complexes. Nature 1991;353:355.
13. Glynne R, Powis SH, Beck S, Kelly A, Kerr LA, Trowsdale J. A proteasome-related gene between the two ABC transporter loci in the class II region of the human MHC. Nature 1991;353:667.
14. Martinez CK, Monaco JJ. Homology of proteasome subunits to a major histocompatibility complex-linked LMP gene. Nature 1991;353:664.
15. Croce CM, Linnenbach A, Huebner K, et al. Control of expression of histocompatibility antigens (H-2) and $\beta_2$-microglobulin in F9 teratocarcinoma stem cells. Proc Natl Acad Sci USA 1981;78:5754.
16. Morello D, Daniel F, Baldacci P, Cayre Y, Gachelin G, Kourilsky P. Absence of significant H-2 and $\beta_2$-microglobulin mRNA expression by mouse embryonal carcinoma cells. Nature 1982;296:260–2.
17. David-Watine B, Israel A, Kourilsky P. The regulation and expression of MHC class I genes. Immunol Today 1990;11:286–92.
18. Jaffe L, Jeannotte L, Bikoff EK, Robertson EJ. Analysis of $\beta_2$-microglobulin gene expression in the developing mouse embryo and placenta. J Immunol 1990;145:3474–82.
19. Jaffe L, Robertson EJ, Bikoff EK. Distinct patterns of expression of MHC class I and $\beta_2$-microglobulin transcripts at early stages of mouse development. J Immunol 147:2740–50.
20. Robertson E, Bradley A, Kuehn M, Evans M. Germline transmission of gene sequences introduced into cultured pluripotential cells by a retroviral vector. Nature 1986;323:445–8.
21. Bernstine EG, Hooper ML, Grandchamp S, Ephrussi B. Alkaline phosphatase activity in mouse teratoma. Proc Natl Acad Sci USA 1973; 70:3899–903.

22. Silverman T, Rein A, Orrison B, et al. Establishment of cell lines from somite stage mouse embryos and expression of major histocompatibility class I genes in these cells. J Immunol 1988;140:4378–87.
23. Bikoff EK, Otten GR, Robertson EJ. Defective assembly of class I major histocompatibility complex molecules in an embryonic cell line. Eur J Immunol 1991;21:1997–2004.
24. Bikoff EK, Jaffe L, Ribaudo RK, Otten GR, Germain RN, Robertson EJ. MHC class I surface expression in embryo-derived cell lines inducible with peptide or interferon. Nature 1991;354:235–8.
25. Jaffe L, Robertson EJ, Bikoff EK. Developmental failure of chimeric embryos expressing high levels of H-2D$^{d}$ transplantation antigens. Proc Natl Acad Sci USA 1992;89:5927–31.

15

# Overexpression of Class I MHC in Murine Trophoblast and Increased Rates of Spontaneous Abortion

JOSEPH R. VOLAND, CHRISTOPHER BECKER,
AND FARIDEH HOOSHMAND

The ability of the placenta to avoid rejection by the maternal immune system has remained a fascinating paradox for the transplantation immunologist. The placenta contains tissue-specific fetal antigens and paternally derived histocompatibility antigens that are potentially antigenic for the maternal immune system. Nevertheless, during the course of a normal pregnancy, deleterious maternal immune responses are not generated.

One of the explanations put forth for the survival of the placenta is that the trophoblast populations, which are in direct contact with maternal blood and tissues, express little or no classical class I or class II *major histocompatibility complex* (MHC) antigen. Consequently, trophoblast may be an immunologically privileged tissue that is incapable of eliciting a classical T cell-mediated immunological rejection response.

We have begun to test this hypothesis in a transgenic mouse model. We have generated a transgenic mouse carrying the classical class I MHC gene $D^d$ under the control of the c-*fos* promoter. This chapter presents data to show that this gene is expressed in the murine placenta and that expression of this gene is correlated with higher rates of spontaneous abortion. Moreover, the rate of transmission of the transgene drops with increasing parity in the mothers, suggesting that we can "immunize" these mothers against pregnancy. We believe these data support the hypothesis that down-regulation of class I MHC on murine trophoblast is essential for survival of the mammalian embryo.

## Historical Background

### *Class I MHC Expression on Trophoblast*

Work from many laboratories over several years has demonstrated the relative lack of class I MHC expression by trophoblast. Similar patterns of class I expression are seen in the human, the mouse, and the rat.

The data from the human system are perhaps the clearest, partially because the anatomy of the human placenta is more thoroughly studied, but also because reagents to human class I antigens were more widely available and better characterized. The syncytiotrophoblast, which lines the chorionic villi and is in direct contact with the maternal blood, expresses no detectable class I MHC protein (1, 2). In situ hybridization studies also show no evidence of mRNA for class I MHC mRNA in the syncytiotrophoblast (3, 4).

The status of MHC gene expression in the cytotrophoblast is very different. Sunderland et al. first showed that the cytotrophoblast of the chorionic shell reacted with the monoclonal antibody W6/32 (5). This antibody recognizes the α3 domain of the class I heavy chain gene when it is complexed with β2 microglobulin. Interestingly, these same cells did not react with antibodies directed against polymorphic determinants of the α1 and α2 domains of the MHC. Moreover, the MHC expressed by cytotrophoblast was found to be truncated with a molecular weight of 39,000–40,000 compared to 45,000 for classical class I molecules (6). The mRNA for class I gene products has also been detected in the cytotrophoblast of first-trimester chorionic villi and in the extravillous cytotrophoblast of the chorionic shell (4). It has now been demonstrated that the MHC molecule expressed by cytotrophoblast is a nonclassical class I molecule, HLA-G (7–10). Expression of HLA-G is highest in the first trimester when cytotrophoblast is more abundant (11), and the expression of this gene declines as pregnancy progresses. HLA-G expression appears confined to extraembryonic tissues, although a recent report indicates that it may also be expressed in mature sperm (12). Interestingly, even though first-trimester villous cytotrophoblast cells express mRNA for HLA-G, no class I protein can be detected. This observation suggests that control of class I expression in trophoblast may occur at both the transcriptional and translational levels.

The function of HLA-G is currently unknown. It has been suggested that its expression on extravillous trophoblast may confer a status of nonpolymorphic *self* on trophoblast, protecting it from *natural killer* (NK) cell-mediated cell lysis. It has also been suggested that the expression of a monomorphic class I MHC may be necessary for nonimmunological cell development or differentiation (13). To date, no definitive experimental data have been provided to support or refute these hypotheses.

The murine placenta demonstrates a similar pattern of MHC expression. The labyrinthine trophoblast, which is the site of fetomaternal exchange, expresses little if any class I MHC (14, 15). The spongiotrophoblast layer of the definitive murine placenta does appear to express paternally derived class I MHC. Redline and Lu (16) demonstrated cells expressing class I MHC that were interstitial in location with a vacuolated appearance perhaps corresponding to *glycogen cells*. These cells were most abundant in the decidua basalis, although small nests of these class I-positive cells could be found in both the spongiotrophoblast and the labyrinthine trophoblast areas of the mature placenta. However, trophoblast in direct contact with maternal blood, whether in the labyrinthine or in the spongiotrophoblast layer, were devoid of class I antigens. In addition, the giant trophoblast cells juxtaposed between the spongiotrophoblast and the decidua basalis were also class I negative.

In situ hybridization studies have confirmed these immunohistochemical findings. Message for class I gene products can be detected in day 9.5 placentas. Interestingly, the message appears to be confined to the trophoblast giant cells (17). By day 12.5 message can be detected in the spongiotrophoblast, and low levels of message can be detected in the labyrinthine trophoblast. When message for β2 microglobulin was analyzed, it was found that the giant cell trophoblast did not express message for β2 microglobulin; however, the remaining layers of the placenta were positive. By day 13.5 the highest levels of β2 microglobulin message were detected in the spongiotrophoblast (18). These results suggest that one mechanism of control for class I gene expression in trophoblast may be the noncoordinate expression of class I message and β2 microglobulin mRNA. These results also indicate that β2 microglobulin message and, presumably, protein are available for assembly with class I heavy chains beginning at day 9.5.

It should be noted that although both serology and in situ hybridization suggest that classical class I MHC is expressed by some populations of murine trophoblast, the possibility that the trophoblast is expressing a cross-reactive nonclassical class I molecule has not been completely excluded. Two lines of evidence suggest that either the class I on murine trophoblast is nonclassical or that it is somehow made inaccessible. First, up-regulation of MHC on spongiotrophoblast can occur in cells that are treated with *interferon α* (IFNα)/*interferon β* (IFNβ) (19). This up-regulation is seen both in vivo and in vitro. Nevertheless, in vivo up-regulation of class I MHC does not significantly compromise the pregnancy (20). Secondly, isolated trophoblast cannot be lysed by allospecific cytotoxic effector cells, even if class I MHC is up-regulated by IFN (21). These results suggest that either the class I MHC expressed by trophoblast is an ineffective target for the T cell receptor of *cytotoxic T lymphocytes* (CTL) or that trophoblast has some mechanism to prevent cell-mediated lysis.

Studies in the rat also point to the possibility that mouse trophoblast may express some form of nonclassical class I MHC. The rat placenta has a similar anatomy to the mouse placenta. Again, the labyrinthine trophoblast is devoid of class I MHC (22). The spongiotrophoblast does express a class I molecule, but in parallel with the human case, the molecule expressed is a nonclassical monomorphic molecule termed *Pa* (23, 24). Unlike the human, these rat trophoblast cells also express classical paternally derived MHC molecules. However, these antigens do not seem to be expressed on the cell surface during an allogeneic pregnancy (25).

In summary, data from both the human system and animal models indicate that trophoblast populations in direct contact with the maternal circulation are devoid of class I MHC molecules. Interstitial trophoblast in contact with maternal tissues express class I MHC, but at least in the case of the human and the rat, these molecules have been shown to be monomorphic nonclassical class I antigens.

Because classical class I MHC molecules are known to be the targets of alloreactive CTL in rejection responses, it has been hypothesized that the lack of these molecules on trophoblast could explain the lack of immunological rejection of the placenta. Definitive proof of this hypothesis requires the up-regulation of class I MHC on trophoblast. To date, such up-regulation has not been effective, primarily because of the tight control of MHC expression exhibited by trophoblast (26).

We have attempted to circumvent this problem by generating a transgenic mouse model in which a classical MHC gene is under the control of a foreign promoter. We chose the c-*fos* promoter because much is known about its regulation (27, 28) and because trophoblast express high levels of c-*fos* during development (29–31). Consequently, a gene driven by the c-*fos* promoter would be expected to be highly expressed during development. We chose the $D^d$ gene because a convenient restriction site exists that allowed us to remove all of the 5′ upstream portion of the gene (32). It is clear from many studies that the 5′ upstream region of the class I genes carries multiple regulatory sites (33, 34). It has also been shown in another transgenic system that mice carrying a human HLA gene B27 containing the 5′ upstream regulatory regions fail to express the gene in the placenta (35). These data suggest that whatever the mechanism controlling class I expression in the placenta, it is probably regulated in this upstream region. By placing our gene totally under the control of the c-*fos* promoter, we hoped to circumvent any transcriptional regulation of class I MHC in trophoblast. Our data indicate that this hybrid gene can be expressed in the placenta and that its expression adversely affects pregnancy.

## *Animal Models of Spontaneous Abortion*

Other investigators have attempted to look at immunologically mediated spontaneous abortion in animal models. Two models have been described in which the immune system clearly plays a role.

The first model is the *Mus musculus*, *Mus caroli* system. *Mus caroli* is a separate mouse species from southeast Asia. When *Mus caroli* embryos are transferred into *Mus musculus* mothers, the pregnancies undergo spontaneous abortion (36). Histologically, there is an inflammatory infiltrate and resorption of the pregnancy. If *Mus caroli* embryos are placed in a trophoblast shell derived from *Mus musculis*, then the pregnancy survives (37). It was originally believed that the abortion of the pregnancy was the result of maternal immune attack against the foreign *Mus caroli* tissues, either because of tissue incompatibility or because of the failure of the foreign trophoblast to properly recruit immunoregulatory cells to the pregnant uterus (38). It was subsequently shown that the failure of the pregnancies was not due to either T cell-mediated or NK cell-mediated cytolytic mechanisms, although the immune system was shown to be involved in the resorption process (39).

The second model of immunologically mediated abortion was the CBA × DBA system. If CBA/J females are bred with DBA/2 males, there is a high rate of spontaneous abortion (40). The abortions can be prevented by immunization of the mothers with third-party haplotype-matched lymphocytes (41, 42) or by complete Freund's adjuvant (43). The pregnancies can be further compromised by injection of double-stranded RNA (poly[I]·poly[C]). It was ultimately shown that the abortions in this model were being mediated by activated NK cells (44) and that treatment with poly(I)·poly(C) was stimulating NK activity, probably through IFN (45, 46) . In support of this model is the observation that trophoblast in vitro, although resistant to T cell-mediated cytolysis, can be effectively lysed by activated NK cells (47).

To date, no model has been described in which immunologically mediated abortion is mediated by $CD8^+$ T cells in a classical graft rejection. Our model suggests that up-regulation of class I MHC may be necessary for T cell-mediated attack. Clearly, the fact that we can only see effective transmission of our transgene in an immunologically compromised β2 microglobulin knockout mouse suggests that $CD8^+$ cells are playing some role in the pregnancy losses we are observing.

# Methods

## *Construction of the Transgene*

The basic structure of the transgene is presented in Figure 15.1. The c-*fos* promoter was excised as a 370-bp *Pst*I-*Pst*I fragment from the full-length c-*fos* gene. This fragment contains the TATA and CATT boxes of the promoter, the serum response enhancer region required for c-*fos* expression (28), and the transcription start site for the c-*fos* message.

The $H\text{-}2D^d$ gene was obtained from a partial *BamH*I digest of a cosmid clone. There is a *BamH*I site 9 bp 5′ of the start ATG (32). Consequently, the gene was obtained with essentially no 5′ untranslated region.

The gene and promoter were sequentially cloned into the *BamH*I and *Pst*I sites of pGEM3 (Promega), and correct orientation was determined by restriction digest analysis. The promoter and gene construct was excised from the plasmid by a *Hind*III-*EcoR*I digest and purified on a

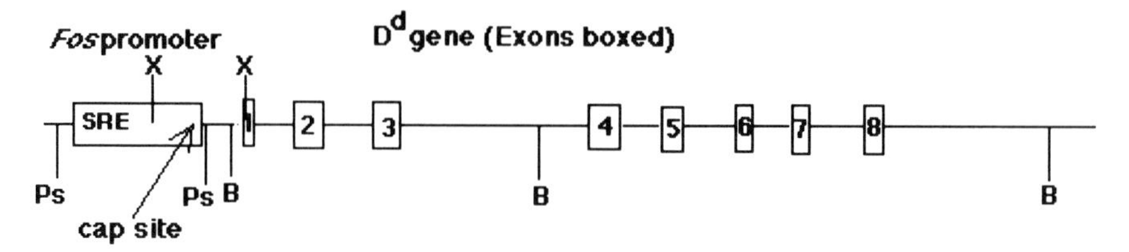


FIGURE 15.1. Structure of the c-*fos*/H-2D$^d$ transgene. The c-*fos* promoter was ligated to the coding region of the H-2D$^d$ gene. The serum response element, cap site, and *Xma*III fragment used for screening are marked. (Ps = *Pst*I site; X = *Xma*III site; B = *BamHI* site; SRE = serum response element.)

sucrose gradient for injection into mouse eggs. The construct was tested by transfection into mouse L cells (H-2$^k$ haplotype). Transfected cells expressed the hybrid gene and reacted with anti-D$^d$ monoclonal antibodies in both FACS analysis and immunohistochemistry (data not shown).

## *Generation of Founder Mice*

Transgenic founders were generated as previously described (48). Briefly, ~400 copies of the transgene were microinjected into fertilized mouse eggs. The eggs were obtained from C57/BL6 × SJL F1 strains. The eggs were implanted into pseudopregnant BDF1 females.

## *Screening of Transgenic Animals*

Tail DNA was obtained from 2-week-old mice and was analyzed by Southern blotting. There is an *Xma*III site in the middle of the c-*fos*

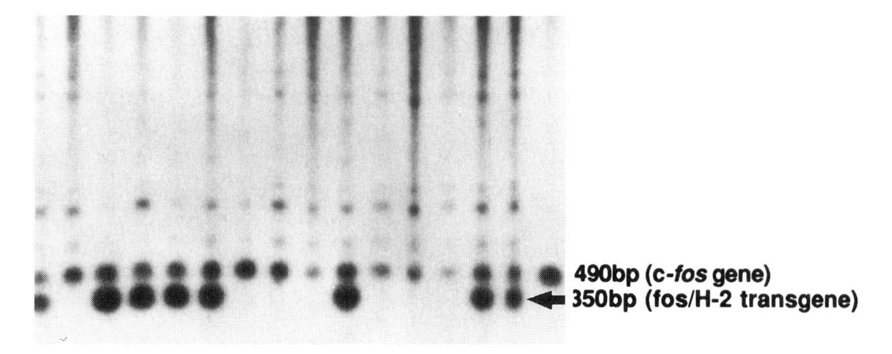


FIGURE 15.2. Southern blot screening of transgenic animals. Mouse tail DNA was digested with *Xma*III and electrophoresed. The resultant fragments were transferred to nylon membrane and probed with the *Xma*III/*Xma*III fragment obtained from the transgene. The transgenic animals were easily identified by a 350-bp band (arrow). Comparison to the 490-bp endogenous c-*fos* promoter gave a measure of a single-copy endogenous gene.

promoter and an *Xma*III site 50 bp from the start ATG. Consequently, cutting the transgene with *Xma*III produces a unique restriction fragment of ~350 bp. In addition, there is a second *Xma*III site in the first exon of the c-*fos* gene that produces a restriction fragment of 490 bp. Blots were probed with the 350-bp *Xma*III/*Xma*III fragment from the transgene. On genomic tail blots both the transgene and the endogenous c-*fos* gene were readily identifiable (Fig. 15.2).

## *Mating Protocols*

Nontransgenic female mice were bred to fos/H-2D$^d$ transgenic males by standard breeding protocols. Virgin C57/BL6 and BALB/c female mice were obtained from Jackson laboratories. Mice were bred starting at 8 weeks of age. The β2 microglobulin knockout mice were maintained in our breeding colony. Virgin females were again bred starting at 8 weeks of age.

Timed pregnancy was dated by the presence of a vaginal plug. The day the plug was detected was considered to be day 0 of pregnancy. Mice were sacrificed at days 8.5 and 17.5 of pregnancy.

## *Immunohistochemistry*

Adult tissue samples were obtained from transgenic mice and were either fixed in 10% buffered formalin or snap-frozen in liquid nitrogen-cooled isopentane as previously described (49). Fetal and placental tissues were obtained from the timed pregnancies. For day 8.5 pregnancies the entire fetal placenta was removed en block, and a small portion was taken for Southern blot analysis. For day 17.5 pregnancies, the placenta was fixed as described while the embryo was utilized for Southern blot analysis.

Fixed tissue was processed by standard methods, and sections were stained with hematoxylin-eosin. Immunohistochemistry was performed on unstained sections using the *monoclonal antibody 34-2-12S* (ATCC) that reacts specifically with the D$^d$ gene product and can be used to stain formalin-fixed tissues. The antibody was directly biotinylated, and the reaction was developed using the avidin-biotin peroxidase method and AEC as a substrate, as previously described (50). Isotype-matched antibodies and antibodies to H2-K$^b$ were utilized as controls. Tissues were also stained from nontransgenic litter mates.

# Results

## *Generation of Transgenic Founders*

A total of 30 eggs were injected, and 8 pups were born. Of these, 2 were transgenic: male 673 and female 678. We believed that the female would be tolerant to her own transgene and that breeding the female to either nontransgenic males or transgenic males would show no obvious pathology. Also, it has been demonstrated that paternal genes and paternal MHC antigens are selectively expressed in the trophoblast (22). Consequently, we used the transgenic males and bred these males to nontransgenic C57/BL6 (allogeneic) and BALB/c (syngeneic) females.

## *Expression of the Transgene in Adult Transgenic Animals*

We chose to look at the expression of the transgene by immunohistochemistry rather than by Northern blot analysis. We felt that it would be difficult to differentiate the message for the transgene from endogenous class I messages. We also believed that immunohistochemistry would allow us to determine more precisely the anatomic locations of transgene expression. Adult transgenic mice bred to a C57/BL6 background were used in these studies. Sections of tissues from adult BALB/c mice were utilized to determine the normal distribution of $D^d$.

TABLE 15.1. Distribution of transgene expression in adult animals.

| Tissue | *fos*/H-2$D^d$ expression |
|---|---|
| Heart | + |
| Lung | |
| Bronchial epithelium | +/− |
| Alveoli | − |
| Capillaries | +/− |
| Liver | |
| Hepatocytes | + |
| Kupffer's cells | + |
| Spleen | |
| Lymphocytes | − |
| Macrophages and connective tissue | ++ |
| Kidneys | |
| Glomeruli | − |
| Convoluted tubules | ++ |
| Collecting tubules | +/− |
| Testis | |
| Spermatogonia | ++ |
| Spermatids | − |
| Leydig cells | − |
| Pancreas | |
| Parenchyma | − |
| Islets | + |

*Note:* Class I MHC expression was detected by immunohistochemistry. Sections were obtained from animals carrying a $D^d$ transgene on a non-H-2$D^d$ background.

The distribution of the transgene is presented in Table 15.1. As can be seen, the transgene was expressed in a variety of tissues. The most interesting areas of expression were in the seminiferous tubules of the testes and the islet cells of the pancreas. There was also prominent staining of the renal tubular epithelium. To date, we have examined only a small number of animals and have been unable to detect any obvious pathology associated with class I expression. Two of the older animals had developed a cystic kidney, and the testes of old males showed less-active spermatogenesis as compared to age-matched nontransgenic litter mates (data not shown). Whether this represents normal variation within the population or is representative of disease secondary to expression of the transgene requires further study.

## *Expression of the Transgene in the Fetoplacental Unit*

We examined the expression of the transgene in day 8.5 and day 17.5 murine placentas. At day 8.5 there was clear staining of the giant cell trophoblast abutting the decidua basalis (Fig. 15.3). The staining appeared to be confined to the cytoplasm of the cell, although at the light microscope level, it is not completely possible to differentiate staining inside and outside the cell. At the 8.5 day stage, there was no obvious inflammation in the placentas.

At day 17.5 of development, the transgenic placentas as compared to nontransgenic controls showed evidence of hemorrhage, widening of the blood sinusoids consistent with edema, and a slight increase in cellularity. However, an overwhelming inflammatory response was not seen (Figs. 15.4A and 15.4B). Immunohistochemical staining revealed clear expression of the transgene in the labyrinthine trophoblast (Figs. 15.4C and 15.4D). In particular, there appeared to be clear expression of the $D^d$ gene in the endovascular trophoblast lining the maternal vascular sinuses. In at least 2 mothers, clear evidence of resorption was also identified that involved at least 3 implantation sites.

## *Reduced Transmission of the Transgene in Both Allogeneic and Syngeneic Immunocompetent Mothers*

The founder transgenic male was bred to either C57/BL6 females or BALB/c females. As can be seen in Table 15.2, while one or two mothers seemed to transmit the gene with almost normal frequency, other mothers had either low or no transmission of the transgene. Averaging the transmission of the transgene over all of the mothers showed that the overall number of transgenic pups produced by these matings was below the expected value of 40%–50%. The allogeneic C57/BL6 mothers also tended to produce slightly fewer transgenic pups as compared to the

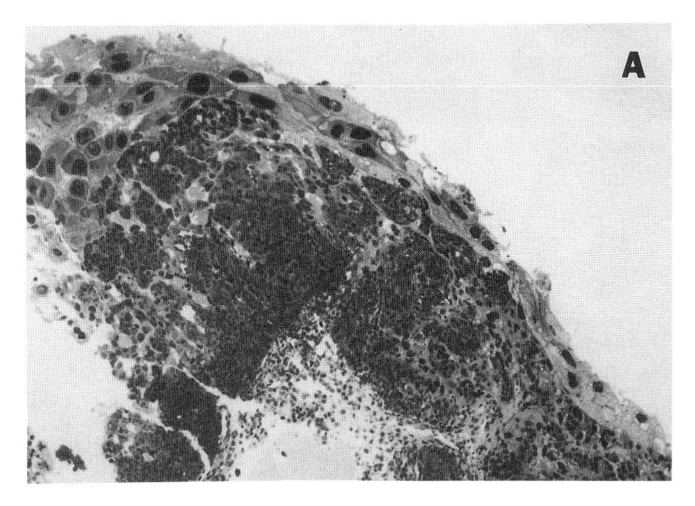


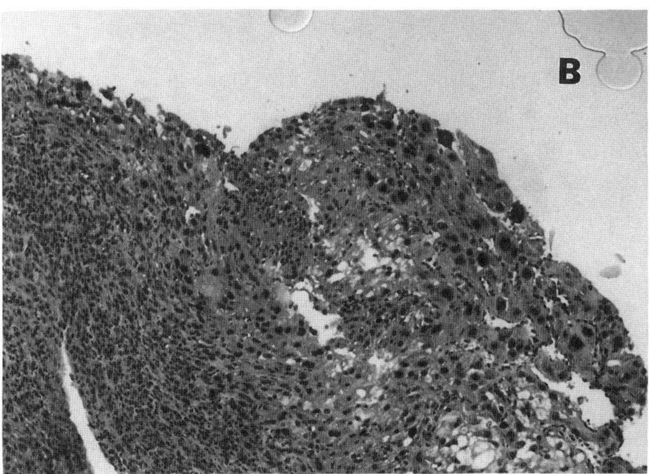


FIGURE 15.3. Transgene expression in day 8.5 gestations. A C57BL/6 mouse that previously had not been producing transgenic pups was bred to a transgenic male and sacrificed at day 8.5 of pregnancy. Immunohistochemical staining of the decidua and placenta showed both nontransgenic (*A*) and transgenic (*B*) trophoblast. Evidence of transgene expression was seen in 5/12 implantation sites. Sections were stained with ATCC, developed with AEC, and counterstained with hematoxylin (×250).

syngeneic BALB/c mothers (Fig. 15.5). Nevertheless, transgenic animals were obtained from both sets of mothers, indicating that expression of the transgene did not always compromise viability. It was unclear whether the differences in the percentage of transmission of the transgene were due to individual animal variation or were the result of immunologic factors. We

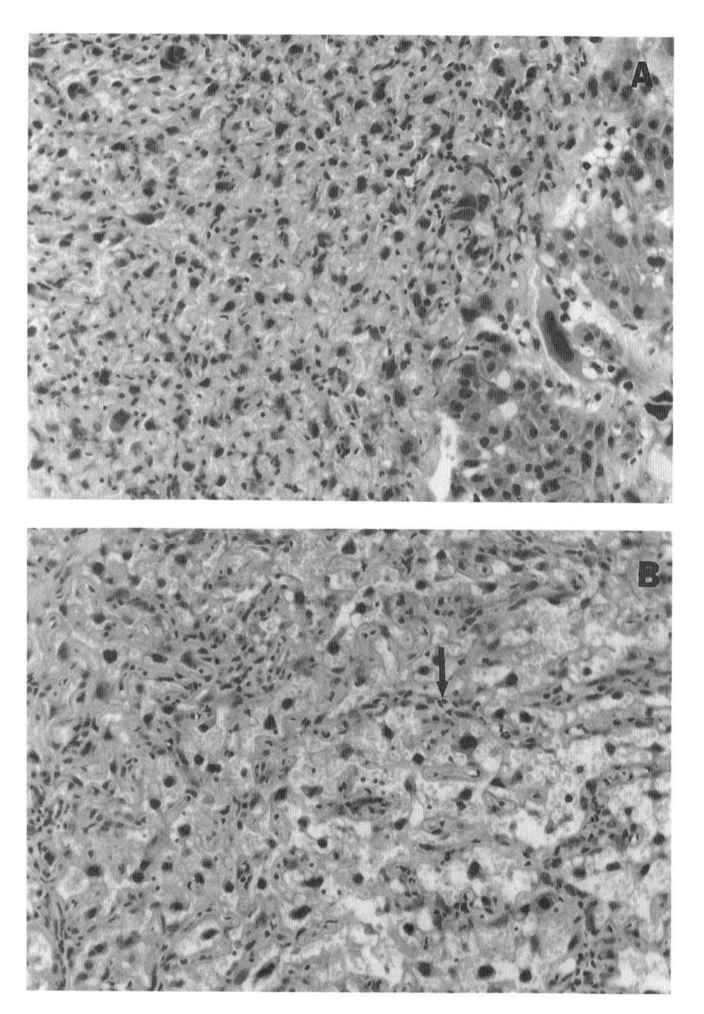


FIGURE 15.4. Transgene expression in day 17.5 placentas. *A:* H&E sections of a nontransgenic placenta demonstrate a normal architecture (×250). *B:* H&E sections of a transgenic placenta demonstrate increased cellularity in the labyrinthine trophoblast (arrow), as well as widening of the blood sinusoids consistent with edema. *C:* Immunohistochemical staining of a nontransgenic placenta is shown. A nontransgenic litter mate was stained with ATCC and developed with AEC. No antigen-specific staining could be detected. *D:* Immunohistochemical staining of a transgenic placenta is shown. Staining of a transgenic placenta with ATCC demonstrates clear expression of the transgene in the labyrinthine trophoblast. In particular, staining is clearly seen in the cells lining the blood sinusoids (arrow).

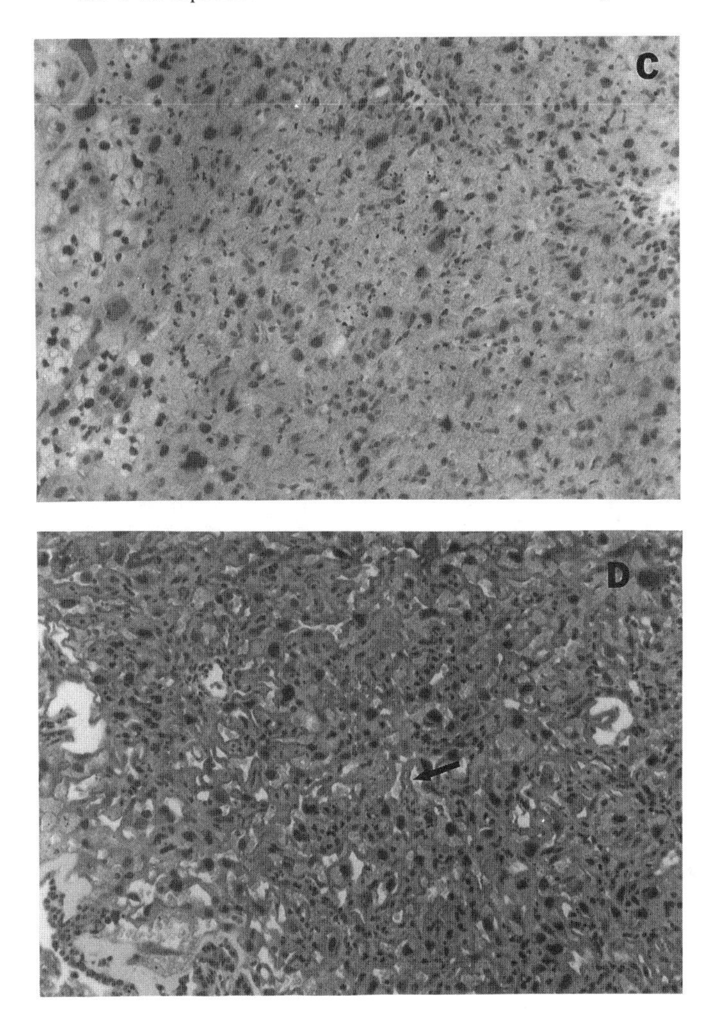


FIGURE 15.4. *Continued*

therefore looked at the rate of transmission of the transgene in immunoincompetent mothers as a function of maternal parity in an attempt to resolve these questions.

## *Enhanced Transmission of the Transgene in Immunoincompetent β2 Microglobulin Knockout Mothers*

It appeared that the rate of transmission of the transgene was reduced—although not abolished—in immunocompetent mothers. We tested the hypothesis that these mothers were failing to produce transgenic offspring

TABLE 15.2. Transmission of $D^d$ transgene in immunocompetent mothers.

| Mother | Number of pups | Number of transgenic pups | Transmission of transgene (%) |
|---|---|---|---|
| C57/BL6 | 9 | 2 | 22.0 |
| C57/BL6 | 7 | 1 | 14.0 |
| C57/BL6 | 9 | 4 | 44.0 |
| C57/BL6 | 7 | 0 | 0.0 |
| C57/BL6 | 10 | 4 | 40.0 |
| | | | Average transmission = 24.0 |
| BALB/c | 7 | 3 | 42.0 |
| BALB/c | 7 | 4 | 57.0 |
| BALB/c | 3 | 0 | 0.0 |
| BALB/c | 8 | 1 | 12.5 |
| BALB/c | 4 | 0 | 0.0 |
| BALB/c | 5 | 4 | 80.0 |
| | | | Average transmission = 27.0 |

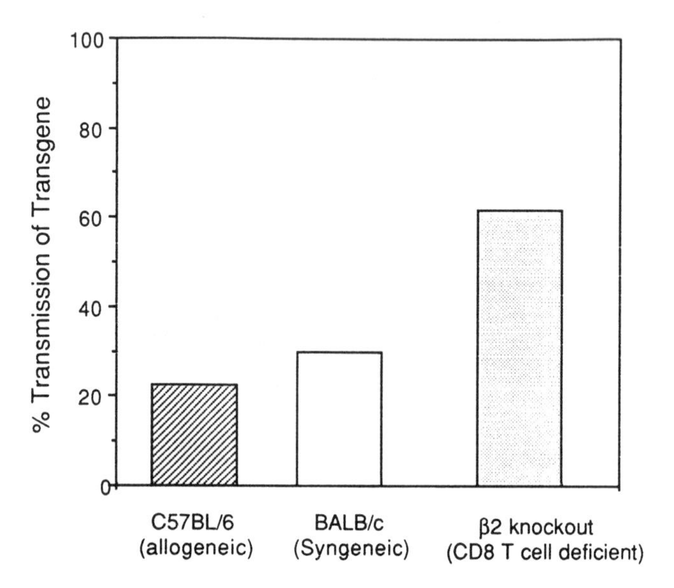


FIGURE 15.5. Transmission of the transgene in immunocompetent vs. immunoincompetent mothers. When all of the breeding data from the initial matings were averaged, the rate of transmission of the transgene in both allogeneic and syngeneic mothers was decreased. The rate of transmission in an immunoincompetent mother was at expected levels.

TABLE 15.3. Transmission of transgene in β2 microglobulin knockout immunoincompetent mothers.

| Mother | Number of pups | Number of transgenic pups | Transmission of transgene (%) |
|---|---|---|---|
| β2 knockout | 5 | 3 | 60 |
| β2 knockout | 8 | 4 | 50 |
| β2 knockout | 5 | 3 | 60 |
| β2 knockout | 6 | 2 | 25 |
| β2 knockout | 8 | 7 | 87 |
| β2 knockout | 7 | 6 | 85 |
| β2 knockout | 3 | 1 | 33 |
| | | | Average transmission = 57 |

because of immunological factors by breeding transgenic males to β2 microglobulin knockout females. These animals lack expression of classical class I MHC and, consequently, fail to positively select $CD8^+$ T cells (51). It would be expected that $CD8^+$ T cells would mediate an immunologic rejection of allotypic class I-bearing tissues. We believed that if the decreased transmission of the transgene was due to immunologic factors, then β2 microglobulin knockout mice—even though they are allogeneic with respect to $D^d$—would not show decreased transmission of the transgene. As can be seen in Table 15.3, we observed an overall increased rate of transmission of the transgene in these immunoincompetent animals. Moreover, the rate of transmission of the transgene was more uniform between mothers.

## *Decreased Transmission of the Transgene with Increasing Parity*

Because the transgene was transmitted with greater frequency in immunoincompetent mothers, we felt that the decreased transmission of the transgene was probably due to immunologic factors. Nevertheless, we did obtain transgenic pups from both C57/BL6 and BALB/c mothers, indicating that even in the case of an allogeneic mother, transmission of the transgene was not an all-or-nothing phenomenon.

We decided to see if a multiparous mother would have greater difficulty in transmitting the transgene. If the mothers were mounting an immunologic response to the transgene, it was possible that in a first pregnancy a sufficient immunologic response would not occur quickly enough to adversely affect the pregnancies. Such a situation would be particularly likely if the class I molecule was not expressed immediately. Because β2 microglobulin does not appear to be transcribed in most of the placenta

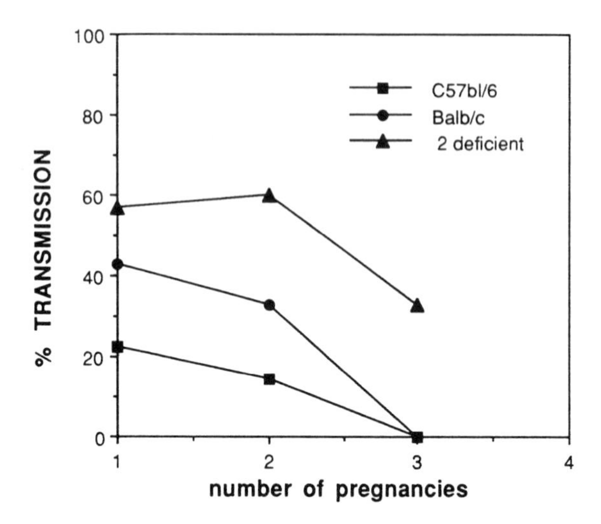


FIGURE 15.6. Decreased transmission of the transgene with increasing parity. Virgin mice were bred multiple times to transgenic males and the number of transgenic pups per litter determined. The rate of transmission of the transgene substantially decreased as a function of parity in immunocompetent mothers. Immunoincompetent mothers showed only slight decreases in transmission. (2 deficient = β2 microglobulin knockout.)

until approximately day 9 of pregnancy (18), it is likely that the $D^d$ gene product is not expressed at the cell surface until that time.

Virgin female mice were repeatedly bred to transgenic males, and the number of transgenic mice produced in each pregnancy was examined. As can be seen in Figure 15.6, the percentage of transgenic animals dropped with every subsequent pregnancy in immunocompetent mothers; eventually, these multiparous mothers produced no transgenic animals at all. By contrast, the pregnancy rate in the β2 microglobulin knockout mice decreased only slightly, and these mothers continued to produce transgenic pups.

## Discussion

We have generated a transgenic mouse model in which the $D^d$ gene under the control of the c-*fos* promoter is expressed in the murine placenta. We believe that the results presented support the hypothesis that the aberrant expression of classical class I MHC leads to immunologically mediated rejection of the placenta by the maternal immune system and, consequently, that the down-regulation of class I MHC by trophoblast is essential for the survival of the murine pregnancy.

## *Expression of Class I MHC in Trophoblast*

Our immunohistochemistry data clearly indicate that the transgene was expressed in multiple trophoblast populations and, in particular, was expressed in the endovascular trophoblast lining the maternal blood sinuses in the labyrinth. It is probable that we are seeing the effects we have reported because we have expressed a class I gene in endovascular trophoblast. Other studies that have attempted to up-regulate class I expression generally found increased expression of class I MHC in the spongiotrophoblast, where it may not be as accessible to the maternal immune system (20). Also, we have introduced the expression of a known classical class I gene that clearly does act as a restricting element for cytotoxic T cells, making it more likely that an immunologic response might occur.

The fact that we obtained expression of our transgene in trophoblast indicates that trophoblast is quite capable of expressing classical class I genes. Moreover, our data, together with work from other transgenic models (35), would suggest that the factors that suppress class I expression in trophoblast must act on regions in the 5′ untranslated portion of the mRNA or in the 5′ upstream regions of the class I gene. These were the only two regions of the $D^d$ gene removed in our construct. The *BamH*I fragment of the $D^d$ gene that we utilized contains an additional 1000 bp of 3′ untranslated and downstream sequences, suggesting that silencer regions are not contained in this portion of the $D^d$ gene or, if present, require additional 5′ upstream sequences to be active.

## *Both Syngeneic and Allogeneic Mothers Appear to Abort Transgenic Pups*

The most surprising finding in our study was that we not only lost the ability to transmit the transgene in allogeneic mothers, but that we also lost the ability to transmit the transgene in syngeneic mothers. This observation suggests that the down-regulation of class I MHC on trophoblast may be necessary for more than the problem of allorecognition.

It is possible that a BALB/c mother would not be tolerant to the transgene. It is possible that trophoblast either splices the gene transcript differently or glycosylates the molecule in a different manner, resulting in a slightly altered $D^d$ that may appear sufficiently foreign to break tolerance. If this is the case, however, then we would predict that multiparous BALB/c mothers should develop cytotoxic effector cells capable of lysing either $D^d$ targets or cell lines transfected with the transgene. However, it would seem reasonable that if tolerance were broken to self-antigen, these animals would be at high risk for developing autoimmune disease. We plan to examine these animals for manifestations of autoimmune illness.

We believe a more likely explanation for the loss of transmission of the transgene in syngeneic mothers is that the presence of a classical class I gene on trophoblast allows the presentation of tissue-specific peptides and the generation of cytotoxic T cells to trophoblast-specific antigens, resulting in a breakdown of tolerance to trophoblast. Of course, the mechanisms responsible for the establishment and maintenance of peripheral tolerance are not yet fully understood. However, evidence has been generated to suggest that the establishment of tolerance may depend on exposure to the antigen at critical periods in development (52) and that maintenance of memory may depend on the persistence of the antigen (53). If both ideas are correct, it is possible that a tolerance to trophoblast does not develop because trophoblast antigens may not have access to the fetal circulation at the critical periods or, more likely, that tolerance is not maintained because the placenta is a transitory organ, and the adult animal is not continually exposed to these antigens. The loss of the transgene in our syngeneic mothers may provide us with a model to study the nature of tolerance to trophoblast antigens.

## *Is the Loss of Transgenic Pups the Result of Immunologically Mediated Abortion?*

The data we have presented strongly suggest that the failure to transmit the transgene is the result of immunologic phenomena. However, other transgenic models that had been reported to be models of immunologically mediated autoimmunity were subsequently shown to be nonimmunologic (54–57). These models most clearly demonstrated that high overexpression of class I MHC was directly toxic to cells and that the cells expressing the transgene underwent nonimmunologically mediated cell death. The mechanism underlying this phenomenon is not known.

We believe that several observations in our system suggest that our loss of transmission of the transgene is not due to nonimmunologic destruction of the trophoblast by overexpression of class I. First, in many of the previous models, several hundred copies of the transgene had been incorporated into the genome. It is possible that so many gene copies simply overwhelm the synthetic capacity of the cell, resulting in cell death. Alternatively, multiple tandem copies of the transgene could disrupt other critical genes or regulatory elements in the genome. An examination of our transgenic blots indicates that by comparison with the c-*fos* endogenous gene, relatively few copies of the transgene have been incorporated into the genome (Fig. 15.2). Consequently, the cell may be able to manage the product of the transgene more efficiently. A similar observation that low copy number results in more viable cellular expression of a transgene has been reported in other systems (58).

The use of the β2 microglobulin knockout mice provides the second line of evidence that we have developed an immunologically mediated model of abortion. These mice are bred onto a C57/BL6 background and in our particular colony are backcrossed 4 times (~95% BL6). The remainder of the genome is contributed by 129 strain mice that were the source of the embryonic stem cells used to create the knockout mutation (51). Consequently, the β2 microglobulin knockouts should behave essentially like a C57/BL6 mouse. In our studies the C57/BL6 mice eventually lost the ability to transmit the transgene. However, the β2 microglobulin knockout mice successfully transmitted the transgene even after multiple pregnancies. If expression of the transgene were nonspecifically killing trophoblast or if there were some genetic factor that resulted in nonimmunologic abortion, then one would also expect the β2 microglobulin knockout mice to fail to transmit the transgene. As the only known defect in the β2 microglobulin mice is the lack of functional $CD8^+$ T cells, our results strongly suggest that the failure to transmit the transgene in C57/BL6 mothers is due to the presence of $CD8^+$ T cells.

Finally, if the loss of the transgene were due to a nonspecific effect of the transgene on development or were due to a direct toxic effect on trophoblast, then one would not expect to see a slow decrease of transmission of the transgene with increasing parity; rather, we should be unable to transmit the transgene initially. The fact that we slowly lose transmission of the transgene is consistent with the idea that we are immunizing mothers against pregnancy.

Ultimately, definitive proof that the loss of the transgene is the result of immunologically mediated abortion will require the demonstration of cytotoxic effector cells in the aborting mothers. Also, it should be possible to restore the ability to transmit the transgene in aborting mothers by the depletion of $CD8^+$ cells before or during pregnancy. These experiments are currently under way.

Although it is reasonable to assume that the abortions are being mediated by cytotoxic $CD8^+$ effector cells directly lysing class I-bearing trophoblast, it is known that trophoblast is highly resistant to such cytotoxic effectors (59). We plan to test the ability of cytotoxic effectors to lyse our transgenic trophoblast in vitro. We also plan to look at the cytokines at the fetomaternal interface. Several lines of investigation now indicate that cytokines may have a beneficial or an adverse effect on pregnancy outcome (44, 60). The possibility exists that the abortions may be mediated by $CD8^+$ cells secreting IFNγ or other deleterious lymphokines, rather than by direct cytotoxic effects.

## Summary

We have generated a transgenic mouse model in which the $D^d$ gene has been placed under the control of the c-*fos* promoter. The gene is expressed in numerous adult tissues, but is also expressed in the placenta, particularly in the endovascular trophoblast of the labyrinthine region. Expression of the transgene results in increased levels of spontaneous abortion. The transmission of the transgene is decreased in immunologically competent—but not in immunologically incompetent—mothers, and the loss of the ability to transmit the transgene increases with parity of the mother. These data suggest that increased expression of classical class I MHC on murine trophoblast, particularly on those populations in direct contact with the maternal blood, compromises pregnancy and that down-regulation of MHC by trophoblast is necessary to prevent immunologically mediated rejection of the fetoplacental unit.

*Acknowledgments.* The authors would like to thank Dr. Stephen Hedrick, who directs the transgenic facility at the UCSD Cancer Center, for allowing us use of it. We would also like to thank Dr. Richard Dutton for continued support and encouragement on this project. The authors thank Dr. Beverly Koller for originally supplying the β2 microglobulin knockout mice, Dr. Eillen Adamson for supplying the c-*fos* gene, and Dr. Michael Steinmetz for providing the $D^d$ gene. We also thank Edda Roberts for technical assistance in maintaining the breeding colonies. This work was supported by NIH Grants AI-31870 and AI-33204. The β2 microglobulin breeding colony is supported by NIH Grant AI-23287. The transgenic facility is supported by NCI Cancer Center Core Grant CA-23100-12.

## *References*

1. Faulk WP, Temple A. Distribution of beta-2 microglobulin and HLA in chorionic villi of human placentae. Nature 1976;262:799–802.
2. Hunt JS, Orr HT. HLA and maternal-fetal recognition. FASEB J 1992;6:2344–8.
3. Hunt JS, Fishback JL, Andrews GK, Wood GW. Expression of class I HLA genes by trophoblast cells. Analysis by in situ hybridization. J Immunol 1988;140:1293–9.
4. Hunt JS, Fishback JL, Chumbley G, Loke YW. Identification of class I MHC mRNA in human first trimester trophoblast cells by in situ hybridization. J Immunol 1990;144:4420–5.
5. Sunderland CA, Redman CWG, Stirrat GM. HLA-A,B,C antigens are expressed on nonvillous trophoblasts of early human placenta. J Immunol 1981;127:2614–5.
6. Redmann CW, McMichael AJ, Stirrat GM, Sunderland CA, Ting A. Class 1 major histocompatibility complex antigens on human extra-villous trophoblast. Immunology 1984;52:457–68.
7. Heinrichs H, Orr HT. HLA non-A,B,C class I genes: their structure and expression. Immunol Res 1990;9:265–74.
8. Wei XH, Orr HT. Differential expression of HLA-E, HLA-F, and HLA-G transcripts in human tissue. Hum Immunol 1990;29:131–42.

9. Ellis SA, Sargent IL, Redman CW, McMichael AJ. Evidence for a novel HLA antigen found on human extravillous trophoblast and a choriocarcinoma cell line. Immunology 1986;59:595–601.
10. Kovats S, Main EK, Librach C, Stubblebine M, Fisher SJ, DeMars R. A class I antigen, HLA-G, expressed in human trophoblasts. Science 1990;248:220–3.
11. Yelavarthi KK, Fishback JL, Hunt JS. Analysis of HLA-G mRNA in human placental and extraplacental membrane cells by in situ hybridization. J Immunol 1991;146:2847–54.
12. Chiang MH, Steuerwald N, Steinleitner LA, Main EK. Detection of HLA-G mRNA in human sperm using RT-PCR [Abstract]. J Immunol 1993;150:283A.
13. Colburn GT, Main EK. Immunology of the maternal-fetal interface in normal pregnancy. Semin Perinatol 1991;15:196–205.
14. Philpott KL, Rastan S, Brown S, Mellor AL. Expression of H-2 class I genes in murine extra-embryonic tissues. Immunology 1988;64:479–85.
15. Head JR, Drake BL, Zuckermann FA. Major histocompatibility antigens on trophoblast and their regulation: implications in the maternal-fetal relationship. Am J Reprod Immunol Microbiol 1987;15:12–8.
16. Redline RW, Lu CY. Localization of fetal major histocompatibility complex antigens and maternal leukocytes in murine placenta. Lab Invest 1989;61: 27–36.
17. Jaffe L, Robertson EJ, Bikoff EK. Distinct patterns of expression of MHC class I and $\beta_2$-microglobulin transcripts at early stages of mouse development. J Immunol 1991;147:2740–50.
18. Jaffe L, Jeannotte L, Bikoff EK, Robertson EJ. Analysis of $\beta_2$-microglobulin gene expression in the developing mouse embryo and placenta. J Immunol 1990;145:3474–82.
19. Zuckermann FA, Head JR. Expression of MHC antigens on murine trophoblast and their modulation by interferon. J Immunol 1986;137:846–53.
20. Mattsson R, Holmdahl R, Scheynius A, Bernadotte F, Mattsson A, Van der Meide PH. Placental MHC class I antigen expression is induced in mice following in vivo treatment with recombinant interferon-gamma. J Reprod Immunol 1991;19:115–29.
21. Zuckermann FA, Head JR. Murine trophoblast resists cell-mediated lysis, I. Resistance to allospecific cytotoxic T lymphocytes. J Immunol 1987;139: 2856–64.
22. Kanbour-Shakir A, Zhang X, Rouleau A, et al. Gene imprinting and major histocompatibility complex class I antigen expression in the rat placenta. Proc Natl Acad Sci USA 1990;87:444–8.
23. Macpherson TA, Ho HN, Kunz HW, Gill TJ III. Localization of the Pa antigen on the placenta of the rat. Transplantation 1986;41:392–4.
24. Misra DN, Kunz HW, Gill TJ III. MHC class I antigens in rat pregnancy: biochemical comparison between the pregnancy-associated (Pa) antigen and the classic class I MHC antigen RT1.Aa in the rat. Transplant Proc 1989; 21:3271–2.
25. Kanbour A, Ho HN, Misra DN, MacPherson TA, Kunz HW, Gill TJ III. Differential expression of MHC class I antigens on the placenta of the rat. A mechanism for the survival of the fetal allograft. J Exp Med 1987;166: 1861–82.

26. Hunt JS, Andrews GK, Wood GW. Normal trophoblasts resist induction of class I HLA. J Immunol 1987;138:2481–7.
27. Deschamps J, Meijlink F, Verma IM. Identification of a transcriptional enhancer element upstream from the proto-oncogene fos. Science 1985;230: 1174–7.
28. Treisman R. Transient accumulation of c-*fos* RNA following serum stimulation requires a conserved 5′ element and c-*fos* 3′ sequences. Cell 1985;42: 889–902.
29. Deschamps J, Mitchell RL, Meijlink F, Kruijer W, Schubert D, Verma IM. Proto-oncogene fos is expressed during development, differentiation, and growth. Cold Spring Harb Symp Quant Biol 1985;50:733–45.
30. Adamson ED, Meek J, Edwards SA. Product of the cellular oncogene, c-*fos*, observed in mouse and human tissues using an antibody to a synthetic peptide. EMBO J 1985;4:941–7.
31. Adamson ED. Expression of proto-oncogenes in the placenta. Placenta 1987; 8:449–66.
32. Sher BT, Nairn R, Coligan JE, Hood LE. DNA sequence of the mouse H-2D$^{d}$ transplantation antigen. Proc Natl Acad Sci USA 1985;82:1175–9.
33. Kimura A, Israel A, Le Bail O, Kourilsky P. Detailed analysis of the mouse H2-K$^{b}$ promoter: enhancer-like sequences and their role in the regulation of class I gene expression. Cell 1986;44:261–72.
34. Ganguly S, Vasavada HA, Weissman SM. Multiple enhancer-like sequences in the HLA-B7 gene. Proc Nat Acad Sci USA 1989;86:5247–51.
35. Oudejans CBM, Krimpenfort P, Ploegh HL, Meijer CJLM. Lack of expression of HLA-B27 gene in transgenic mouse trophoblast. J Exp Med 1989;169: 447–56.
36. Rossant J. The mechanism of survival of the fetal allograft. Ann Immunol (Paris) 1984;135D:312–5.
37. Rossant J, Mauro VM, Croy BA. Importance of trophoblast genotype for survival of interspecific murine chimaeras. J Embryol Exp Morphol 1982;69: 141–9.
38. Croy BA, Rossant J, Clark DA. Histological and immunological studies of postimplantation death of *Mus caroli* embryos in the *Mus musculus* uterus. J Reprod Immunol 1982;4:277–93.
39. Croy BA, Rossant J, Clark DA. Effects of alterations in the immunocompetent status of *Mus musculus* females on the survival of transferred *Mus caroli* embryos. J Reprod Fertil 1985;74:479–89.
40. Chaouat G, Kiger N, Wegmann TG. Vaccination against spontaneous abortion in mice. J Reprod Immunol 1983;5:389–92.
41. Chaouat G, Kolb JP, Kiger N, Stanislawski M, Wegmann TG. Immunologic consequences of vaccination against abortion in mice. J Immunol 1985;134: 1594–8.
42. Kiger N, Chaouat G, Kolb JP, Wegmann TG, Guenet JL. Immunogenetic studies of spontaneous abortion in mice. Preimmunization of females with allogeneic cells. J Immunol 1985;134:2966–70.
43. Szekeres-Bartho J, Kinsky R, Kapovic M, Chaouat G. Complete Freund adjuvant treatment of pregnant females influences resorption rates in CBA/J × DBA/2 matings via progesterone-mediated immunomodulation. Am J Reprod Immunol 1991;26:82–3.

44. Gendron R, Baines M. Infiltrating decidual natural killer cells are associated with spontaneous abortion in mice. Cell Immunol 1988;113:261–7.
45. Chaouat G, Menu E, Clark DA, Dy M, Minkowski M, Wegmann TG. Control of fetal survival in CBA × DBA/2 mice by lymphokine therapy. J Reprod Fertil 1990;89:447–58.
46. Kinsky R, Delage G, Rosin N, Thang MN, Hoffmann M, Chaouat G. A murine model of NK cell mediated resorption. Am J Reprod Immunol 1990; 23:73–7.
47. Drake BL, Head JR. Murine trophoblast can be killed by lymphokine-activated killer cells. J Immunol 1989;143:9–14.
48. Kaye J, Hsu M-L, Sauron ME, Jameson S, Gascoigne N, Hedrick SM. Selective development of CD4+ T cells in transgenic mice expressing a class II MHC-restricted antigen receptor. Nature 1989;341:746–9.
49. Scully PA, Steinman HK, Kennedy C, Trueblood K, Frisman DM, Voland JR. AIDS-related Kaposi's sarcoma displays differential expression of endothelial surface antigens. Am J Pathol 1988;130:244–51.
50. Voland JR, Frisman DM, Baird SM. Presence of an endothelial antigen on the syncytiotrophoblast of human chorionic villi: detection by a monoclonal antibody. Am J Reprod Immunol Microbiol 1986;11:24–30.
51. Koller BH, Marrack P, Kappler JW, Smithies O. Normal development of mice deficient in beta 2 microglobulin, MHC class I proteins, and CD8+ T cells. Science 1990;248:1227–30.
52. Juretic A, Knowles BB. Frequency of SV40-specific cytotoxic T-lymphocyte precursors in two SV40 T-antigen transgenic mouse lines. APMIS 1991;99: 213–8.
53. Gray D, Skarvall H. B-cell memory is short-lived in the absence of antigen. Nature 1988;336:70–3.
54. Allison J, Campbell IL, Morahan G, Mandel TE, Harrison LC, Miller JF. Diabetes in transgenic mice resulting from over-expression of class I histocompatibility molecules in pancreatic beta cells. Nature 1988;333:529–33.
55. Allison J, Malcolm L, Culvenor J, Bartholomeusz RK, Holmberg K, Miller JF. Overexpression of beta 2–microglobulin in transgenic mouse islet beta cells results in defective insulin secretion. Proc Natl Acad Sci USA 1991;88: 2070–4.
56. Harrison LC, Campbell IL, Allison J, Miller JF. MHC molecules and beta-cell destruction: immune and nonimmune mechanisms. Diabetes 1989;38: 815–8.
57. Miller JF, Morahan G, Slattery R, Allison J. Transgenic models of T-cell self-tolerance and autoimmunity. Immunol Rev 1990;118:21–35.
58. Woodward JG, Martin WD, Stevens JL, Egan RM. A transgenic mouse model for tolerance in the eye: an immunologically privileged site [Abstract]. J Immunol 1993;150:5A.
59. Zuckermann FA, Head JR. Possible mechanism of non-rejection of the feto-placental allograft: trophoblast resistance to lysis by cellular immune effectors. Transplant Proc 1987;19:554–6.
60. Chaouat G, Menu E, Athanassakis I, Wegmann TG. Maternal T cells regulate placental size and fetal survival. Reg Immunol 1988;1:143–8.

# Part VI

# Immunological Aspects of Human Infertility

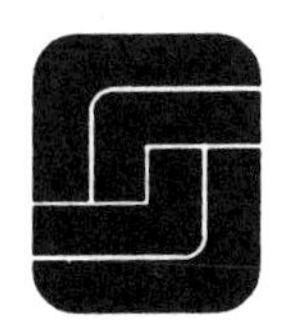

# 16

# Mechanism of Self-Tolerance and Autoimmune Disease Pathogenesis: Analysis Based on Ovarian Autoimmune Models

KENNETH S.K. TUNG, YA-HUAN LOU, AND HEDY SMITH

*Premature ovarian failure* (POF) is a human disease characterized by amenorrhea before the age of 40 (or 35), regardless of whether the disease occurs before or after puberty (1). The known causes of POF have included genetic abnormalities, viral infections, and chemical and drug toxicity. However, these factors account for only a small subset of POF patients. For the majority of POF patients, the cause of the disease is unknown, and an immunologic basis has been considered. A recent review estimated that POF occurs in 0.3% of U.S. women of reproductive age, and about 14,000 women are at risk for autoimmune ovarian disease (1). This is likely to be an overestimate since the two classical criteria of organ-specific autoimmunity—serum autoantibodies and inflamed ovaries (oophoritis)—have rarely been documented in POF patients (reviewed in 2). Notwithstanding the uncertain incidence of immunologic POF, the immunologic basis for POF rests on solid grounds. In POF patients antiovarian autoantibodies and oophoritis have been detected; moreover, autoimmune oophoritis occurs as part of the polyendocrine autoimmunity syndromes, along with thyroiditis, insulitis, adrenalitis, myasthenia gravis, and others (1). In addition, autoimmune oophoritis can be readily induced in experimental animals.

Experimental disease of the ovary (autoimmune oophoritis) was induced in rats immunized with heterologous or homologous ovarian antigens in complete Freund's adjuvant (3, 4). Contraceptive vaccine research in the 1980s then described ovarian failure in several animal species immunized with heterologous zona pellucida or ZP3 (5–8), and in 1992 autoimmune mechanism was established as the basis for the ovarian failure since oophoritis and ovarian failure develop in mice immunized with a novel peptide from murine ZP3 (9).

Meanwhile, a completely different experimental approach to autoimmune oophoritis was explored systematically by Nishizuka and his colleagues (10–15). Their first report in 1967 described ovarian failure in mice thymectomized soon after birth, usually on day 3 (D3TX) (10). Subsequent reports documented the autoimmune basis for the ovarian failure and further indicated that several other manipulations of the immune system could also lead to a high incidence of autoimmune oophoritis (reviewed in 16). We recently discovered that the mere transfer of T cells from normal mice into syngeneic athymic *nu/nu* recipients also elicits organ-specific autoimmune diseases (17).

In the past few years, we have exploited models of autoimmune oophoritis to investigate the fundamental problems of self-tolerance and autoimmune disease pathogenesis. Our findings suggest that self-tolerance is maintained through the balance between pathogenic effector T cells and regulatory (suppressor) T cells. When the homeostasis is perturbed, either by depletion of the regulatory cells or activation of the pathogenic effector cells, autoimmune diseases follow.

## Mechanisms That Maintain Self-Tolerance to Ovarian and Gastric Autoantigens

The highly exciting research on self-tolerance in the past decade has employed several experimental approaches (18–30). One approach examines the fate of *T cell receptor* (TCR) Vβ segment-specific T cells that recognize endogenous superantigens; a second relies on analysis of the phenotype and function of the T cell population in mice with a self-reactive transgenic TCR. A third approach analyzes the T cell response and pathology in mice with transgenic antigens or cytokines expressed in a unique extrathymic site; and a fourth studies the biochemical events of signal transduction that lead to T cell anergy, predominantly in vitro.

The conclusion drawn from this large body of work indicates that multiple mechanisms may be operative to maintain tolerance, and these mechanisms include clonal deletion of the self-reactive T cells through apoptosis in the thymus or in the periphery and induction of T cell anergy (unresponsiveness) by suboptimal stimulation by antigen and/or antigen-presenting cells. In contrast, these studies did not address the mechanism of regulation of effector T cells by suppressor T cells. We have recently dissected the possible mechanisms of tolerance in a simple animal model in which we studied self-antigens expressed in physiological quantities and locations, used disease pathology and autoantibodies as experimental end points, and evaluated the tolerance state of normal T cells, rather than transgenic T cells (17). The key findings of our study are summarized in Table 16.1 and are detailed below.

TABLE 16.1. Capacity of normal murine splenocytes and thymocytes to induce ovarian and gastric autoimmunity in syngeneic BALB/c *nu/nu* or BALB/c *SCID* mice.

| Cells transferred (no. × $10^{-6}$/ mouse) | Pathology | | | | | | | | Antibody | |
|---|---|---|---|---|---|---|---|---|---|---|
| | Oophoritis | | | | Gastritis | | | | | |
| | | Severity | | | | Severity | | | | |
| | Incidence (%) | 1 | 2 | 3 | Incidence (%) | 1 | 2 | 3 | Ovary | Stomach |
| A. Normal BALB/c *nu*/+ mice | | | | | | | | | | |
| None | 0/6 (0) | | | | 0/6 (0) | | | | 0/6 | 0/6 |
| B. From BALB/c *nu*/+ mice to BALB/c *nu/nu* mice | | | | | | | | | | |
| Day 3 spleen cells | | | | | | | | | | |
| Untreated (20) | 11/15 (73) | 1 | 2 | 8 | 8/15 (53) | 0 | 3 | 5 | 8/11 | 1/5 |
| CD4-depleted (20) | 0/6 (0) | | | | 0/6 (0) | | | | 0/6 | 0/6 |
| CD8-depleted (20) | 4/5 (80) | 0 | 1 | 3 | 4/5 (80) | 0 | 1 | 3 | 3/4 | 4/5 |
| Day 7 spleen cells | | | | | | | | | | |
| Untreated (30) | 5/8 (63) | 2 | 2 | 1 | 4/8 (50) | 0 | 4 | 0 | 1/3 | 5/8 |
| Adult spleen cells | | | | | | | | | | |
| Untreated (25) | 0/8 (0) | | | | 0/8 (0) | | | | 0/8 | 0/8 |
| CD5-depleted (20) | 14/19 (74) | 2 | 6 | 6 | 6/19 (32) | 3 | 2 | 1 | 1/4 | 0/4 |
| Day 3 (20) and adult (20) spleen cells | | | | | | | | | | |
| Untreated | 0/7 (0) | | | | 1/7 (14) | 0 | 0 | 1 | 0/7 | 0/7 |
| Day 7 (30) and adult (30) spleen cells | | | | | | | | | | |
| Untreated | 1/15 (7) | 0 | 0 | 1 | 0/15 (0) | | | | 0/15 | 0/15 |
| Day 3 thymocytes | | | | | | | | | | |
| Untreated (50) | 7/9 (78) | 0 | 2 | 5 | 6/9 (66) | 1 | 2 | 3 | 3/5 | 2/5 |

TABLE 16.1. *Continued*

| Cells transferred (no. × $10^{-6}$/mouse) | Pathology | | | | | | | | Antibody | |
|---|---|---|---|---|---|---|---|---|---|---|
| | Oophoritis | | | | Gastritis | | | | | |
| | | Severity | | | | Severity | | | | |
| | Incidence (%) | 1 | 2 | 3 | Incidence (%) | 1 | 2 | 3 | Ovary | Stomach |
| Adult thymocytes | | | | | | | | | | |
| Untreated (30) | 8/11 (73) | 6 | 1 | 1 | 8/11 (73) | 2 | 3 | 3 | 3/9 | 7/9 |
| J11D-depleted (7) | 6/8 (75) | 5 | 1 | 0 | 6/8 (75) | 2 | 2 | 2 | 4/5 | 4/5 |
| CD8-depleted (20) | 5/8 (63) | 4 | 1 | 0 | 8/8 (100) | 0 | 2 | 6 | 1/8 | 8/8 |
| CD4-depleted (20) | 1/9 (11) | 0 | 1 | 0 | 0/9 (0) | | | | 0/9 | 0/9 |
| C. From BALB/c CB17 mice to *SCID* mice | | | | | | | | | | |
| Adult spleen cells | | | | | | | | | | |
| Untreated (25) | 0/9 (0) | | | | 0/9 (0) | | | | 0/9 | 0/9 |
| CD5-depleted (25) | 4/6 (67) | 1 | 1 | 2 | 3/6 (50) | 1 | 2 | 0 | NS | NS |

The results are reproducible; data of each group were pooled from at least 2–3 independent experiments. Significant differences in disease prevalence among experimental groups were determined by $x^2$ analysis. Recipients of adult splenocytes had less disease than recipients of 3-day-old splenocytes (oophoritis, $P = 0.04$; gastritis, $P = 0.03$) or 7-day-old splenocytes (oophoritis, $P = 0.03$; gastritis, $P = 0.08$). CD4-depleted 3-day-old splenocytes transferred less disease than CD8-depleted 3-day-old splenocytes (oophoritis, $P = 0.03$; gastritis, $P = 0.03$). Recipients of CD5-depleted adult spleen cells had more disease than those of untreated adult spleen cells in BALB/c *nu/nu* recipients (oophoritis, $P = 0.002$; gastritis $P = 0.2$) or BALB/c.*SCID* recipients (oophoritis, $P = 0.02$; gastritis, $P = 0.08$). Recipients of untreated adult thymocytes had a similar frequency of disease as recipients of J11D antibody-depleted adult thymocytes (oophoritis, $P = 0.7$; gastritis, $P = 0.7$) or CD8-depleted adult thymocytes (oophoritis, $P = 0.98$; gastritis, $P = 0.3$), whereas CD4-depleted adult thymocytes had less disease (oophoritis, $P = 0.02$; gastritis, $P = 0.004$). Recipients of mixed 3-day-old and adult spleen cells had less disease than recipients of 3-day-old spleen cells alone (oophoritis, $P = 0.006$; gastritis, $P = 0.2$). Recipients of mixed 7-day-old and adult spleen cells had less disease than recipients of 7-day-old spleen cells alone (oophoritis, $P = 0.0002$; gastritis, $P = 0.003$).

*Source:* Reprinted with permission from Smith, Lou, Lacy, and Tung (17), Copyright 1992, The Journal of Immunology.

First, we transferred $CD4^+$ $CD8^-$ thymocytes from normal sex-matched adult BALB/c *nu*/+ mice to athymic BALB/c *nu*/*nu* mice. Two months later, 75% of the recipients developed significant autoimmune oophoritis and gastritis. Tissue pathology occurrence was associated with the detection of serum autoantibodies to the respective organ-specific antigens, including oocyte antigen and the gastric parietal cell $H^+K^+$-dependent ATPase. Since $CD4^+$ $CD8^-$ thymocytes represent a subpopulation of thymoctyes that have matured beyond the known stage of clonal deletion, we conclude that pathogenic self-reactive T cells can mature in the adult thymus.

Second, we discovered that neonatal splenic $CD4^+$ $CD8^-$ T cells readily transferred similar autoimmune diseases to *nu*/*nu* syngeneic recipients, whereas T cells from adult spleens did not. In an earlier study based on $V\beta11^+$ T cells that recognize self-IE and endogenous retroviral peptides, we showed that $V\beta11^+$ T cells are not deleted in the neonatal thymus and that the T cell repertoire in the neonatal spleen is enriched in $V\beta11^+$ T cells (31). These two studies collectively indicate that tolerance for self-antigens, including those relevant in autoimmune diseases, is ontogenetically regulated. Specifically, the neonatal repertoire is enriched in self-reactive T cells that for reasons that remain unclear, do not require regulation. Since thymectomy soon after birth should skew the T cell repertoire to one enriched in self-reactive T cells, this scenario could explain why autoimmune diseases might develop spontaneously in the D3TX mice. Indeed, $V\beta11^+$ self-reactive T cells are enriched in adult D3TX mice (31). As discussed below, this is only one of the possible mechanisms for autoimmune disease occurrence in the D3TX mice.

Third, although adult spleen T cells do not transfer autoimmune oophoritis and gastritis to athymic *nu*/*nu* recipients, the fraction of adult splenic $CD4^+$ T cells expressing a low level of cell surface CD5 readily transfer severe autoimmune diseases and autoantibodies. In this experiment we followed the protocol established by Sagaguchi et al. (14) by treating the splenic T cells with CD5 antibody and complement, then injecting the residual $CD5^{low}$ T cells into *nu*/*nu* recipients. Thus, self-reactive T cells are not deleted in the peripheral repertoire of adult mice. Instead, they appear to have a unique phenotype that includes the expression of a low level of surface CD5, a known T cell activation marker. In a similar study on normal rats, pathogenic self-reactive $CD4^+$ T cells were reported to have the $T_{H1}$ subset, as they produce *interleukin-2* (IL-2) and not *interleukin-4* (IL-4) upon activation (32).

Fourth, to investigate the existence of regulatory T cells that may down-regulate self-reactive T cells, we cotransferred adult splenic T cells and neonatal T cells into *nu*/*nu* recipients. It was found that autoimmune diseases that would be induced by neonatal T cells were abrogated by the adult T cells (Table 16.1). Therefore, pathogenic autoreactive T cells exist in normal adult mice, but they are normally nonfunctional. It is possible that the suppressor T cells render the self-reactive T cells nonfunctional. However, since function returns when the suppressor cells are removed, the nonfunctional state of self-reactive T cells is not permanent. Additional experimental findings that support the existence of suppressor T cells and analyze their phenotype are described below under the D3TX autoimmune disease model.

## Potential Mechanisms of Autoimmune Disease Induction

The analysis of normal T cell subpopulations, described above, indicates that pathogenic self-reactive T cells exist in normal mice and that they are regulated by suppressor T cells. The balance of these two cell populations ensures that tolerance is maintained in normal adult mice. When the balance is tipped in favor of effector T cells, autoimmune diseases could ensue in the genetically susceptible host. This scenario is illustrated by two experimental models of autoimmune oophoritis: (i) depletion of suppressor cells in D3TX mice and (ii) activation of effector cells through molecular mimicry at the level of an ovarian T cell peptide.

## Autoimmune Oophoritis Due to Regulatory T Cell Depletion

Normal (C57BL/6 × A/J)F1 mice thymectomized between days 1 and 4 after birth develop autoimmune oophoritis spontaneously. The disease can be readily transferred to young recipients by $CD4^+$, but not $CD8^+$, T cells (33). Of even greater significance, D3TX-induced autoimmune oophoritis is prevented if the D3TX mice are infused with T cells from normal adult mice, as long as the cells are given to D3TX mice that are younger than 10–12 days of age. Analysis of the normal adult T cells that are responsible for disease suppression indicates that they also are $CD4^+$ $CD8^-$ (33). Moreover, they express high levels of CD5, in contrast with the $CD5^{low}$ pathogenic T cells in adults (see above). The finding on disease suppression by normal T cells suggests that autoimmune diseases occur because of deprivation of regulatory T cells by D3TX. This conclusion is further substantiated by two findings described earlier: (i) Adult T cells transfer disease after $CD5^{high}$ T cells have been depleted; and (ii) cotransfer with normal adult spleen T cells abrogates disease transfer by normal neonatal T cells (17). Therefore, as the pathogenic basis of D3TX autoimmunity, two mechanisms are invoked: (i) the enrichment and expansion of the self-reactive pathogenic T cells of the neonatal repertoire, as discussed earlier, and (ii) the deprivation of suppressor T cells.

## Autoimmune Oophoritis Due to Activation of Pathogenic T Cells by Molecular Mimicry

Autoimmune oophoritis can also develop if the balance is tipped in favor of an expansion of pathogenic T cells. This occurs when effector T cells are expanded and become activated through molecular mimicry at the level of T cell peptide recognition. This has been demonstrated in a new model of autoimmune oophoritis induced in mice by immunization with a unique ovarian peptide in adjuvant.

*Zona pellucida* (ZP) is the acellular matrix that surrounds the developing and ovulated oocytes. It contains three major proteins: ZP1, ZP2, and ZP3 (34). ZP3 is a glycoprotein with a polypeptide backbone of 424 amino acids and polysaccharides that account for 40% of its molecular mass. Certain O-linked glycoconjugates are known to function as the sperm-binding structure during fertilization (35). In a study on murine ZP3 as an experimental contraceptive vaccine antigen, a 15-mer peptide of murine ZP3 (328–342) was generated that contained the known B cell epitope of ZP3 (336–342) (36). Subsequently, when we immunized mice with ZP3 (328–342) in complete Freund's adjuvant, with or without the carrier protein, the mice developed a high incidence of severe autoimmune oophoritis (9). In addition, their T cells proliferated in response to the peptide, and antibodies to ZP3 were detected in their sera and bound to the ovarian ZP. By studying the truncated ZP3 (328–342) peptides, we were able to assign the shortest oophoritogenic peptide to the 8-mer sequence, ZP3 (330–337). This peptide overlaps by 2 residues with the known 7-mer B cell epitope (336–342) established earlier (9, 36).

That the autoimmune disease of the ovary is mediated by $CD4^+$ T cells is demonstrated by adoptive transfer experiments—initially, with bulk lymph node $CD4^+$ T cells; later, by ZP3 peptide-specific T cell lines and T cell clones (9, 37). The oophoritogenic $CD4^+$ T cell lines and T cell clones produce IL-2, *interferon* $\gamma$ (IFN$\gamma$), and *tumor necrosis factor* (TNF), but not IL-4. Moreover, oophoritis transfer was significantly attenuated by cotransfer of antibody to TNF, but not by cotransfer of antibody to IFN$\gamma$.

The opportunity to study an autoimmune disease elicited by the well-defined ZP3 self-peptide led to our recent investigation on molecular mimicry at the level of the T cell peptide. We asked whether nonovarian peptides could be found that mimic the ZP3 peptide with respect to specificity of T cell recognition and immunopathological consequence of the immune response. And, if we were successful, what would be the mechanism behind peptide mimicry? These studies were timely in view of the current research interest on the immunochemistry of nominal peptides as T cell epitopes (reviewed in 38).

We searched the protein sequence library for peptides with amino acid sequences that simulate the nanomer ZP3 peptide 330–339 and uncovered

TABLE 16.2. Cross-reaction between murine ZP3 peptide and murine AChR δ-chain peptide for induction of autoimmune oophoritis and ovary-bound anti-ZP antibody response and for stimulation of an oophoritogenic T cell clone.

| | | | Oophoritis | | | | | | |
|---|---|---|---|---|---|---|---|---|---|
| | | | Incidence | Severity | | | | | Proliferative responses of J3 |
| Adjuvant | Peptide | Sequence | (%) | 1 | 2 | 3 | 4 | ZP-bound IgG | clone (Δ cpm) |
| CFA | None | | 0/5 | — | — | — | — | 0/5 | 302 |
| CFA | ZP3 330–338 | *N*SSSS*QFQI* | 5/5 (100) | 0 | 2 | 3 | 0 | 5/5 | 121,222 |
| CFA | mACRδ 120–128 | *N*NNDGS*FQI* | 3/5 (60) | 1 | 2 | 0 | 0 | 3/5 | 95,659 |
| CFA | mACRγ 117–124 | *N*NVDGV*F*EV | 0/5 | — | — | — | — | 0/5 | −2,097 |
| CFA | None | | 0/5 | — | — | — | — | 0/5 | 243 |
| CFA | ZP3 330–338 | *N*SSSSQ*FQI* | 4/8 (80) | 1 | 1 | 2 | 0 | 5/5 | 51,364 |
| CFA | mACRδ 120–128 | *N*NNDGS*FQI* | 2/5 (40) | 1 | 1 | 0 | 0 | 3/5 | 15,397 |
| CFA | mACRγ 117–124 | *N*NVDGV*F*EV | 0/5 | — | — | — | — | 0/5 | −2,734 |
| ICFA | None | | 0/5 | — | — | — | — | 0/5 | 276 |
| ICFA | ZP3 330–338 | *N*SSSSQ*FQI* | 4/4 (100) | 1 | 1 | 2 | 0 | 4/4 | 249,314 |
| ICFA | mACRδ 120–128 | *N*NNDGS*FQI* | 3/4 (75) | 1 | 1 | 1 | 0 | 3/4 | 22,021 |
| ICFA | mACRγ 117–124 | *N*NVDGV*F*EV | 0/5 | — | — | — | — | 0/5 | −2,683 |

In 3 independent experiments adult female B6AF1 mice immunized with 50 μg of the peptide in adjuvant, or adjuvant alone, were studied 14 days later. Antibody to ZP is detected by direct immunofluorescence as intense IgG bound to ovarian ZP. The proliferative responses of the ZP3 (330–342)-specific oophoritogenic T cell clone, J3, to the peptides (30 μM) in presence of mitomycin-treated L cells transfected with Ia-$\alpha^k\beta^b$ are expressed as Δ cpm.

*Source:* Reprinted from Luo, Garza, Hung, and Tung (37), by copyright permission of the American Society for Clinical Investigation.

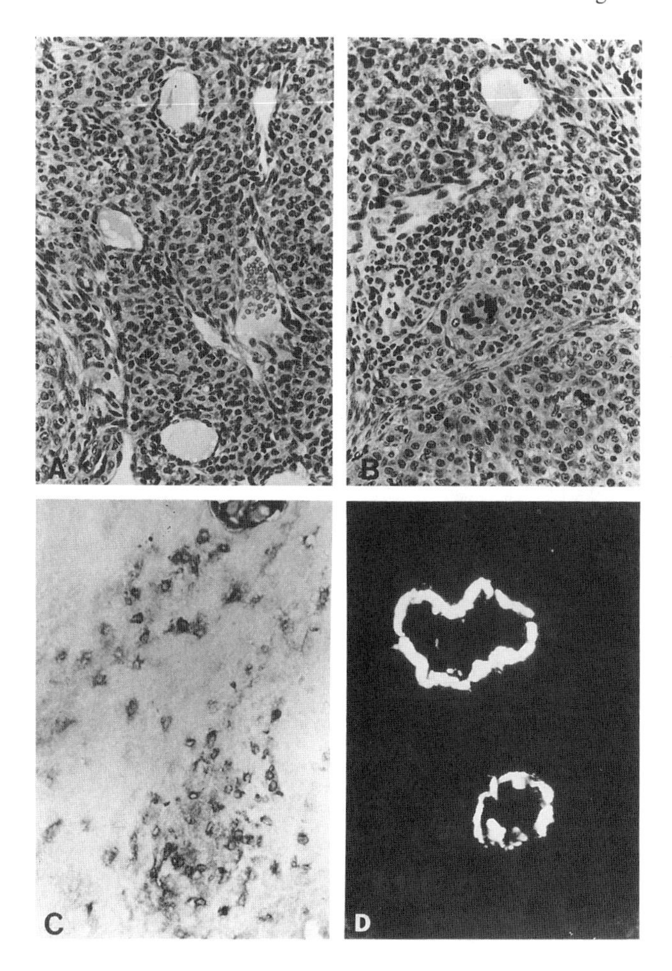


FIGURE 16.1. Ovarian immunopathology of a mouse immunized with the murine nicotinic AChR peptide, ACRδ (120–128). Compared with normal mouse ovary (*A*), which is free of identifiable inflammatory leukocytes, the ovarian interstitial space of a mouse immunized with ACRδ (120–128) in CFA contains infiltrates of mononuclear inflammatory cells that include a multinucleated giant cell (*B*). In a mouse immunized with the ACRδ (120–128) in CFA, T ($CD3^+$) cells identified by immunoperoxidase are found in the ovarian inflammatory infiltrate (*C*), and intense staining of mouse IgG (ZP antibody) is detectable by direct immunofluorescence in the ovarian ZP (*D*) (*A* and *B*, ×100; *C* and *D*, ×400). Reprinted from Luo, Garza, Hunt, and Tung (37), by copyright permission of the American Society for Clinical Investigation.

TABLE 16.3. Mapping the amino acid residues in ZP3 (330–338) important for induction of autoimmune oophoritis and anti-ZP antibodies and for stimulation of an oophoritogenic T cell clone.

| Study | Peptide sequence | Oophoritis: Incidence (%) | Severity 1 | Severity 2 | Severity 3 | Severity 4 | ZP-bound IgG | Proliferation of J3 clone (Δ cpm): 0.1 (Peptide, μM) | 1 | 10 | 100 |
|---|---|---|---|---|---|---|---|---|---|---|---|
| 1 | NSSSSQFQI | 9/10 (90) | 1 | 3 | 5 | 0 | 10/10 | 169 | 12,216 | 32,874 | 249,315 |
| | *A*SSSSQFQI | 0/9 (0) | 0 | 0 | 0 | 0 | 0/9 | −185 | 518 | 600 | 46,739 |
| | N*A*SSSQFQI | 8/9 (90) | 2 | 3 | 2 | 1 | 6/9 | −56 | 25,924 | 51,738 | 236,450 |
| | NS*A*SSQFQI | 10/10 (100) | 2 | 2 | 6 | 0 | 4/5 | 1,756 | 34,433 | 85,815 | 219,133 |
| | NSS*A*SQFQI | 5/5 (100) | 0 | 1 | 3 | 1 | 5/5 | 835 | 44,120 | 79,919 | 167,197 |
| | NSSS*A*QFQI | 6/10 (60) | 2 | 2 | 2 | 0 | 3/5 | −167 | 1,046 | 34,234 | 141,256 |
| | NSSSS*A*FQI | 0/9 (0) | 0 | 0 | 0 | 0 | 0/5 | −194 | −206 | 637 | 15,810 |
| | NSSSSQ*A*QI | 0/9 (0) | 0 | 0 | 0 | 0 | 0/9 | −218 | −225 | 86 | 398 |
| | NSSSSQF*A*I | 4/10 (40) | 3 | 1 | 0 | 0 | 2/5 | −211 | −225 | −33 | 7,048 |
| | NSSSSQFQ*A* | 5/5 (100) | 0 | 1 | 2 | 2 | 4/5 | 36,211 | 37,337 | 221,861 | 241,213 |
| 2 | *Q*SSSSQFQI | 0/5 (0) | — | — | — | — | 0/5 | −163 | 3 | 259 | 997 |
| | NSSSS*N*FQI | 0/5 (0) | — | — | — | — | 0/5 | −3 | 348 | 964 | 2,142 |
| | NSSSSQ*Y*QI | 3/5 (60) | 0 | 1 | 1 | 1 | 4/5 | 6,262 | 28,758 | 197,601 | 310,881 |
| | NSSSSQF*N*I | 0/5 (0) | — | — | — | — | 0/5 | 25 | 1,663 | 1,663 | 3,082 |
| 3 | *NAAAAQFQA* | 8/10 (80) | 3 | 3 | 2 | 0 | 8/10 | 27,284 | 109,783 | 158,031 | 89,808 |
| | *NAAAAAFQI* | 13/15 (87) | 3 | 7 | 3 | 0 | 15/15 | 30,769 | 64,960 | 124,713 | 121,605 |
| | NAAAAAFAA | 0/10 (0) | — | — | — | — | 0/7 | −2,615 | −2,039 | 1,295 | 11,456 |

The results of the in vivo experiments in studies 1 and 3 are pooled from 2 independent experiments; those of study 2 are from 1 experiment. The proliferative responses of the J3 clone to the various peptides are highly reproducible, and the result of a representative experiment is shown.
*Source:* Reprinted from Luo, Garza, Hunt, and Tung (37), by copyright permission of the American Society for Clinical Investigation.

one from the δ-chain and one from the γ-chain of the murine *acetylcholine receptor* (AChR) (37). Despite the limited number of residues shared between these peptides and ZP3 (330–339), we investigated the AChR peptides because of the known clinical association between POF and myasthenia gravis (1).

Autoimmune oophoritis was found to occur in mice immunized with the AChR δ-peptide that shares 4 amino acid residues with the ZP3 peptide (Table 16.2 and Fig. 16.1). In contrast, mice immunized with the AChR γ-peptide, which shares only 2 common residues with the ZP3 peptide, do not develop ovarian pathology. Since both the ZP3 peptide and the AChR δ-peptide are recognized by a ZP3-specific T cell clone and the responses in both cases are restricted to the same *major histocompatibility complex* (MHC) molecule, Ia-$\alpha^k\beta^b$, the ZP3 peptide and AChR δ-peptide must be cross-reactive at the level of the T cell receptor.

We next investigated the mechanism of antigen mimicry between these T cell peptides by determining the residues in the ZP3 peptide and AChR δ-peptide critical for disease induction and for recognition by pathogenic T cells. Of the 9 residues in the ZP3 peptide, 4 are critical, and of the 4, 3 are shared between the ZP3 peptide and the AChR δ-peptide (Table 16.3). To provide direct evidence for molecular mimicry at these residues, we studied nanomer polyalanine peptides into which different residues of the ZP3 peptide and/or the AChR δ-peptide have been inserted. The peptide with the 4 critical residues of the ZP3 peptide and the peptide with the common residues shared between ZP3 peptide and AChR δ-peptide both elicit severe oophoritis, and they stimulate the T cell clone to proliferate (Table 16.3) (37).

Thus, nonovarian peptides that share sufficient critical residues with the self-peptide of ZP3 have the potential to stimulate ZP3 peptide-specific T cell clones. Through the mechanism of molecular mimicry, a nonovarian peptide can elicit autoimmune oophoritis by clonal expansion and activation of ZP3-specific pathogenic T cells. Our ongoing research will attempt to identify foreign peptides that also mimic the function of the ZP3 peptide.

## A Novel Mechanism of Autoantibody Induction in Autoimmune Disease

The clinicopathological picture of an organ-specific autoimmune disease, such as POF, is thought to include (i) T cell response to the self-peptide, (ii) immunopathology in the target organ, (iii) serum autoantibodies to the antigens in the target organ, and (iv) binding of autoantibodies to the target organ. So far, we have shown that autoimmune oophoritis induced by the ZP3 peptide is associated with T cell response to the peptide and the development of oophoritis. Our next discovery was that activation of effector T cells can also lead to the spontaneous occurrence of autoantibodies against the native ZP3 protein (39).

The production of autoantibody in response to a pure T cell epitope was initially suggested by the detection of antibody to the ZP in mice immunized with (i) the truncated ZP3 peptide 330–338 lacking the intact B epitope 336–342 (Table 16.2), (ii) the AChR δ-peptide that mimics the ZP3 peptide (Table 16.2), and (iii) the polyalanine peptide that contains either the critical residues of ZP3 peptide or those of the AChR δ-chain peptide (Table 16.3). That an exclusive T cell epitope of murine ZP3 elicits autoantibodies against ZP3 outside the T cell peptide was confirmed by the following experimental findings (39). First, the truncated ZP3 peptides we investigated in Table 16.3 do not contain any additional B cell epitopes that cross-react with native ZP3. Second, since the zona-bound IgG is 400-fold enriched in antibody activity against the ZP, we conclude that the zona-bound IgG truly represents tissue-bound antibodies. Third, the antibodies produced by mice immunized with a T cell epitope can react with ZP3 by immunoblot, and they bind to native ZP in vivo. Fourth, and perhaps most important, endogenous ovarian antigens are required for antibody induction since autoantibodies to the ZP are not detected in ovariectomized mice immunized with the ZP3 peptides that lack the B epitope 336–342.

Our very recent study has further documented that ovarian pathology is not a requirement for induction of antibody to the ZP (Lou, Tung, unpublished observations). In other words, the antibody response is not strictly a sequelae of immune response to antigens released from diseased ovaries. In contrast, the finding strongly implies that in the normal mouse ovarian antigens, including ZP3, can reach extraovarian sites (including the regional lymphoid tissues). In this location and in the presence of activated, ZP3-specific helper T cells, the ovarian antigens would stimulate B cells to produce antibody to the ZP.

Our investigation on the phenomenon of autoantibody amplification has therefore uncovered a novel mechanism of autoantibody induction. This finding will impact on our understanding of autoantibodies in autoimmune diseases. In addition, it provides an approach to addressing the nature of physiological tolerance of autoreactive B cells.

## Conclusions

Based on our recent studies on experimental autoimmune oophoritis, we have obtained strong experimental evidence for the notion that autoimmune diseases could occur in situations where regulatory or suppressor T cells become defective or where pathogenic T cells are clonally expanded through molecular mimicry at the level of TCR. Moreover, T cell response triggered by a pure self-, or cross-reactive nonself, T cell epitope can spontaneously elicit a concomitant autoantibody response. The implication is that while the central event of autoimmune disease induction depends on mechanisms capable of driving a strong and persistent T cell response against self-peptides, once initiated, the T cell response itself is sufficient to bring about tissue pathology, B cell response, and autoantibody production.

## References


1. LaBarbera AR, Miller MM, Ober C, Rebar RW. Autoimmune etiology in premature ovarian failure. Am J Reprod Immunol Microbiol 1988;16:115–22.
2. Tung KSK, Lu CY. In: Kraus FT, Damjanov I, Kaufmann N, eds. Pathology of reproductive failure. Baltimore: Williams and Wilkins, 1991:308–33.
3. Jankovic BD, Markovic BM, Petrovic S, Isakovic K. Experimental autoimmune oophoritis in the rat. Eur J Immunol 1973;3:375–7.
4. Vajnstangl M, Petrovic S, Jankovic BD. Autoimmune oophoritis in the rat induced with isologous ovarian tissue. Periodicum Biologorum 1979;81: 249–51.
5. Wood DM, Liu C, Dunbar BS. Effect of alloimmunization and heterimmunization with zonae pellucidae on fertility in rabbits. Eur J Immunol 1981; 25:439–50.
6. Mahi-Brown CA, Huang TTF, Yanagimachi R. Infertility in bitches induced by active immunization with porcine zonae pellucidae. J Exp Zool 1982;222: 89–95.
7. Sacco AG, Pierce DL, Subramanian MG, Yurewicz EC, Dukelow WR. Ovaries remain functional in squirrel monkeys (*Samiri sciureus*) immunized with porcine zona pellucida 55,000 macromolecule. Biol Reprod 1987;36: 484–90.
8. Gulyas BJ, Gwatkin RBL, Yuan LC. Active immunization of cynomolgus monkeys (*Macaca fascicularis*) with porcine zonae pellucidae. Gamete Res 1983;4:299–307.
9. Rhim SH, Millar SE, Robey F, et al. Autoimmune diseases of the ovary induced by a ZP3 peptide from the mouse zona pellucida. J Clin Invest 1992;89:28–35.
10. Nishizuka Y, Sakakura T. Thymus and reproduction: sex linked dysgenesis of the gonad after neonatal thymectomy. Science 1969;166:753–5.
11. Taguchi O, Takahashi T, Masao S, Namikawa R, Matsuyama M, Nishizuka Y. Development of multiple organ localized autoimmune disease in nude mice after reconstitution of T cell function by rat fetal thymus graft. J Exp Med 1986;164:60–71.
12. Taguchi O, Nishizuka Y. Self-tolerance and localized autoimmunity: mouse models of autoimmune disease that suggest that tissue-specific suppressor T-cells are involved in tolerance. J Exp Med 1987;165:146–56.
13. Kojima A, Prehn RT. Genetic susceptibility to post-thymectomy autoimmune disease in mice. Immunogenetics 1981;14:15–27.
14. Sakaguchi SK, Fukuma K, Kuribayashi T, Masuca. Organ-specific autoimmune diseases induced in mice by elimination of T cell subset, I. Evidence for

the active participation of T cells in natural self-tolerance: deficit of a T cell subset as a possible cause of autoimmune disease. J Exp Med 1985;161:72–87.
15. Sakaguchi S, Sakaguchi N. Organ-specific autoimmune disease induced in mice by elimination of T cell subsets, V. Neonatal administration of cyclosporin A causes autoimmune disease. J Immunol 1989;142:471–80.
16. Tung KSK, Taguchi O, Teuscher C. Testicular and ovarian autoimmune diseases. In: Cohen IR, Miller A, eds. Guidebook to animal models for autoimmune diseases. New York: Academic Press, 1993.
17. Smith H, Lou Y-H, Lacy P, Tung KSK. Tolerance mechanism in experimental ovarian and gastric autoimmune diseases. J Immunol 1992;149:2212–8.
18. Kappler JW, Roehm N, Marrack P. T cell tolerance by clonal deletion in the thymus. Cell 1987;49:273–80.
19. MacDonald HR, Schneider R, Lees RK, et al. T cell receptor Vβ use predicts reactivity and tolerance to Mls-encoded antigens. Nature 1988;332:40–5.
20. Kisielow P, Bluthmann H, Staerz UD, Steinmetz M, Von Boehmer H. Tolerance in T cell receptor transgenic mice involves deletion of non-mature CD4+CD8+ thymocytes. Nature 1987;333:742–6.
21. Von Boehmer H, Kisielow P. Self-nonself discrimination by T cells. Science 1990;248:1369–73.
22. Murphy KM, Heimberger AG, Lou DY. Induction by antigen of intrathymic apoptosis of CD4+CD8+ $TC^{lo}$ thymocytes in vivo. Science 1990;1720–3.
23. Lo D, Burkey LC, Widera G, et al. Diabetes and tolerance in transgenic mice expressing class II MHC molecules in pancreatic beta cells. Cell 1988;53: 159–68.
24. Morahan G, Brennan FE, Bhathal PS, Allison J, Cox KO, Miller JFAP. Expression in transgenic mice of class 1 histocompatibility antigens controlled by the metallothionine promoter. Proc Natl Acad Sci USA 1989;86:3782–6.
25. Wietes K, Hammer RE, Jones-Youngblood S, Forman J. Peripheral tolerance in mice expressing a liver specific class I molecule: inactivation/deletion of a T cell subpopulation. Proc Natl Acad Sci USA 1990;87:6604–8.
26. Ohashi PS, Oehen S, Buerki K, et al. Ablation of "tolerance" and induction of diabetes by virus infection in viral antigen trangenic mice. Cell 1991;65: 305–17.
27. Oldstone MBA, Nerenberg M, Southern P, Price J, Lewicki H. Virus infection triggers insulin-dependent diabetes mellitus in a transgenic model: role of anti-self (virus) immune response. Cell 1991;65:319–31.
28. Lamb J, Skidmore BJ, Green N, Chiller JM, Feldmann M. Induction of tolerance in influenza virus-immune T lymphocyte clones with synthetic peptides of influenza hemagglutinin. J Exp Med 1983;157:1434–47.
29. Jenkins K, Schwartz RH. Antigen presentation by chemically modified splenocytes induces antigen-specific T cell unresponsiveness in vitro and in vivo. J Exp Med 1987;165:302–19.
30. Arnold B, Schonrich G, Hammerling G. Extrathymic T-cell selection. Curr Opin Immunol 1992;2:166–70.
31. Smith H, Chen I-M, Kubo R, Tung KSK. Neonatal thymectomy results in a repertoire enriched in T cells deleted in adult thymus. Science 1989;245: 749–52.
32. Fowell D, McKnight AJ, Powrie F, Dyke R, Mason D. Subsets of CD4 T cells and their roles in the induction and prevention of autoimmunity. Immunol Rev 1991;123:37–64.

33. Smith H, Sakamoto Y, Kasai K, Tung KSK. Effector and regulatory cells in autoimmune oophoritis elicited by neonatal thymectomy. J Immunol 1991; 147:2928–33.
34. Dean J. Biology of mammalian fertilization: role of the zona pellucida. J Clin Invest 1992;89:1055–9.
35. Bleil JD, Wassarman PM. Structure and function of the zona pellucida: identification and characterization of the proteins of the mouse oocyte's zona pellucida. Dev Biol 1980;76:185–202.
36. Millar SE, Chamow SM, Baur AW, Oliver C, Robey F, Dean J. Vaccination with a synthetic zona pellucida peptide produces long-term contraception in female mice. Science 1989;246:935–8.
37. Luo A-M, Garza KM, Hunt D, Tung KSK. Antigen mimicry in autoimmune disease: sharing of amino acid residues critical for pathogenic T cell activation. J Clin Invest 1993.
38. Sette A, Grey HM. Chemistry of peptide interactions with MHC proteins. Curr Opin Immunol 1992;4:79–86.
39. Lou Y-H, Tung KSK. T cell peptide of a self-protein elicits autoantibody to the protein antigen: implications in mechanism and specificity of autoantibodies. J Immunol 1993.

# 17

# Immunobiological Effects of Vasectomy and Vasovasostomy in the Rat Model

JOHN C. HERR, CHARLES J. FLICKINGER, AND STUART S. HOWARDS

Vasectomy is known to induce increases in serum antisperm antibodies in most species that have been studied (1, 2). As a result, there has been considerable interest in the possibility of systemic effects as a consequence of the immune response. Concerns were heightened by reports of glomerulonephritis in rabbits (3), tumors in mice (4), and an increased incidence of atherosclerosis in monkeys (5, 6) after vasectomy. However, epidemiologic studies have not revealed an increased incidence of any serious diseases in vasectomized men (7–10).

Recently, attention has turned to the potential reversibility of the effects of vasectomy because a significant number of vasectomized men now seek a subsequent vasovasostomy (11–14). Techniques of vasovasostomy have been improved (20) and now permit restoration of anatomic continuity of the vas deferens in up to 90% of the cases. However, the proportion of men that become fertile is less, ranging between 40% and 70% in various studies (11–14). This suggests that some alterations after vasectomy, either local or systemic, may persist even after reconnection of the vas deferens.

We have been studying the effects of vasectomy and vasovasostomy in the Lewis and Sprague-Dawley strains of rats. This species is convenient for laboratory studies, and the Lewis rat strain was selected because it is prone to develop antisperm antibodies and testicular alterations after vasectomy (3, 24, 25).

## Methods, Results, and Discussion

### *Serum Antibody Responses in Lewis Rats to Vasectomy*

Using an ELISA assay with $10^5$ intact sperm plated as an assay target, the profile of serum antibody (diluted 1/45) arising in response to a

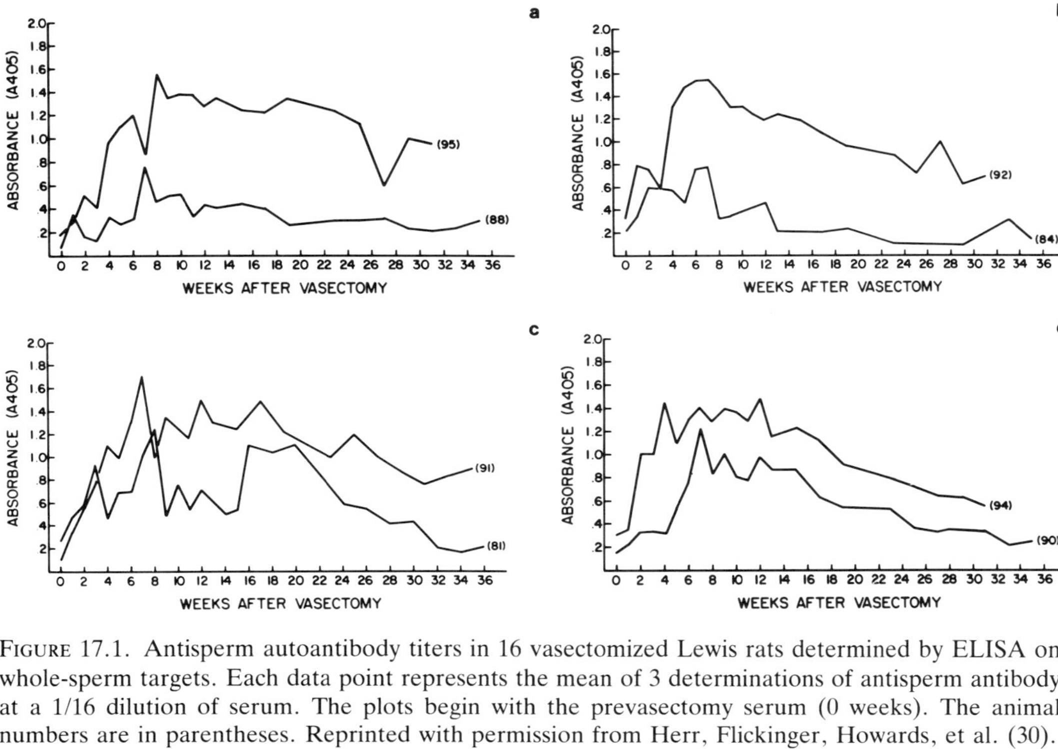


FIGURE 17.1. Antisperm autoantibody titers in 16 vasectomized Lewis rats determined by ELISA on whole-sperm targets. Each data point represents the mean of 3 determinations of antisperm antibody at a 1/16 dilution of serum. The plots begin with the prevasectomy serum (0 weeks). The animal numbers are in parentheses. Reprinted with permission from Herr, Flickinger, Howards, et al. (30).

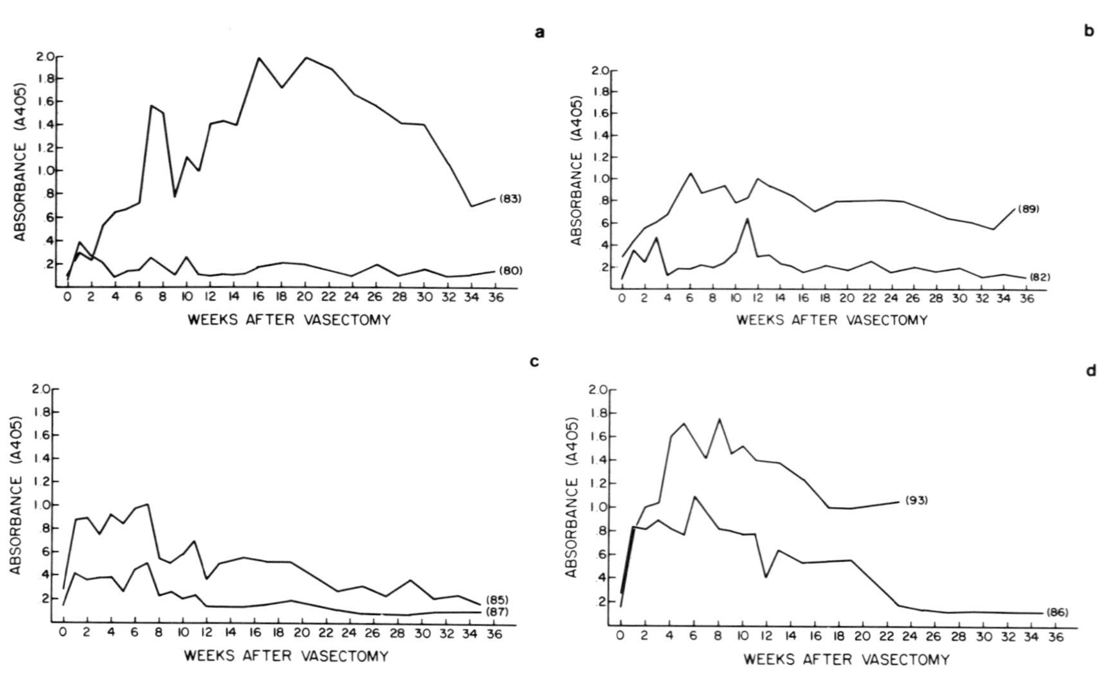


FIGURE 17.2. Antisperm autoantibody titers in 16 vasectomized Lewis rats determined by ELISA on whole-sperm targets. Each data point represents the mean of 3 determinations of antisperm antibody at a 1/16 dilution of serum. The plots begin with the prevasectomy serum (0 weeks). The animal numbers are in parentheses. Reprinted with permission from Herr, Flickinger, Howards, et al. (30).

vasectomy can be seen in Figures 17.1 and 17.2 (30). There was considerable variation in the temporal appearance and magnitude, as well as in the persistence and decline, of antisperm autoantibodies among the 16 Lewis rats tested. Several of the animals demonstrated pronounced and sustained antisperm antibody responses (animals 95, 92, 91, 94, 83, and 93), whereas other animals were less responsive (88, 80, 82, and 87). Prevasectomy or "natural" antisperm antibody showed a 5.5-fold difference (0.061 to 0.337 absorbance units).

All animals demonstrated a positive antisperm antibody response (defined as 1.96 standard deviations above the mean of the preimmune serum [15]) at some point during the postvasectomy period. When the results were expressed as percentage that were positive responders at each time point, 50% of the animals had positive responses by the end of week 1 (Fig. 17.3), and 81% of the animals had positive responses by the end of week 2. The greatest percentage of animals showing a positive response over the course of the study was found at the end of week 7 postvasectomy (88%). The 2 animals that were not positive for antisperm antibodies at week 7 (animals 80 and 82) demonstrated low positive responses during the first few weeks of the study. Between postoperative weeks 8 and 17, the percentage of animals showing a positive response fluctuated from 75% to 81% (Fig. 17.3).

A gradual decline in the percentage of animals positive for antisperm antibody began at week 13. At postoperative week 23, 56% of animals had positive responses, and by week 35 only 25% had positive antisperm responses. In general, the return, with time, to values comparable to the preimmune level correlated with lower overall responsiveness to sperm autoantigens. Calculations made of the areas under the absorbance curves

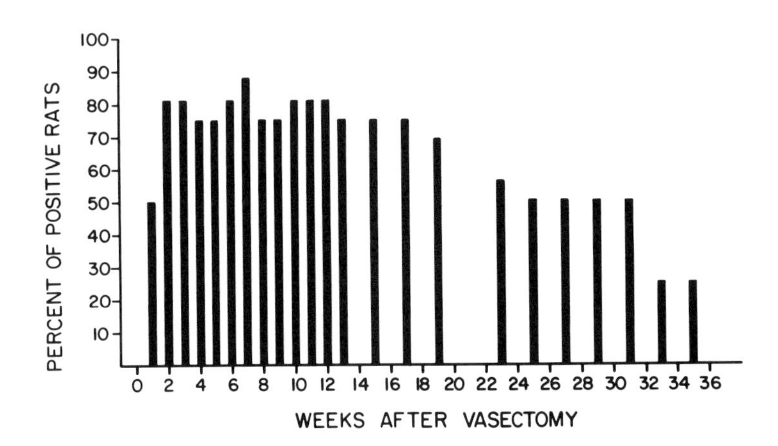


FIGURE 17.3. Percentage of Lewis rats demonstrating antisperm autoantibodies over time as a consequence of vasectomy. Reprinted with permission from Herr, Flickinger, Howards, et al. (30).

of each animal for the first 15 weeks of the study and during weeks 15–30 showed that the magnitude of the response of the group as a whole was somewhat less during weeks 15–30 than in the initial weeks. Examination of the individual curves indicated that those animals with high end point absorbance values had generally higher responses during the initial weeks of the study. Conversely, animals that returned to preimmune values generally had lower areas under the absorbance curves during the first 15 weeks of the study. In addition, pronounced fluctuations in antibody response over time were noted on inspection of the absorbance curves (Figs. 17.1 and 17.2) for certain animals (e.g., 83, 95, 81, 90, and 91), whereas other animals exhibited less fluctuation (92, 80, and 87).

In order to analyze the antisperm autoantibody responsiveness of the 16 animals as a group, calculations were made of the mean absorbance values for all 16 animals at intervals after vasectomy (Fig. 17.4). This demonstrated that in the group as a whole, antisperm antibody rose for the first 7 weeks, when the highest mean absorbance value was obtained (Fig. 17.4). From week 7 to week 19, mean absorbance values of the group remained relatively constant. By week 23 mean absorbance values declined to ~0.5 OD units and declined again to ~0.3 units by week 31.

The results indicate that immunologic challenge with spermatozoa or sperm antigens occurs relatively quickly, within days after vasectomy. The elevations in antisperm antibodies following vasectomy are similar

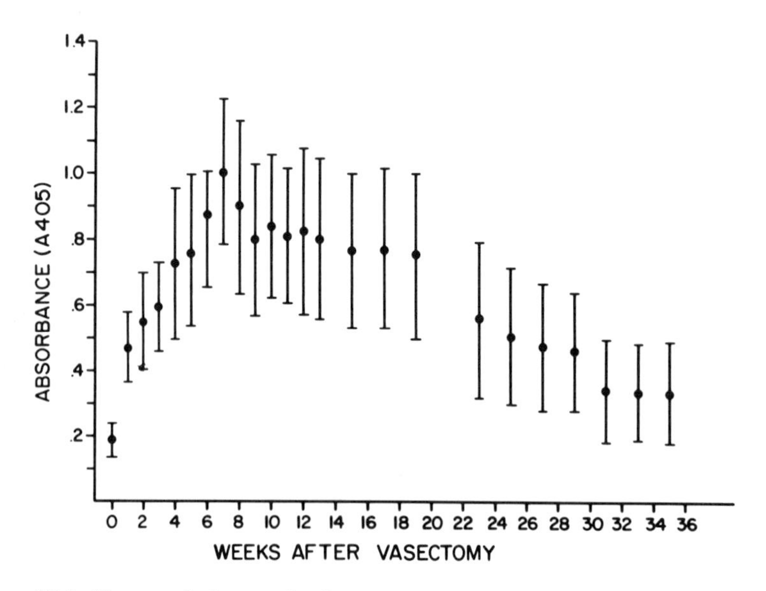


FIGURE 17.4. Temporal changes in the mean group absorbance values on ELISA as a measure of antisperm autoantibodies appearing in the sera of vasectomized Lewis rats. Absorbance values represent mean ± SEM. Reprinted with permission from Herr, Flickinger, Howards, et al. (30).

in temporal appearance to antibody responses to inoculation with spermatozoa or bacteria (16, 17). The mode of this challenge is presently unclear, but most likely involves release of antigenic components from the surgical site, sites of granulomas (see below), or possible alterations in the blood-testis or blood/epididymal barriers. The elevations in detectable antisperm autoantibody during the 2 weeks following vasectomy point to a critical window following the surgical procedure, during which autoantibody-producing cells are induced, antibody is synthesized, and serum autoantibody levels rise.

## *Correlation Between Antisperm Antibodies and Changes in Testis Weight and in the Seminiferous Epithelium in Lewis Rats That Have Been Vasectomized and Vasovasostomized*

The relationship between antisperm antibodies as determined by ELISA and the occurrence of alterations in testicular weight and histology were studied following vasectomy in Lewis rats (29). The effects of vasovasostomy on antisperm antibody levels were also examined. A bilateral vasectomy was performed on rats in the vasectomy group. Animals in the vasovasostomy group received a bilateral vasectomy followed 3 months later by a bilateral vasovasostomy. Rats in the group designated *sham* were subjected to a sham vasectomy and, 3 months later, to a sham vasovasostomy. Thirteen vasectomy, 25 vasovasostomy, and 14 sham-operated rats were killed for study at intervals of 3, 4, and 7 months. The *testicular biopsy score count* (TBSC), a semiquantitative method, was employed for assessing the extent of changes in the seminiferous epithelium (18).

The weights of testes in vasectomized and vasovasostomized rats fell into two groups. The group of small testes showed severe microscopic alterations consisting of depletion of germ cells (Fig. 17.5b). Many seminiferous tubules lacked all stages of germ cells except for a few spermatogonia and, thus, were composed almost entirely of Sertoli cells. Others contained varying numbers of spermatocytes, but spermatids were uncommon. In contrast, the normal-sized testes in the vasectomy and vasovasostomy groups resembled those of the shams (Fig. 17.5a). These qualitative observations of testis morphology were matched by quantitative differences shown by the TBSC; small testes had very low scores, indicative of severe depletion of germ cells, whereas normal-sized testes had scores of 9–10 that reflect normal testis morphology (19).

For purposes of comparison with respect to serum antisperm antibodies, animals in the vasectomy and vasovasostomy groups were divided into those possessing or lacking testicular alterations. Altered testes were defined as those with a weight less than 0.185 g per 100 g of body weight;

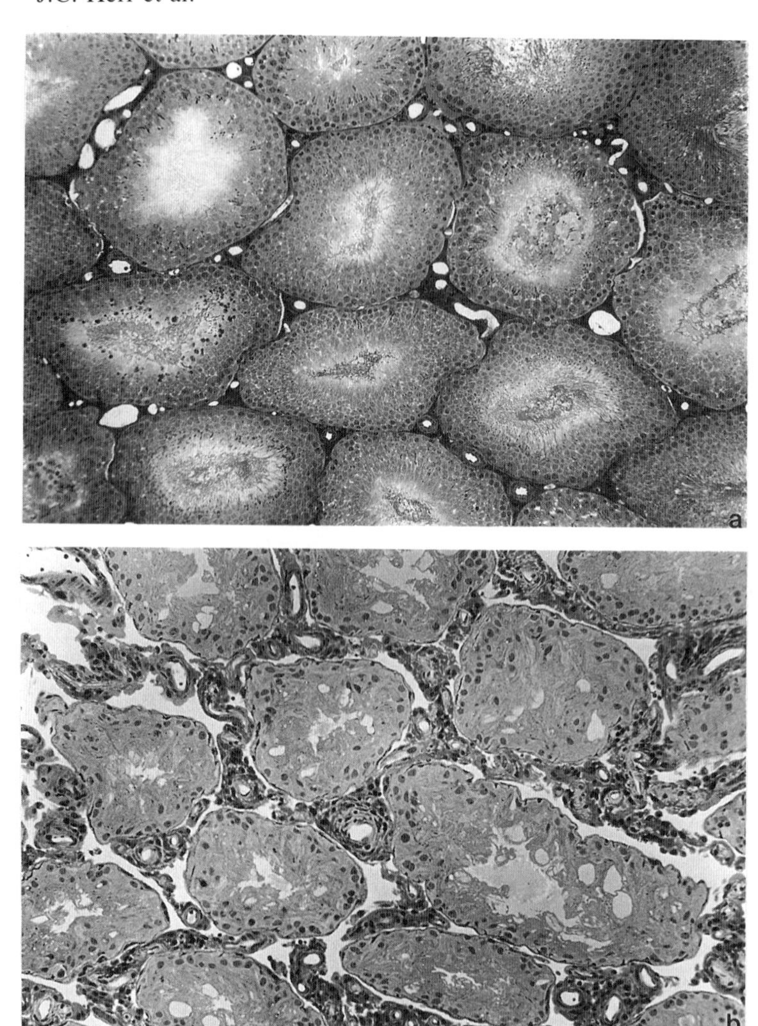


FIGURE 17.5. Cross section of seminiferous tubules from (*a*) a testis showing normal morphology and (*b*) a testis from a Lewis rat showing altered testicular morphology, including a lack of spermatids and spermatocytes (this testis received a TBSC of 3.1) (×150). Reprinted with permission from Herr, Flickinger, Howards, et al. (29).

these small testes also had very low TBSC. The category with testicular alterations included animals with bilateral and unilateral testicular changes. Sera from animals with and without testicular alterations were analyzed by ELISA for antisperm antibodies, and the results were compared with those from sham-operated rats (Table 17.1).

TABLE 17.1. Absorbance value from antisperm antibody ELISA using sera from rats in vasectomy and vasovasostomy groups with or without testicular alterations and from sham-operated animals.

| | Testicular alterations | | |
|---|---|---|---|
| Month | Yes | No | Sham |
| 1[a] | 1.072 ± 0.113 | 0.791 ± 0.081 | 0.501 ± |
| 0.051 | (16) | (21) | (14) |
| 3[a] | 1.284 ± 0.108 | 0.868 ± 0.082 | 0.503 ± |
| 0.045 | (16) | (21) | (14) |
| 4[a] | 1.341 ± 0.122 | 1.035 ± 0.101 | 0.488 ± |
| 0.086 | (13) | (19) | (7) |
| 7[b] | 0.984 ± 0.152 | 0.843 ± 0.132 | 0.417 ± |
| 0.025 | (4) | (7) | (3) |
| pre[c] | 0.491 ± 0.056 | 0.427 ± 0.034 | 0.462 ± |
| 0.042 | (16) | (21) | (14) |

*Note:* The mean absorbance ± SEM at intervals after vasectomy are shown.
[a] The 3 values are significantly different from one another ($P < 0.05$).
[b] No significant difference.
[c] Pre = prevasectomy values ($n$).

At 1-, 3-, and 4-month intervals, the mean absorbance for sera from animals with altered testes was significantly greater than that from rats lacking testicular alterations, and the mean absorbances for operated animals with or without testicular alteration exceeded that for the sham group ($P < 0.05$). At the 7-month interval the mean absorbances lay in the same order, but the differences were not statistically significant; this may reflect the smaller number of animals at the 7-month interval. To examine an individual animal's responses more closely, a positive antisperm antibody response was considered an absorbance exceeding the mean of the prevasectomy absorbance values + 1.96 SD; that is, an absorbance in excess of 95% of the prevasectomy absorbances. At 3 months postvasectomy, for example, 68% of the animals displayed a positive antisperm response (Table 17.2). It can be seen from Table 17.2 that animals with positive and negative responses were present among those with testis changes as well as in those lacking testicular alterations at each interval after operation.

The observation that mean antisperm antibody levels were higher in rats that exhibited testicular changes than in those in which the testes remained normal suggests that the generation of antisperm antibodies may play a role in the development of testicular lesions in the Lewis rat after vasectomy. However, further work is necessary to define the nature of the association between antisperm antibodies and testicular alterations

TABLE 17.2. Numbers of rats with positive antisperm antibody responses (absorbance in ELISA >0.808) in relation to testicular alterations.

| | Testicular alterations | | |
|---|---|---|---|
| Months | Yes | No | All animals |
| 1 | 12:4 (75) | 8:13 (38) | 20:17 (54) |
| 3 | 14:2 (88) | 11:10 (52) | 25:12 (68) |
| 4 | 11:2 (85) | 11:8 (58) | 22:10 (69) |
| 7 | 3:1 (75) | 3:4 (43) | 6:5 (55) |

*Note:* The ratios of positive responders:nonresponders are shown. The percentage of positive responders is in parentheses. The significance of the difference in the proportion of positive responders between groups with testicular alterations or no testicular changes was marginal at 1 and 3 months ($P = 0.0576$; $P = 0.0566$).

in this model. For example, antisperm antibodies have not been localized within the adluminal compartment of the seminiferous tubules. The possibility cannot be excluded that antisperm antibodies rise secondarily to testicular damage. A further possibility is that antisperm antibodies act in concert with cell-mediated processes to effect testicular damage.

Although animals with testicular changes clearly have higher mean antisperm antibody levels than those that lack alterations, there are individual exceptions within both groups. That is, some animals with testicular alterations do not show a positive antisperm antibody response, and, conversely, some rats that lack testicular alterations develop high levels of antisperm antibodies.

## *Influence of Vasovasostomy on Antisperm Antibodies in Lewis Rats*

Mean absorbance values determined by ELISA for the sera of animals that gave positive antibody responses in the sham, vasectomy, and vasovasostomy groups were compared (Fig. 17.6) (19). For preimmune sera harvested prior to any manipulation, the mean absorbances for the three groups were similar. In the case of the sham-operated animals, the mean absorbances at all of the subsequent intervals did not change significantly from the preimmune value. In contrast, at each interval studied antisperm antibody levels in the vasectomy and vasovasostomy groups were greater than those in the sham group. When the Newman-Keuls test for sig-

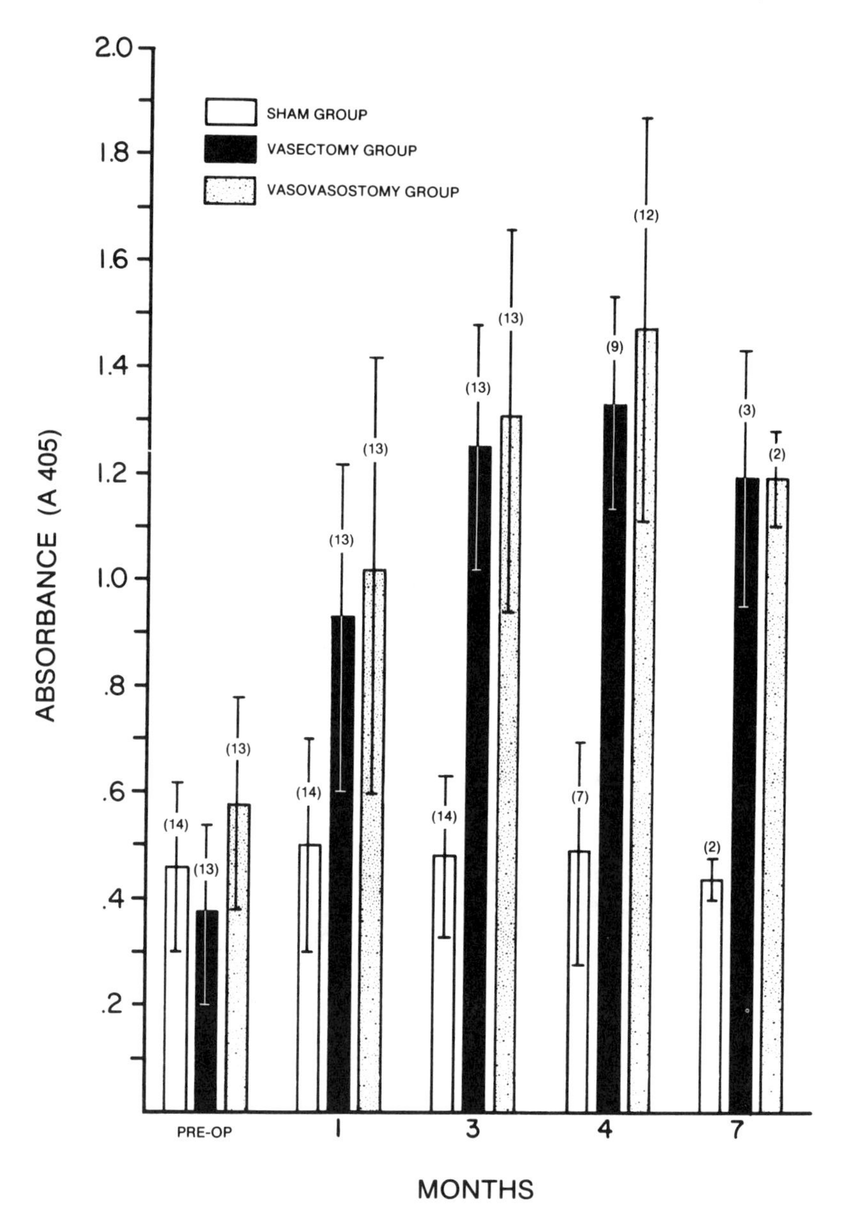


FIGURE 17.6. Mean antisperm antibody absorbance values on ELISA of sera from sham, vasectomized, and vasovasostomized Lewis rats assayed at various intervals following vasectomy. The standard deviation is indicated by error bars. The number of experimental animals is in parentheses. Reprinted with permission from Flickinger, Herr, Howards, et al. (19).

nificance was applied, vasectomy and vasovasostomy values differed significantly from sham values at all intervals except at 7 months. This lack of significance at 7 months was due to a low number of samples, although values differed by approximately 0.75 OD units. However, vasectomy and vasovasostomy mean antisperm antibody values were not significantly different from each other at any interval in the Lewis rats.

The results indicate that vasovasostomy did not cause a significant decrease in antisperm antibody in positive responders 1 or 4 months after reanastomosis as compared to levels in responders with a persisting vasectomy. This was not due to continuing occlusion of the vas deferens in the animals that underwent a vasovasostomy (19) since we studied fluid flows in the vas and demonstrated patency. We also observed that while vasovasostomy prevented the progression of testicular alterations with time, the operation did not appear to reverse testicular changes (19). Although these results in the rat cannot be generalized directly to other species, it is of interest to note that the proportion of men that become fertile after vasovasostomy ranges between 40% and 70% (11–14), even though restoration of anatomic continuity of the vas deferens is possible in up to 90% of the cases (20). It may be speculated that either the persistence of antisperm antibodies, or testicular alterations, or both, may play roles in limiting the restoration of fertility after vasovasostomy.

## *Influence of Vasovasostomy on Antisperm Antibodies in Sprague-Dawley Rats*

Serum antisperm antibodies were also studied in Sprague-Dawley rats after vasectomy and vasovasostomy (27). In general, Sprague-Dawley rats are immunologically less responsive to vasectomy than Lewis rats (3, 24). Animals in the vasectomy group received a bilateral vasectomy, those in the vasovasostomy group received a bilateral vasectomy followed 3 months later by a bilateral vasovasostomy, and sham operations were performed on rats in the sham group. Blood samples were obtained at 1, 3, 4, and 7 months, and antisperm antibodies were assayed by an ELISA.

### Antisperm Antibodies After Vasectomy or Vasectomy Followed by Vasovasostomy

The mean serum antisperm antibody levels in the vasectomy and vasovasostomy groups were higher than those calculated for sham-operated animals 1 month after the operation (Fig. 17.7), but the differences were not statistically significant. At 3 months the presence of a significant difference among the three groups was indicated by ANOVA ($P < 0.05$), but the Newman-Keuls test was not capable of indicating unambiguously where the difference lay (21). It may be noted, however, that antibody

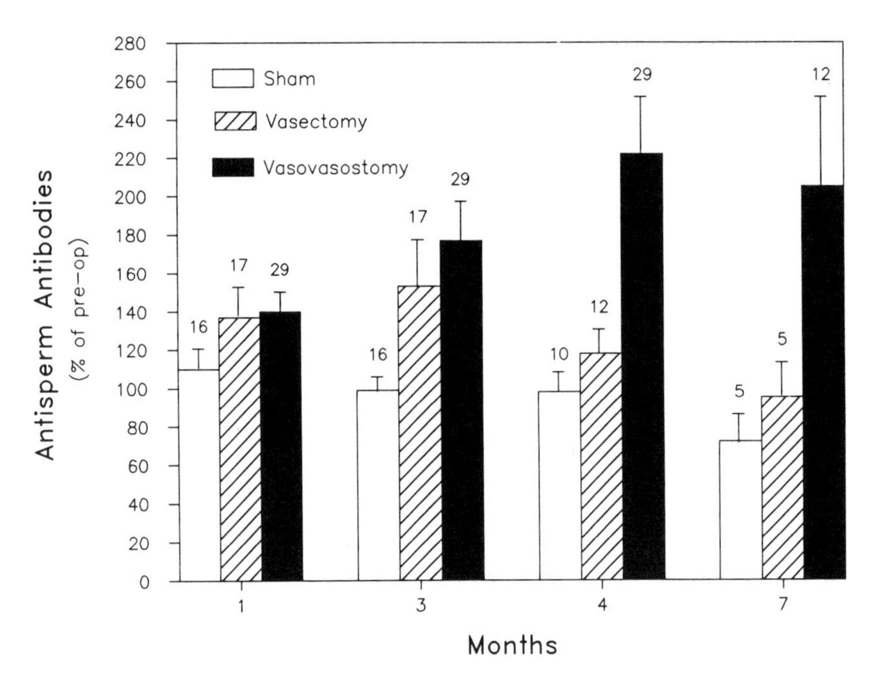


FIGURE 17.7. Bar graph showing serum antisperm antibodies in Sprague-Dawley rats as a percentage of preoperative values for rats in vasectomy, vasovasostomy, and sham-operated groups at various intervals. The mean and SEM for each group are shown; the number of animals in each sample appears above the error bar. The vasovasostomy group was significantly different from the vasectomy and sham-operated groups at the 4-month interval. Reprinted with permission from Flickinger, Howards, Carey, et al. (27), © American Urological Assn.

values were higher for the vasectomy and vasovasostomy groups than for the sham-operated group at 3 months.

Vasovasostomies were performed at 3 months on animals in the vasovasostomy group. Subsequently, at the 4-month interval (1 month after vasovasostomy), the groups differed significantly ($P < 0.01$ by ANOVA), antisperm antibodies in the vasovasostomy group being significantly elevated in comparison with vasectomized or sham-operated animals ($P < 0.05$ in the Newman-Keuls test). At 7 months the value for rats in the vasovasostomy group was approximately double that for the vasectomy group, but the difference was not significant, perhaps due to the fact that the number of animals was smaller at 7 months than at the earlier intervals. There was no difference among the three groups in mean absorbance of preoperative sera in the ELISA (0.377 ± 0.033, 0.359 ± 0.023, and 0.385 ± 0.028, mean ± SE for vasovasostomy, vasectomy, and sham groups, respectively).

TABLE 17.3. Antisperm antibodies in rats with or without spermatic granulomas after vasovasostomy at 3 months.

| Months | Granuloma absent | Granuloma(s) present |
|---|---|---|
| 1 | 127[a] ± 23<br>(7)[b] | 144 ± 10<br>(22) |
| 3 | 145 ± 44<br>(7) | 187 ± 20<br>(22) |
| 4 | 142[c] ± 26<br>(7) | 247[c] ± 36<br>(22) |
| 7 | 103 ± 30<br>(4) | 257 ± 60<br>(8) |

[a] Absorbencies on ELISA are expressed as a percentage of the preoperative value (i.e., the absorbance of a sample was divided by the absorbance of the preoperative sample from the same animal ×100). Means ± SEM are shown.
[b] Numbers of animals are shown in parentheses.
[c] Means are significantly different ($P < 0.05$).

## Influence of Spermatic Granulomas

Spermatic granulomas occurred bilaterally around the proximal cut and ligated end of the vas deferens in the majority (16/18, 89%) of the Sprague-Dawley rats in the vasectomy group. Spermatic granulomas were also observed at the site of the anastomosis in vasovasostomized rats, and some animals had spermatic granulomas of similar appearance adherent to the cauda epididymis. A total of 22 of 29 (76%) (Table 17.3) vasovasostomized animals had one or more granulomas that were distributed in the following manner. Eight rats had granulomas located bilaterally at the site of anastomosis on the vas deferens, and 9 rats had a unilateral granuloma of the vas deferens (4 of these also had an epididymal granuloma or a large vas granuloma that extended into the epididymal region); 5 rats possessed only epididymal granulomas. In gross view a granuloma was a spherical or irregularly shaped yellowish mass of tissue with a flexible shell surrounding a semisolid core. Study of microscopic sections showed that the wall of the granuloma was composed of fibrous connective tissue surrounding epithelioid cells (Fig. 17.8). The core was occupied by many sperm in different stages of disintegration. Plasma cells and lymphocytes were also present in the connective tissue of the granuloma wall, predominantly between the fibrous tissue and the epithelioid cells (Figs. 17.8 and 17.9). In some granulomas, macrophages and neutrophils containing parts of phagocytosed sperm were observed at the margins of the core close to the epithelioid cells.

The results demonstrated that vasovasostomized Sprague-Dawley rats had higher mean antisperm antibody levels than animals with a persisting

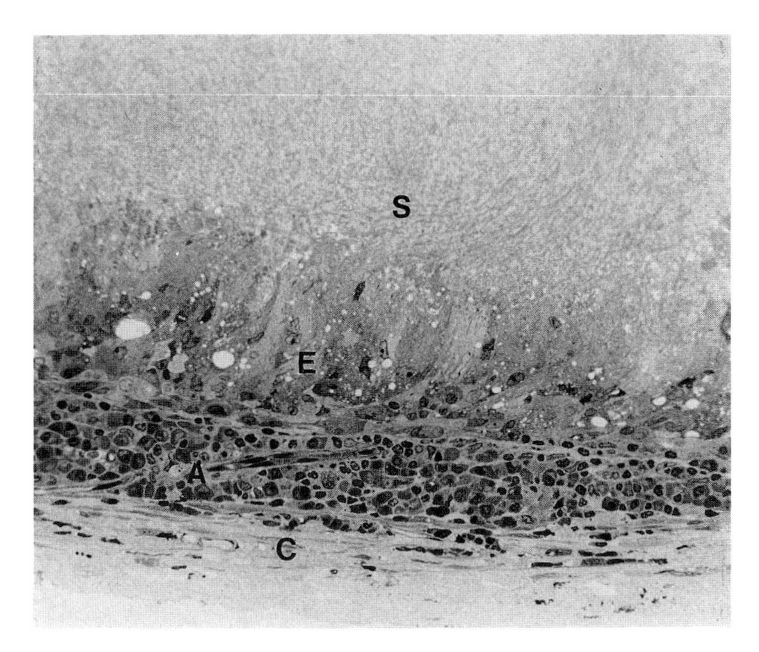


FIGURE 17.8. Light micrograph of a portion of a spermatic granuloma of the vas deferens from a Sprague-Dawley rat 4 months after vasovasostomy (7-month interval). Proceeding toward the center of the granuloma, the wall is composed of connective tissue (C), a cellular layer that includes plasma cells and lymphocytes (A), vacuolated epithelioid cells (E), and a core containing many sperm (S). (×360). Reprinted with permission from Biology of Reproduction 1989;40:353–60.

vasectomy or a sham operation. This suggests that under certain conditions vasovasostomy may stimulate an antibody response to sperm rather than a reduction in antibody levels, as might be anticipated after removal of the obstruction of the duct system.

The mechanism by which vasovasostomy led to increased antisperm antibodies is uncertain, but the observation that many vasovasostomized rats developed spermatic granulomas at the site of the reanastomosis may be pertinent. Spermatic granulomas commonly occur after vasectomy in both Lewis and Sprague-Dawley rats as the result of leakage of sperm from the excurrent duct system. Although the site and time of appearance of granulomas vary with the species—perhaps depending on the physical characteristics of the reproductive ducts—a very high proportion of rats develop a granuloma around the proximal ligated end of the vas deferens within a few weeks of vasectomy. In our studies these vas granulomas were excised at the time of vasovasostomy, so the granulomas observed near the vas deferens in vasovasostomized animals at the time they were killed must have formed de novo after vasovasostomy. It should be noted

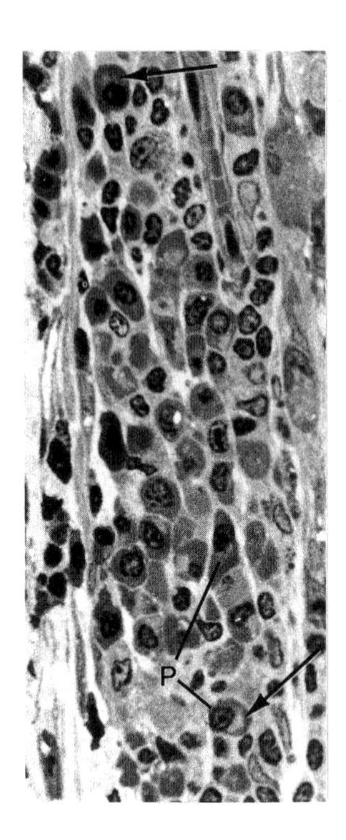


FIGURE 17.9. A portion of Figure 17.8 and an adjacent field shown at higher magnification. In the cellular layer (A in Fig. 17.8), many plasma cells (P) can be identified on the basis of their basophilic cytoplasm polygonal shape and eccentrically located nucleus with prominent clumps of heterochromatin. In several instances the negative image of the Golgi apparatus (arrow) is visible near the nucleus (×730). Reprinted with permission from Biology of Reproduction 1989;40:353–60.

that in this system the occurrence of granulomas after vasovasostomy did not occlude the vas deferens, as shown by studies of fluid flow through the vas (22).

The presence of plasma cells, lymphocytes, and a variety of other cells in the walls of the spermatic granulomas indicates that the granulomas were potential sites of antigen processing, antigen presentation, and immune responsiveness. The reoccurrence of granulomas after vasovasostomy suggests that these structures serve as sites for continued antigenic challenge.

The increase in antisperm antibodies after vasovasostomy in Sprague-Dawley rats may be related to their relatively low immunologic responsiveness to vasectomy since a second major immunologic challenge (vasovasostomy with secondary granuloma formation) may be needed to stimulate antisperm antibody production in this strain. Conversely, in the more-responsive Lewis strain, a maximal response may be achieved as a result of the vasectomy alone so that the subsequent vasovasostomy leads to no further increase in antibodies. The results in the two strains suggest that contrary to expectation, vasovasostomy may not bring about a decrease in antisperm antibodies and may even result in an increase in such antibodies depending on the immunological responsiveness of the individual. Paradoxically, an increase in antibodies appears to be more likely to occur after vasovasostomy in less-responsive rats.

## *Inflammatory Changes in the Epididymis After Vasectomy in the Lewis Rat*

The histology of epididymides of Lewis rats was studied at intervals up to 7 months after vasectomy, vasectomy followed 3 months later by vasovasostomy, or sham operations (33). Epididymal histology was related to testicular alterations and to serum antisperm antibodies as determined by ELISA.

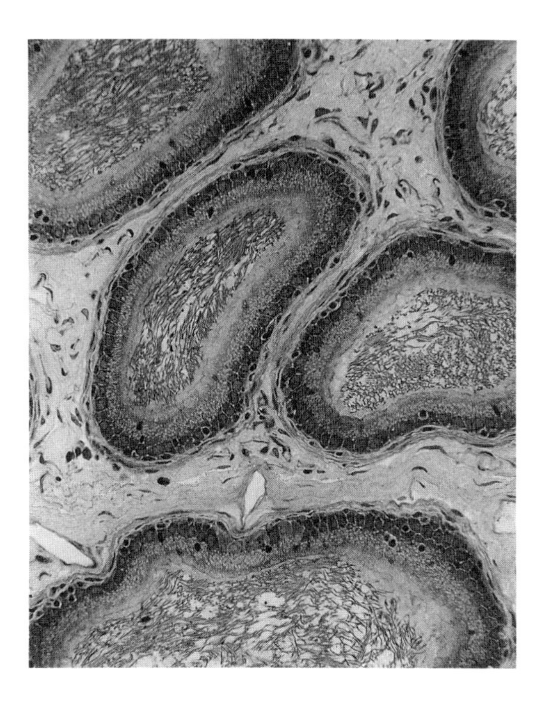

FIGURE 17.10. Light micrograph of the caput epididymidis of a sham-operated rat at 7 months. The pseudostratified columnar epithelium of uniform height surrounds a lumen containing spermatozoa. The interstitial loose connective tissue is composed mainly of collagen fibers, ground substance, and fibroblasts (×260). Reprinted with permission from Flickinger, Herr, Howards, et al. (33).

### Sham-Operated Rats

The epididymides of rats that underwent a sham operation had a normal histological appearance (Figs. 17.10 and 17.11). The pseudostratified columnar epididymal epithelium decreased in height along the length of the duct. As in normal rats, it consisted of abundant columnar principal cells with stereocilia on their apical surfaces, small oval or triangular basal cells, round halo cells, clear (or light) cells with vacuoles and granules in the cytoplasm, and (in the proximal part) narrow cells. The lumen of the epididymis contained numerous spermatozoa that were closely packed in the wide lumen of the cauda epididymides. The interstitial tissue of the epididymis was composed of loose connective tissue that contained fibroblasts, collagen fibers, abundant ground substance, and small blood vessels. Mast cells contained granules that stained metachromatically with toluidine blue, and small cells with round heterochromatic nuclei, which may represent lymphocytes, were widely scattered in the interstitium.

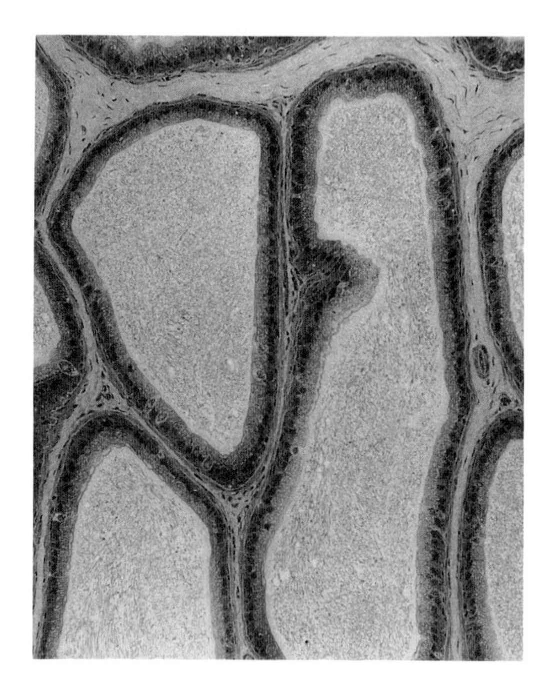

FIGURE 17.11. Cauda epididymidis of a 7-month sham-operated rat. The wide epididymal lumen contains numerous sperm surrounded by a smoothly contoured epithelium. Loose connective tissue underlies the epithelium (×150). Reprinted with permission from Flickinger, Herr, Howards, et al. (33).

## Epididymides of Rats with Testicular Alterations

Of 33 rats in the *vasectomy* (Vx) and *vasovasostomy* (Vv) groups, 13 had testicular alterations (5 Vx and 8 Vv). Of these 13 animals, severe epididymal lesions that included pronounced interstitial changes were present in 7 (2 Vx and 5 Vv). All 13 rats with testicular alterations, including the 6 without interstitial epididymal modification (3 Vx and 3 Vv), showed reduced numbers of epididymal sperm. In contrast, there were no instances of severe epididymal changes among 20 animals that lacked testicular alterations (6 Vx and 14 Vv). Study of specimens from animals at the 3-, 4-, and 7-month intervals did not reveal any obvious temporal pattern of epididymal changes. Since there were no qualitative differences

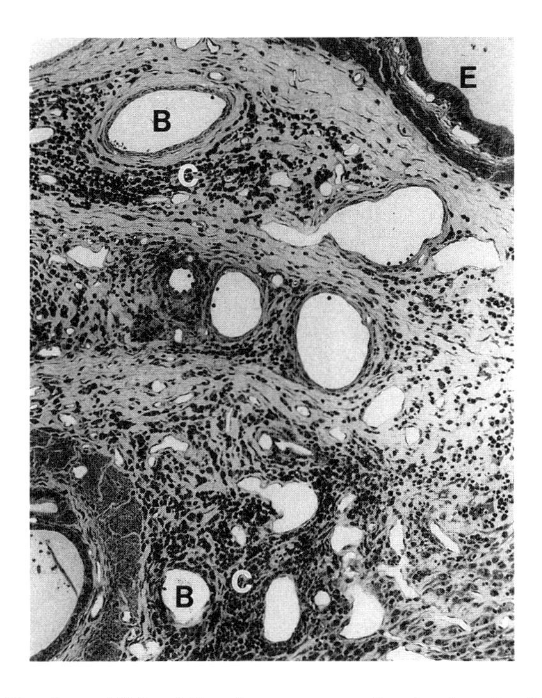


FIGURE 17.12. Cauda epididymidis of a vasovasostomized animal at 7 months. The low-power micrograph illustrates infiltrates of mononuclear cells (C) in the interstitial tissue of the epididymis of a rat that had severe testicular alterations. The cells are visible at this magnification by virtue of their densely staining nuclei; similar examples are shown at higher magnification in Figures 17.5a and 17.5b. They are concentrated around blood vessels (B) in the upper and lower parts of the picture. The vessels have clear lumina because the blood was washed out in the course of perfusion fixation. Small portions of the epididymal duct (E) are visible in corners of the field (×150). Reprinted with permission from Flickinger, Herr, Howards, et al. (33).

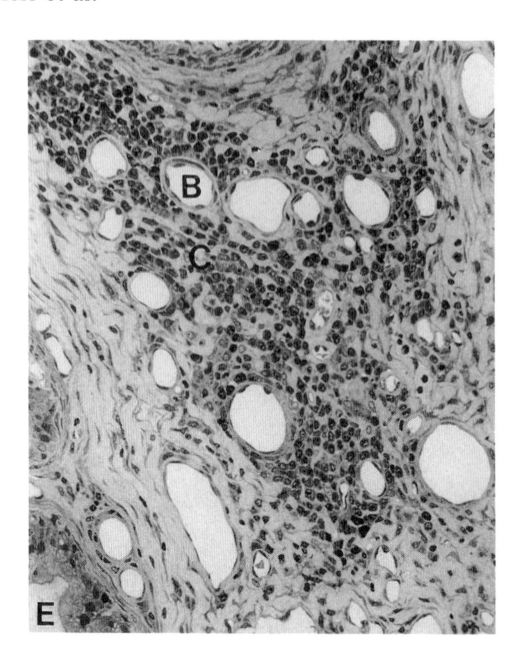


FIGURE 17.13. Vasovasostomy group, 4 months, cauda epididymidis. Numerous cells (C) are visible in the epididymal interstitial tissue around and between small blood vessels (B) (×270). (E = epididymal lumen.) Reprinted with permission from Flickinger, Herr, Howards, et al. (33).

in epididymal histology between the two experimental groups, specimens from vasectomized and vasovasostomized animals are described together.

Each of the 7 animals with severe interstitial changes in the epididymis (Figs. 17.12 through 17.14) displayed aggregates of mononuclear cells in the connective tissue between many small blood vessels (Figs. 17.12 and 17.13). The cellular composition of these aggregates differed from one animal to another, but consisted of varying numbers of cells identified as lymphocytes, plasma cells, and macrophages (Fig. 17.14). Lymphocytes were small round cells that stained intensely with toluidine blue and had a high nucleus: cytoplasm ratio. Plasma cells displayed the classical features of polygonal shape, basophilic cytoplasm, and an eccentrically located nucleus with prominent clumps of chromatin disposed about the inner aspect of the nuclear envelope. In some of the plasma cells, a clear area next to the nucleus, believed to be occupied by the Golgi apparatus, was distinctly visible. Macrophages were large oval cells that had a "foamy" or granulated cytoplasm, imparting to them a distended appearance. In a few instances lymphocytes or plasma cells surrounded individual blood vessels (Fig. 17.12); more commonly, they formed larger masses inter-

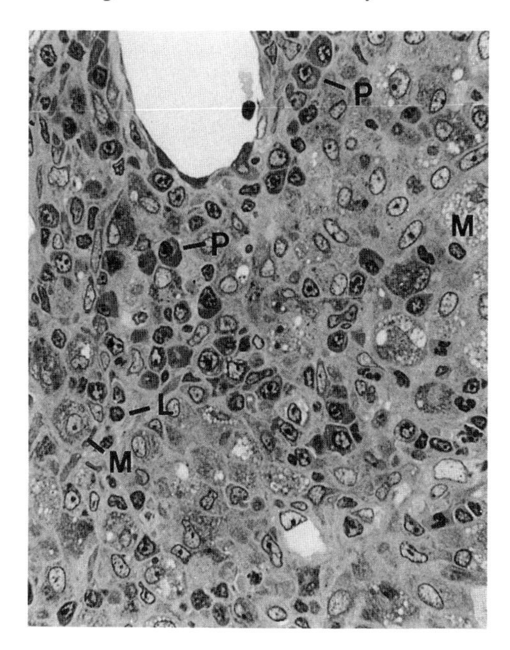


FIGURE 17.14. Vasovasostomy, 4 months, cauda epididymidis. The epididymal interstitium contains aggregates of cells of different types. Some, with polygonal shape, basophilic cytoplasm, and eccentric nucleus, are clearly plasma cells (P). Others are identified as macrophages on the basis of their large size and foamy or granulated cytoplasm (M). Cells with small heterochromatic nuclei and scant cytoplasm are believed to be lymphocytes (L) (×590). Reprinted with permission from Flickinger, Herr, Howards, et al. (33).

spersed among numerous small vessels (Fig. 17.13). Such aggregations of macrophages, lymphocytes, and plasma cells in the epididymal interstitium are the histopathologic hallmarks of a chronic inflammatory response.

The lumina of the epididymides with prominent interstitial changes contained cells that are not normally found there. Most striking were the polymorphonuclear leukocytes (Fig. 17.15) observed in the lumina of 4 epididymides, in 2 instances in the cauda, and in 2 instances in the caput epididymides. Larger mononuclear cells (Fig. 17.16) were observed in the lumen in 3 cases. These cells were generally round or oval with some irregular surface projections and granulated and/or vacuolated cytoplasm. Thus, they resembled macrophages.

Each of the 13 animals with testicular alterations displayed the following additional epididymal changes whether interstitial alterations were present or not. As anticipated from the depletion of germ cells in the testes, spermatozoa were absent or greatly reduced in number in the

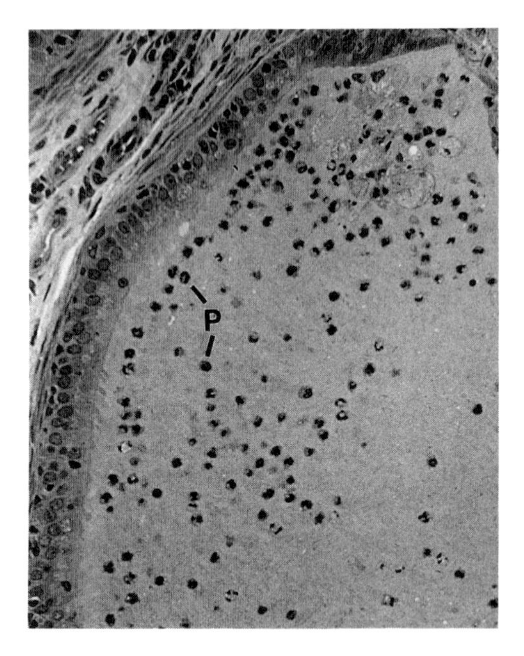


FIGURE 17.15. Four-month vasovasostomized rat. Polymorphonuclear leukocytes (P) with their characteristic multilobed nuclei occupy the lumen of the caput epididymidis of a rat that showed interstitial epididymal changes (×320). Reprinted with permission from Flickinger, Herr, Howards, et al. (33).

lumina of all the caput epididymides that drained an altered testis. In no case were more than a few scattered sperm found in the caput epididymides. The proximal portion of the cauda epididymides was lined by a columnar epithelium and generally had a collapsed lumen devoid of sperm (Fig. 17.17). Some epithelial cells contained vacuoles and/or many cytoplasmic granules that stained deeply with toluidine blue (Fig. 17.18). In 2 epididymides (vasovasostomy, 4 and 7 months), large intraepithelial vacuoles contained elongated, densely staining structures (Fig. 17.19).

## *Antisperm Antibodies in Lewis Rats with Severe Epididymal Changes*

Serum antisperm antibodies measured with ELISA were compared for rats with and without the severe epididymal changes that included interstitial aggregates of mononuclear cells (Fig. 17.20) (33). At the 1- and 3-month intervals, the mean levels of antisperm antibodies for rats in vasectomy and vasovasostomy groups that possessed these epididymal changes were significantly higher than for rats that lacked the epididymal

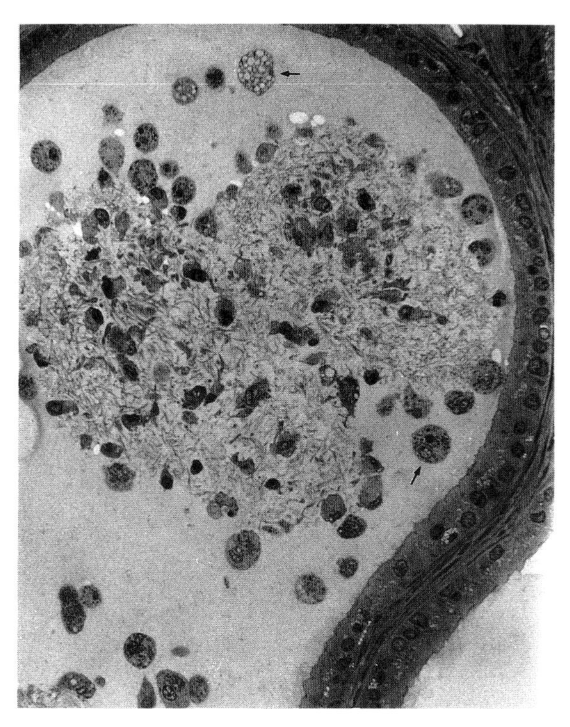

FIGURE 17.16. Four-month vasovasostomized rat. The lumen of the cauda epididymidis contains some sperm accompanied by large cells with granulated and/or foamy cytoplasm (arrows) (×320). Reprinted with permission from Flickinger, Herr, Howards, et al. (33).

changes ($P < 0.05$ and $P < 0.01$, respectively). At 1, 3, and 4 months, the means of antisperm antibodies for rats in the vasectomy and vasovasostomy groups (with or without epididymal changes) were significantly higher than those in the sham group ($P < 0.05$, $P < 0.01$, and $P < 0.01$). There were no significant differences between the mean absorbances on ELISA for serum antisperm antibodies preoperatively (0-month interval) or at the 7-month interval (which included fewer animals).

All the animals with severe epididymal interstitial changes also had testicular alterations. On the other hand, epididymal lesions were not detected in some (6/13) of the rats with altered testes. We did not serially section all of the epididymides, however, and the epididymal interstitial changes consisted of localized foci of inflammation. Thus, it seems likely that some epididymal alterations were missed by the sampling procedure. Alternatively, the epididymitis may be transient in some regions. In any event, epididymal changes were observed with similar frequency in vasectomy (2/11, 18%) and vasovasostomy (5/22, 23%) groups and

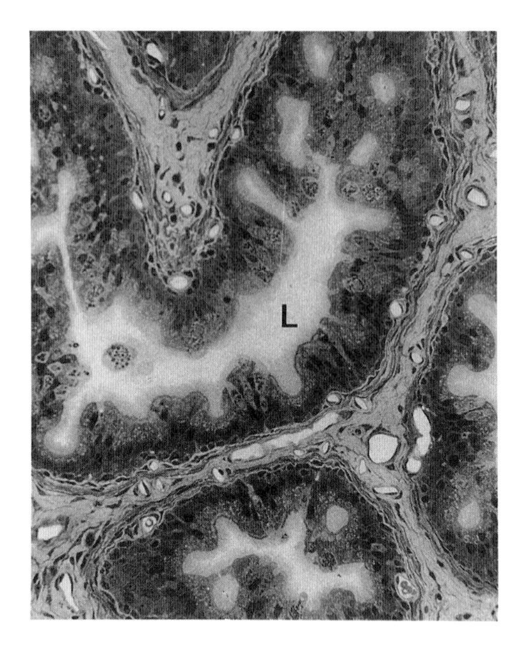


FIGURE 17.17. Vasectomized group, 3 months after operation. The epididymal lumen (L) is collapsed and lacks sperm, and the epithelium that surrounds it has an irregular contour (×250). Reprinted with permission from Flickinger, Herr, Howards, et al. (33).

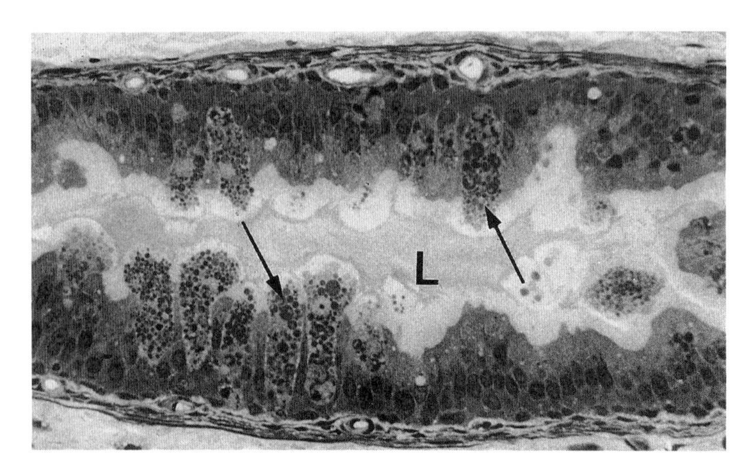


FIGURE 17.18. Seven months after vasectomy. Columnar epithelial cells appear distended with numerous densely staining granules (arrows) (×300). (L = lumen.) Reprinted with permission from Flickinger, Herr, Howards, et al. (33).

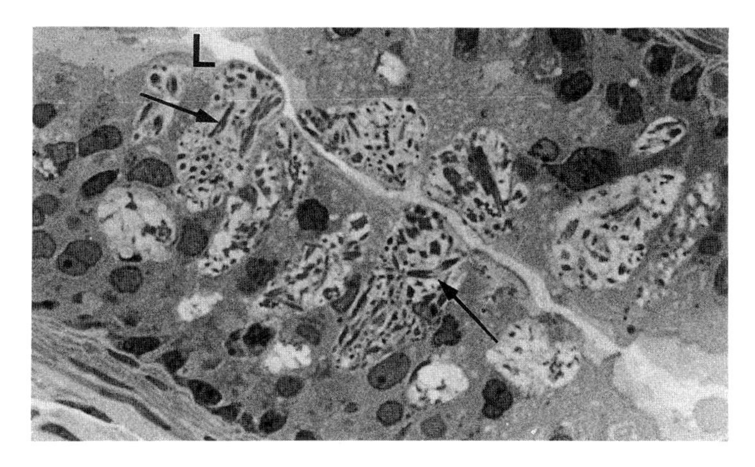


FIGURE 17.19. Vasovasostomy group, 4-month interval. A portion of the epithelium shows large vacuoles (arrows) that contain dense elongated structures (×570). (L = lumen). Reprinted with permission from Flickinger, Herr, Howards, et al. (33).

remained present at least as long as 4 months after vasovasostomy. This suggests that vasovasostomy did not result in an improvement in epididymal alterations over the time period studied. Similarly, in this model vasovasostomy did not appear to result in a reversal of testicular changes, although vasovasostomy did prevent the progression of testicular alterations after vasectomy.

## *Immunodominant Autoantigens Recognized by Postvasectomy Sera*

Protein autoantigens recognized by the majority of postvasectomy Lewis rats were identified on Western blots of reduced Laemmli extract separated by 10% SDS-PAGE (28). Immunoreactivity of a band using postvasectomy antiserum was compared to prevasectomy control serum from the same rat. Immunoblots of serum from 9 vasectomized rats and of hyperimmune antiserum (prepared by immunizing rats with isologous spermatozoa) are shown in Figure 17.21.

Several observations can be made from these blots of sperm peptide autoantigens. First, reduced peptides at 86, 63, 43, and 20 kd were recognized by a majority of the postvasectomy sera screened. The autoantigenic band at 86 kd was diffuse, spreading from ~78 to 89 kd, and in some blots appeared to be a doublet. The broad 86-kd band was recognized by 8 of 9 postvasectomy sera as well as by the positive-control hyperimmune serum. The 63-kd band was prominent at the bottom of a triplet of antigens from 63 to 74 kd. Seven of 9 postvasectomy sera

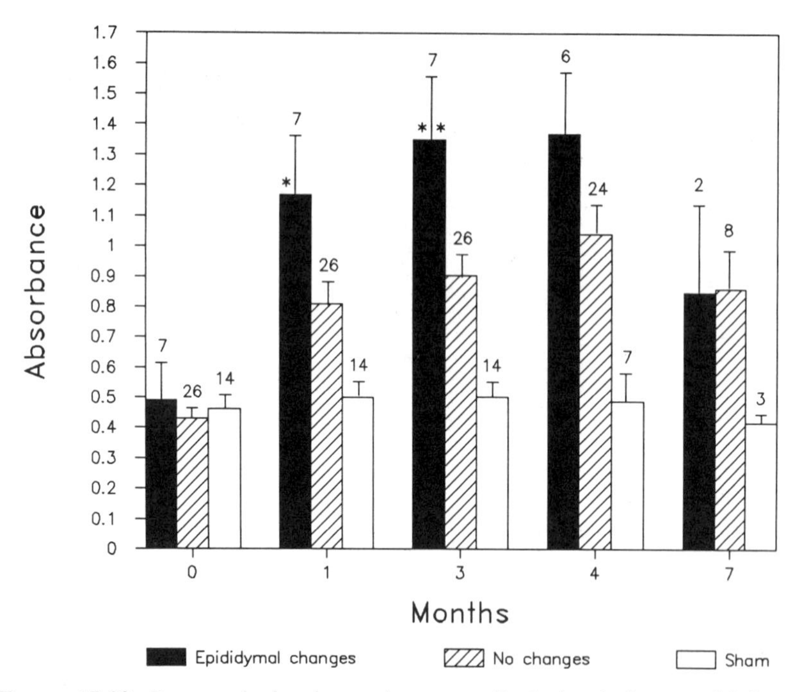


FIGURE 17.20. Bar graph showing antisperm antibody levels for rat with interstitial epididymal changes and without interstitial epididymal changes, along with those for the sham-operated group. The mean absorbance on ELISA and the SEM for each group are shown; the number of animals in each sample appears above the error bar. The animals with epididymal changes were significantly different from those without epididymal changes at 1 and 3 months (* = $P < 0.05$; ** = $P < 0.01$). Antibody levels for rats with or without epididymal changes in vasectomy and vasovasostomy groups were significantly increased over levels in sham-operated animals at all times postvasectomy except at 7 months ($P < 0.05$; $P < 0.01$; $P < 0.01$). Reprinted with permission from Flickinger, Herr, Howards, et al. (33).

recognized the 63-kd band (Fig. 17.21). The majority of individuals (8/9) recognized a band at 43 kd as well. This tight, 43-kd band migrated just below another autoantigen at 44 kd that was recognized by 5 out of 9 animals. The 20-kd band, representing a diffuse group of peptides with molecular masses of between 16 and 21 kd, was bound by 6 out of 9 postvasectomy sera.

Of additional note, 2 autoantigens at 54 and 57 kd were recognized by a majority of individuals in the study. These antigens were only weakly detected at this titer in all animals except animal 1. Autoantigen recognition in the postvasectomy Lewis rat is summarized in Table 17.4, which outlines the repertoire of major and minor Laemmli-soluble autoantigens recognized by each individual.

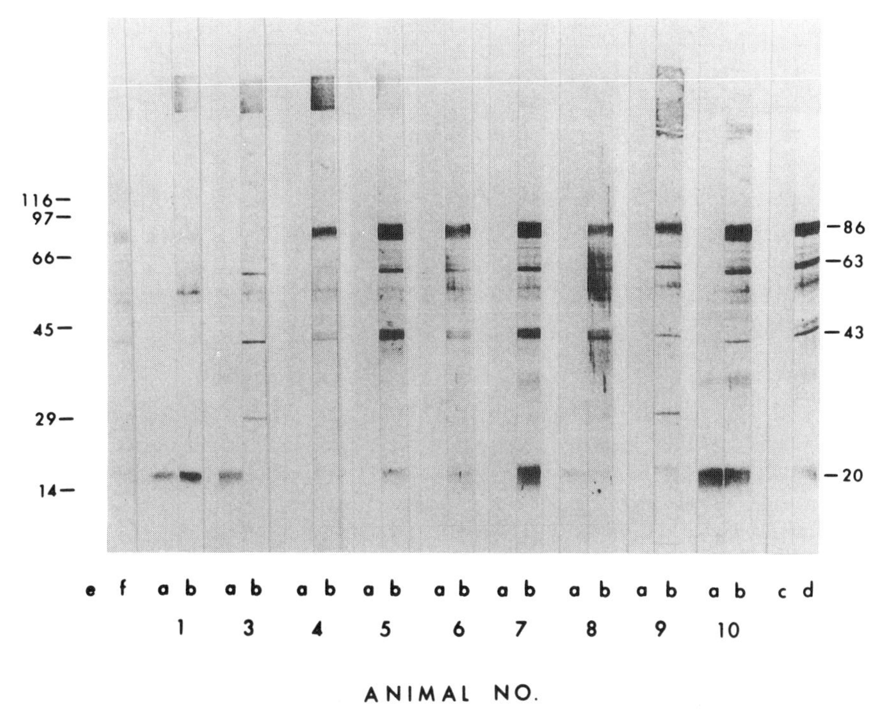


FIGURE 17.21. Comparison of sperm autoantigens bound by the sera of individual vasectomized rats. Lewis rat sperm antigens were solubilized in Laemmli buffer containing 2% 2-β-mercaptoethanol and separated by 10% SDS-PAGE. Transblots were stained with prevasectomy serum (a-lanes), postvasectomy serum (b-lanes, 56 days), pooled normal Lewis rat serum (c-lane), or pooled hyperimmune serum (d-lane). Amido black was used to identify molecular mass standards (e-lane) or total transblotted sperm proteins (f-lane). The majority of postvasectomy sera bound antigens of 86, 63, 43, and 20 kd. Reprinted with permission from Handley, Flickinger, and Herr (28).

## Monospecific Isoantiserum

Antigen-specific isoantisera were purified from the hyperimmune isoantiserum (32) by a modification of the affinity purification method described by Olmsted (26). The specificity of each affinity-purified isoantiserum was assessed on Western blots of SDS-PAGE-separated sperm antigens (Fig. 17.22). All antisera bound the antigens from which they were derived. As shown in Figure 17.22, the anti-63-kd (blot B) isoantiserum and the anti-28-kd (blot D) and the anti-20-kd (blot E) isoantibodies bound only their respective antigens. In contrast, the anti-86-kd (blot A) and the anti-43-kd (blot C) isoantibodies demonstrated significant cross-reaction with each other and with other sperm proteins.

TABLE 17.4. Summary analysis of individual postvasectomy antigen recognition.

| | Animal number | | | | | | | | | | |
|---|---|---|---|---|---|---|---|---|---|---|---|
| kd ($R_f$) | 1 | 3 | 4 | 5 | 6 | 7 | 8 | 9 | 10 | Hypr | +/total |
| 86 (0.30) | − | − | + | + | + | + | + | + | + | + | 8/10 |
| 63 (0.39) | − | + | + | + | + | + | + | + | + | + | 9/10 |
| 57 (0.43) | + | + | + | + | + | + | + | + | + | + | 10/10 |
| 54 (0.46) | + | − | + | + | + | + | + | − | + | + | 8/10 |
| 44 (0.53) | − | − | + | + | + | + | + | − | − | + | 6/10 |
| 43 (0.55) | + | + | + | + | + | + | + | + | + | + | 10/10 |
| 42–38 (0.56–0.63) | − | − | − | + | − | + | + | − | − | + | 4/10 |
| 37 (0.65) | − | − | − | − | − | + | − | − | + | + | 3/10 |
| 31 (0.65) | − | + | − | − | − | − | − | + | − | − | 3/10 |
| 26 (0.81) | − | − | − | − | − | − | ± | − | − | + | 2/10 |
| 20 (0.86–0.92)[a] | − | − | − | + | + | + | − | + | + | + | 6/10 |
| 15 (0.00) | + | + | + | + | − | − | − | + | + | − | 6/10 |

*Note:* Postvasectomy serum produced an immunoreaction product distinctly greater than prevasectomy serum (±), slightly greater than prevasectomy serum (+), or indistinctly different from prevasectomy serum (−). (Hypr = hyperimmune antiserum rat.)

[a] The 20-kd band represents a diffuse group of peptides with molecular masses of between 16 and 21 kd.

## *Localization of Sperm Autoantigens on Sperm and in Testis*

Indirect immunofluorescence with the monospecific isoantisera was utilized to locate the major postvasectomy autoantigens in caudal epididymal spermatozoa (32). The polyclonal antiserum from which the monospecific antisera were produced stained the entire spermatozoon, while monospecific antibodies bound only to the sperm tail, staining the proximal portion (43 and 20 kd), a distal domain (63 kd), or the entire tail (86 kd). Immunohistochemically stained sections of normal rat testes revealed that the 63-, 43-, and 20-kd autoantigens were synchronously expressed in the cytoplasm of spermatids in the apical portions of seminiferous tubules during stages II–VIII in the cycle of the seminiferous epithelium. Figure 17.23 shows the stage-specific expression of the 63-kd monospecific isoantisera.

## Summary and Future Direction

Vasectomy induces serum antisperm antibodies in Lewis rats. In most animals these antibodies arise within a week or two of the operation. The antibodies decline in titer gradually. Some rats develop severe testicular lesions and have reduced testis weights as a result of vasectomy and

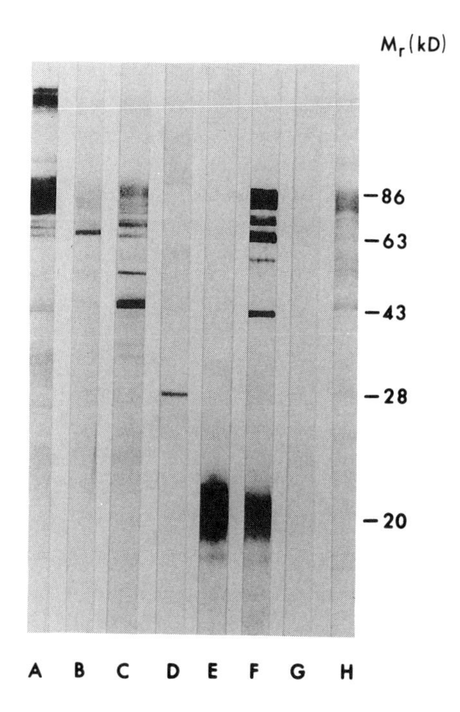


FIGURE 17.22. Western immunoblot showing the specificity of affinity-purified monospecific isoantisera to the sperm antigens at 86 (A), 63 (B), 43 (C), 28 (D), and 20 kd (E). Pooled polyclonal hyperimmune serum (F) and a pooled normal Lewis rat serum (G) were diluted 1:50 before use. Monospecific isoantisera were diluted 1:20 before use. (H = amido black stain of total nitrocellulose bound sperm proteins after separation in the identical preparative gel.) Reprinted with permission from Handley, Herr, and Flickinger (32).

vasovasostomy. Antisperm antibodies are generally higher in animals with altered testes than in animals with normal testes, although there are exceptions in individual animals. Epididymitis occurs in all animals with testicular lesions. Antisperm antibodies are elevated in animals with epididymal lesions.

Vasovasostomy in Lewis rats does not reduce the levels of antisperm antibodies 1 or 4 months after reanastomosis compared to rats with a persisting vasectomy. While vasovasostomy in Lewis rats prevents the progression of testicular alterations with time, the operation did not appear to reverse testicular changes. Surprisingly, vasovasostomy in Sprague-Dawley rats causes an increase in antisperm antibody above that of animals with persisting vasectomy. Granulomas developed as a result of vasovasostomy in many Lewis and Sprague-Dawley rats.

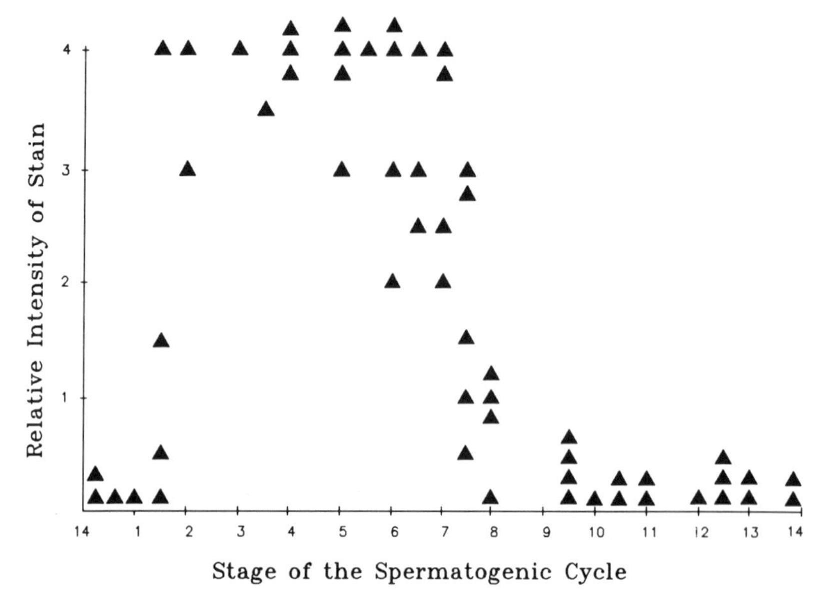


FIGURE 17.23. Plot of the relative intensity of the immunocytochemical reaction of anti-63-kd serum with seminiferous tubules at each stage of spermatogenesis, with each point representing a single observation. Serial sections of the Lewis rat testis were stained alternately with toluidine blue or monospecific isoantibodies to the 63-kd autoantigen. Reprinted with permission from Handley, Herr, and Flickinger (32).

Finally, immunodominant sperm autoantigens of 86, 63, 43, and 20 kd have been identified. The 86-kd antigen is associated with the entire tail, the 43- and 20-kd antigens with the proximal tail, and the 63-kd antigen with the distal tail. Each of these antigens is expressed during stages II–VIII of the seminiferous cycle. The cloning, sequencing, and detailed molecular characterization of the dominant autoantigens recognized during the postvasectomy autoimmune response represent important avenues for future research.

## *References*

1. Alexander NJ, Anderson DJ. Vasectomy—consequences of autoimmunity to sperm antigens. Fertil Steril 1979;32:253–60.
2. Linnet L. Clinical immunology of vasectomy and vasovasostomy. Urology 1983;22:101–14.
3. Bigazzi PE, Kosuda LL, Hsu KC, Andres GA. Immune complex orchitis in vasectomized rabbits. J Exp Med 1976;143:382–404.
4. Anderson DJ, Alexander NJ, Fulgham DL, Palotay JL. Spontaneous tumors in long-term vasectomized mice: increased incidence and association with antisperm immunity. Am J Pathol 1983;111:129–39.

5. Alexander NJ, Clarkson TB. Vasectomy increases the severity of diet-induced atherosclerosis in *Macaca fascicularis*. Science 1978;201:538–41.
6. Clarkson TB, Alexander NJ. Vasectomy: effects on the occurrence and extent of atherosclerosis in rhesus monkeys. J Clin Invest 1980;65:15–25.
7. Goldacre M, Vessey M, Clarke J, Heasman M. Record linkage study of morbidity following vasectomy. In: Lepow IH, Crozier R, eds. Vasectomy: immunologic and pathophysiologic effects in animals and man. New York: Academic Press, 1979:567–75.
8. Walker AM, Jick H, Hunter JR, Danford A, Rathman KJ. Hospitalization rates in vasectomized men. JAMA 1981;245:2315–7.
9. Petitti DB, Klein R, Kipp H, Kahn W, Siegelaub AB, Friedman GD. A survey of personal habits, symptoms of illness, and histories of disease in men with and without vasectomies. Am J Public Health 1982;72:476–80.
10. Massey FJ, Bernstein GS, O'Fallon WM, et al. Vasectomy and health: results from a large cohort study. J Am Med Assoc 1984;252:1023–9.
11. Silber SJ. Vasectomy and vasectomy reversal. Fertil Steril 1978;29:125–40.
12. Lee L, McLoughlin MG. Vasovasostomy—comparison of macroscopic and microscopic techniques at one institution. Fertil Steril 1980;33:54–5.
13. Martin DC. Microsurgical reversal of vasectomy. Am J Surg 1981;142:48–50.
14. Thomas AJ Jr, Pontes JE, Rose NR, Segal S, Pierce JM Jr. Microsurgical vasovasostomy: immunologic consequences and subsequent fertility. Fertil Steril 1981;35:447–50.
15. Herr JC, Flickinger CJ, Howards SS, et al. An enzyme-linked immunosorbent assay for measuring antisperm autoantibodies following vasectomy in Lewis rats. Am J Reprod Immunol 1986;11:75–81.
16. Ielasi A, Kotlarski I. Species variation in antibody response, I. Quantitation of bactericidal antibody production to *S. typhimurium* in rats and mice. Aust J Exp Biol Med Sci 1969;47:689–99.
17. Rumke P, Titus M. Spermagglutinin formation in male rats by subcutaneously injected syngeneic epididymal spermatozoa and by vasoligation or vasectomy. J Reprod Fertil 1970;21:69–70.
18. Johnsen SG. Testicular biopsy score count—a method for registration of spermatogenesis in human testes: normal values and results in 335 hypogonadal males. Hormones 1970;1:2–25.
19. Flickinger CJ, Herr JC, Howards SS, et al. The influence of vasovasostomy on testicular alterations after vasectomy in Lewis rats. Anat Rec 1987;217:137–45.
20. Howards SS. Vasovasostomy. In: Skinner DG, Gloege GM, eds. The craft of urologic surgery. Urol Clin North Am 1980;7:165–9.
21. Zar JH. Biostatistical analysis. Englewood Cliffs, NJ: Prentice-Hall, 1974.
22. Carey PO, Howards SS, Flickinger CJ, et al. Effects of granuloma formation at site of vasovasostomy. J Urol 1988;139:853–6.
23. Sun EL, Flickinger CJ. Development of cell types and regional differences in the rat epididymis. Am J Anat 1979;154:27–56.
24. Neaves WB. The effect of vasectomy on the testes of inbred Lewis rats. J Reprod Fertil 1978;54:405–11.
25. Brannen GE, Coffey DS. Immunologic implications of vasectomy, II. Serum-mediated immunity. Fertil Steril 1974;25:515–20.
26. Olmsted JB. Affinity purification of antibodies from diazotized paper blots of heterogeneous protein samples. J Biol Chem 1981;256:11955–7.

27. Flickinger CJ, Howards SS, Carey PO, et al. Testicular alterations are linked to the presence of elevated antisperm antibodies in Sprague-Dawley rats after vasectomy and vasovasostomy. J Urol 1988;140:627–31.
28. Handley HH, Flickinger CJ, Herr JC. Post-vasectomy sperm autoimmunogens in the Lewis rat. Biol Reprod 1988;39:1239–50.
29. Herr JC, Flickinger CJ, Howards SS, et al. The relation between antisperm antibodies and testicular alterations after vasectomy and vasovasostomy in Lewis rats. Biol Reprod 1987;37:1297–1305.
30. Herr JC, Flickinger CJ, Howards SS, et al. Temporal appearance of antisperm autoantibodies in Lewis rats following vasectomy. J Androl 1987;8:253–8.
31. Flickinger CJ, Howards SS, Herr JC, et al. The incidence of spermatic granulomas and their relation to testis weight after vasectomy and vasovasostomy in Lewis rats. J Androl 1987;7:285–91.
32. Handley HH, Herr JC, Flickinger CJ. Localization of post-vasectomy sperm autoantigens in the Lewis rat. J Reprod Immunol 1991;20:205–20.
33. Flickinger CJ, Herr JC, Howards SS, et al. Inflammatory changes in the epididymis after vasectomy in the Lewis rat. Biol Reprod 1990;43:34–45.

# Antiphospholipid Antibodies and Placental Development

NEAL S. ROTE, TIMOTHY W. LYDEN, ELIZABETH VOGT,
AND AH KAU NG

The *antiphospholipid antibody syndrome* (aPL syndrome) is characterized by elevated levels of *autoantibodies against negatively charged phospholipids* (aPLs) associated with several severe obstetrical complications, including recurrent pregnancy loss, pregnancy-induced hypertension and preeclampsia before 34 weeks' gestation, and intrauterine growth retardation (reviewed in 1, 2). The syndrome appears to be a prothrombotic state in which the patients are at risk for thrombosis at virtually any site. The most thoroughly studied aPLs are the lupus anticoagulant and *antibodies against the phospholipid cardiolipin* (aCLs). It is commonly proposed that aCLs, which are cross-reactive with other phospholipids, are primarily responsible for the thrombosis. The obstetric complications have been attributed to aCL-induced placental damage resulting from placental or decidual thrombosis.

Despite the intensity with which the above hypothesis has been studied, there is extremely little definitive experimental evidence that supports the principle aspects of the theory; that is, that aCL is responsible for the clinical manifestations and that thrombosis is the mechanism by which the placenta is damaged. This chapter discusses a more plausible hypothesis of aPL-mediated placental damage.

Certain observations suggest that the currently widely held hypothesis is incorrect. Although thrombosis is frequently observed in the decidua and placenta of aPL-positive patients, the observation is, in fact, very difficult to confirm or state as specific for the aPL syndrome. The observations on decidual/placental thrombosis have generally been made from the examination of the products of miscarriage, which have been subjected to a long process of tissue destruction and inflammation.

One study of the decidual/placental interface associated with aPLs and a fetal death described maternal spiral arteriolar vasculopathy, including acute atherosis and intraluminal thrombosis (3). Only a single patient with aPLs, however, was investigated. The observation of decidual/placental thrombosis has been confirmed in 14 of 17 patients with aPLs and intrauterine fetal death (4). In that study, however, only 8 specimens, 4 with aPLs and 4 without, had adequate decidual material for evaluation of the maternal vessels. In none of these was atherosis observed. The association between aPLs and placental thrombosis has not been confirmed in a further study in which 5 placentas from patients with *systemic lupus erythematosus* (SLE), aCL, and abnormal fetal heart rate tests (4 of the 5 patients) were examined (5). Three of these placentas were significantly smaller than expected. One resulted in a fetal death. In all cases placental thrombosis was rare and inadequate to explain the clinical symptoms. Placental or decidual thrombosis, although commonly observed in placentas of aPL-positive women, is not present in an adequate number of patients or to a sufficient degree to account for the amount of pregnancy loss associated with this syndrome.

Additionally, the observations of decidual/placental thrombosis are not specific to aPL-positive patients. Thrombosis has also been reported in cases of *intrauterine growth retardation* (IUGR) (6), preeclampsia (7), SLE (8), and fetal death unrelated to aPLs (4). In one of the studies quoted, over half of the aPL-negative patients with fetal deaths also had signs of placental thrombosis (4). Decidual/placental thrombosis, therefore, may be a characteristic of damage to the decidual/placental interface, regardless of the presence or absence of aPLs.

Placental thrombosis is conceptually even less well supported as a cause of aPL-induced pregnancy loss. In order for placental thrombosis to occur, aPLs must cross the placenta. IgG is the only transplacental antibody in humans, yet several patients have been described who have the obstetric complications of the aPL syndrome and yet have only IgM aPLs, which are not transplacental (9–11). In addition, most affected pregnancies are lost in the first trimester when the transplacental passage of IgG is very inefficient (12–15). These observations preclude placental thrombosis as a mechanism of pregnancy loss without preexisting damage to the trophoblastic interface, which would allow free passage of antibody across the placental barrier.

Another aspect of the prevalent hypotheses is that aCL is responsible for the pathology of the syndrome. There are several difficulties with that portion of the hypothesis. In order for pathophysiology to occur, circulating aPLs must react with antigen. There is little indication that the aPL syndrome is an immune complex disease; therefore, the antigen is probably expressed on the surface of affected cells. Platelets and endothelial cells have been studied the most thoroughly and are the most commonly proposed targets for aPLs. *Cardiolipin* (CL), however, is confined to the inner leaflet of mitochondria (16–18). The current dogma is that antibodies do not freely cross the membranes of living cells. If that is correct, CL is not accessible to aPLs. The general response to this observation is that aCLs are cross-reactive and can bind to other phospholipids, particularly those in plasma membranes. The other cross-reactive phospholipids, however, are also inaccessible to aPLs. The predominant phospholipid on the cell surface is *phosphatidylcholine* (PC), which is generally not reactive with aPLs.

The aPLs react primarily with *phosphatidylserine* (PS) and, in some studies, *phosphatidylethanolamine* (PE); these are primarily on the inner side of the plasma membrane and are not accessible to aPLs under normal conditions (19–22). *Phosphatidylinositol* (PI) and *phosphatidylglycerol* (PG) are also candidates for aPL binding. Antibodies against these two phospholipids have been described, but only in a minority of the patients (11, 24–26). Only *antibody against phosphatidylserine* (aPS) is observed at an incidence similar to aCL. Under normal conditions, therefore, aPLs cannot access their target antigens.

Another discrepancy of the current hypothesis is that some patients have significant levels of aPLs, yet have uneventful pregnancies. Most studies document a 60%–90% pregnancy loss rate, a 40%–50% rate of pregnancy-induced hypertension, and a 20%–30% IUGR rate in women with aPLs (1, 2, 15, 27). Thrombosis, which is also observed in these patients, is relatively rare. Only a third of aPL-positive patients have a

TABLE 18.1. Summary of monoclonal aPL reactivities.

| | Monoclonal aPL | | |
|---|---|---|---|
| Reactivity | BA3B5C4 | 3SB9b | D11A4 |
| ELISAs[a] | | | |
| PS | ++ | ++ | − |
| CL | ++ | − | ++ |
| PA, PG, or PI | ++ | ++ | ++ |
| PE or PC | − | − | − |
| Platelets | | | |
| Resting | − | − | − |
| Thrombin-activated | + | + | − |
| Human endothelial cells | | | |
| Unfixed | − | − | − |
| Fixed/cytoskeleton | − | ++ | − |
| Trophoblast | | | |
| Villous cytotrophoblast | ++ | ± | − |
| Syncytiotrophoblast | ± | ++ | − |
| Extravillous cytotrophoblast | + | + | − |
| BeWo (− forskolin) | + | ± | − |
| BeWo (+ forskolin) | ± | ++ | − |
| In vivo mouse model | | | |
| Induction of IUGR | − | + | − |

[a] Reactivity against the phospholipids phosphatidylserine (PS), cardiolipin (CL), phosphatidic acid (PA), phosphatidylglycerol (PG), phosphatidylinositol (PI), phosphatidylethanolamine (PE), and phosphatidylcholine (PC) were tested in ELISAs.

history of thrombosis—commonly, with years separating individual thrombotic events—although most patients seem to have stable levels of circulating aPLs (28, 29). Thrombosis, therefore, does not often occur while aPLs are present, suggesting that other rare events or cofactors are necessary to initiate aPL-related thrombosis. Pregnancy, on the other hand, may be extremely sensitive to the effects of aPLs and may not need other factors. An extremely plausible explanation of this discrepancy is that aPL-associated thrombosis and pregnancy loss are independent events involving different cells.

In order to explain the mechanism of aPL-mediated pregnancy loss, therefore, we must search for a new paradigm. Our data, as well as supporting evidence from others, have resulted in the formulation of a new model. This model suggests that the placental trophoblasts can be directly affected by aPLs and explains many of the discrepancies of previously proposed mechanisms. The crux of the model is based on the following observations. First, aPS is as prevalent, if not more so, as aCL (11, 24–26). Second, PS is a predominant phospholipid on the inner side of plasma membranes (20–22). Third, under certain normal physiologic conditions, PS is externalized and, therefore, exposed to circulating aPLs (20–37). Fourth, throughout normal placental development the trophoblasts externalize PS during a constant and obligatory intercellular fusion process (38, 39, and unpublished data). Fifth, aPLs can bind to externalized PS and may cause incomplete differentiation or directly damage the trophoblast (unpublished data).

## Materials and Methods

### *Monoclonal Antibodies Against Phospholipids*

Production and serologic characterization of monoclonal aPLs have been described elsewhere (38). Each *monoclonal antibody* (mAb) is an IgM and reacted in ELISAs against the negatively charged phospholipids *phosphatidic acid* (PA), PI, and PG and did not react with PE or PC (Table 18.1). Each mAb, however, reacted differently with CL and PS; 3SB9b reacted only with PS ($CL^-/PS^+$), D11A4 reacted only with CL ($CL^+/PS^-$), and BA3B5C4 reacted with both CL and PS ($CL^+/PS^+$). The concentrations of IgM in undiluted tissue culture supernatant were 3SB9b, 80 μg/mL; BA3B5C4, 15 μg/mL; and D11A4, 30 μg/mL.

### *Antibodies from aPL-Positive Patients*

Three patients were selected for study based on positive IgM-aPLs without IgG-aPLs. One patient with the primary aPL syndrome was investigated. Two patients had aPLs secondary to SLE. All patients were tested for aPLs and other autoantibodies at the Foundation for Blood Research in Scarborough, Maine.

Patient 1 had experienced 3 successful uncomplicated pregnancies followed by 3 sequential first-trimester spontaneous pregnancy losses. She had a positive antinuclear antibody test with a homogenous staining pattern and had a diagnosis of SLE. Testing for aPLs revealed a positive lupus anticoagulant, positive IgM antibodies against CL and PS, and negative IgG and IgA antibodies against CL and PS.

Patient 2 had experienced 1 first-trimester pregnancy loss. She was positive for antinuclear antibodies with a homogenous staining pattern and had a diagnosis of SLE. Testing for aPLs revealed a positive lupus anticoagulant, positive IgM antibodies against CL and PS, positive IgG antibody against CL, and negative IgG against PS.

Patient 3 had experienced 2 first-trimester spontaneous pregnancy losses without a successful pregnancy. She was negative for antinuclear antibodies and anti-double-stranded DNA antibodies and had no clinical indication of SLE. Testing for aPLs revealed a positive lupus anticoagulant activity, positive IgM antibodies against CL and PS, and negative IgG and IgA antibodies against CL and PS. After donating serum for this study, patient 3 was successfully treated with low-dose aspirin and heparin and delivered twins on the following pregnancy (40). Controls consisted of female laboratory personnel who were negative on complete aPL testing.

## *Preparation of Human Placentas*

Placentas were collected anonymously at Miami Valley Hospital from normal deliveries or from repeat cesarean sections. Tissue associated with the basal plate in the region of the marginal lake was dissected from freshly delivered normal placentas, cut into sections of approximately 2–3 $cm^2$, and immersed immediately in liquid nitrogen or fixed in 10% formalin for 24 h at room temperature. The frozen tissue was wrapped in aluminum foil and stored at −70°C. The tissue was transferred to a chilled cryostat, mounted to a holder using O.C.T. mounting media, and cut into 6-μm sections.

The formalin-fixed tissue was cut into 1-$cm^2$ pieces and processed for embedding using a Fisher histoprocessor. Sections (6 μm) were cut and dried onto gelatin/chrome-alum-coated slides for 18 h at 42°C. Sections were deparaffinized by immersion in xylene for 15 min, which was repeated twice, and rehydrated through a series of graded alcohols: 100% ETOH for 10 min (×2); 95% ETOH for 10 min (×2); 70% ETOH for 10 min (×2); and PBS (pH 7.3) for 10 min (×2). Endogenous peroxidase activity in the sections was blocked by routine immersion in 9 parts methanol/1 part hydrogen peroxide for 10 min. The sections were washed in PBS (pH 7.3) for 2 min (×3).

## *Immunoperoxidase Labeling of Placental Preparations*

Nonspecific binding was blocked by incubation for 10 min in 10% non-immune rabbit serum. Test antibody (200 μL) was added to the sections following removal of the blocking solution. The sections were incubated for 1 h at room temperature in a moist chamber. Slides were rinsed for 2 min in flowing PBS. Biotinylated polyclonal anti-mouse IgM (200 μL) (Histostain-SP Kit, Zymed Inc.) was placed on each section and incubated for 30 min at room temperature. This was rinsed with flowing PBS for 2 min, followed by a 5 min incubation with avidin-peroxidase conjugate, a rinse in PBS, and incubation for 5–15 min with the kit's substrate-chromogen reagent. The slides were rinsed well with DDH2O and counterstained for 3′ with hematoxylin, washed with DDH2O, and then mounted with cover slips and evaluated.

## *In Vivo Activity in Pregnant Mice*

Female BALB/c mice were mated and injected IP on day 8 of pregnancy with 1 mL of an identical concentration of each monoclonal aPL per mouse with 10 animals per group. Negative controls included an IgM mAb against double-stranded DNA (Ch26-1352, ATCC). On day 15 each of the animals was sacrificed. The uterus was dissected and evaluated for numbers of reabsorption sites, numbers of potentially viable fetuses, and the weights of fetuses and placentas.

# Results

## *3SB9b and BA3B5C4 Reactivity with Placenta*

The most notable reaction with BA3B5C4 ($CL^+/PS^+$) was associated with the villous cytotrophoblastic cell layer (Fig. 18.1A). Formalin-fixed cells reacted very strongly with a clearly cytoplasmic labeling. The syncytiotrophoblastic layer was very poorly labeled in most areas examined, although in some focal areas the syncytiotrophoblastic layer showed an intense pattern of labeling. These particular areas appeared to be devoid of an underlying cytotrophoblast layer, suggesting that the syncytiotrophoblastic staining may be at sites of recent cytotrophoblast fusion.

3SB9b ($CL^-/PS^+$) showed an intense labeling at the apical surface of the syncytiotrophoblastic region (Fig. 18.1B). This labeling was most pronounced on villous surfaces associated with open areas of the maternal blood space. In addition to the apparent surface labeling observed, the syncytiotrophoblast also displayed a strong cytoplasmic labeling that was most intense below the apical surface. Occasional cellular labeling was associated with the stroma of the villi and may involve some villous cytotrophoblast cells.

A

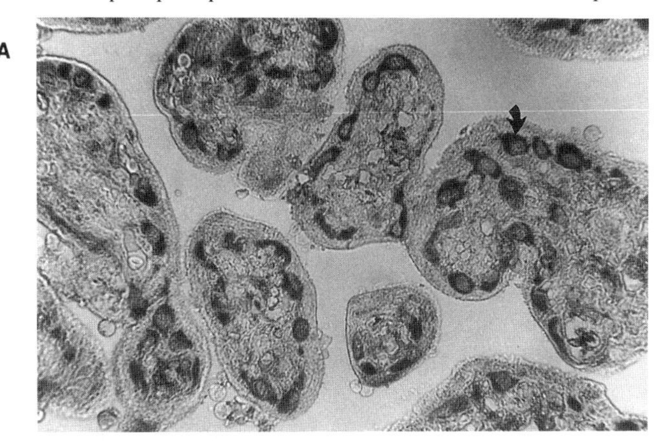

B

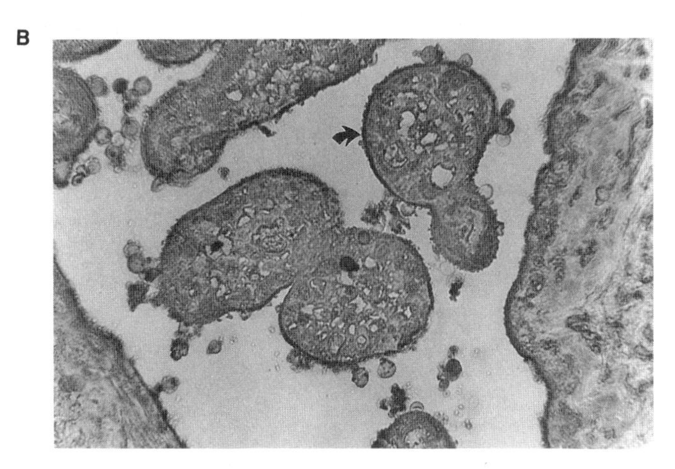

FIGURE 18.1. Indirect immunoperoxidase staining of formalin-fixed placental sections in the chorionic villus region reacted with BA3B5C4 ($CL^+/PS^+$) (*A*) or 3SB9b ($CL^-/PS^+$) (*B*). BA3B5C4 reactivity was localized in the cytotrophoblast (arrow), and 3SB9b reacted in the apical area of the syncytiotrophoblast (arrow) (×33).

Throughout this study the normal PBS controls were negative for labeling. Monoclonal aPL D11A4 effectively served as a negative control for nonspecific mouse IgM binding. Anti-cytokeratin 4.62 was used as a positive control for cytotrophoblastic cell staining and demonstrated expected labeling patterns.

In all cases throughout this study, frozen sections reacted with a similar cellular distribution as that observed in formalin-treated tissue; although cytoplasmic staining could not be observed, the same cells labeled on the cell surface with the same relative intensity.

## *Reactivity of aPL-Positive Patients' Sera with Normal Human Placenta*

In order to make direct comparisons with the IgM monoclonal aPLs, 3 patients with naturally occurring IgM aPLs were tested for reactivity against formalin-fixed normal placenta. Each patient had IgM that reacted with placenta, but with a different pattern of staining. Serum from patient 2 had antibodies that reacted with both the syncytium and the villous

A

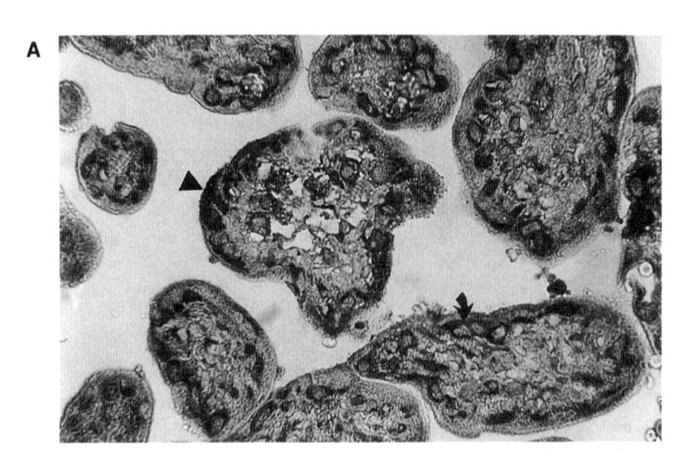

B

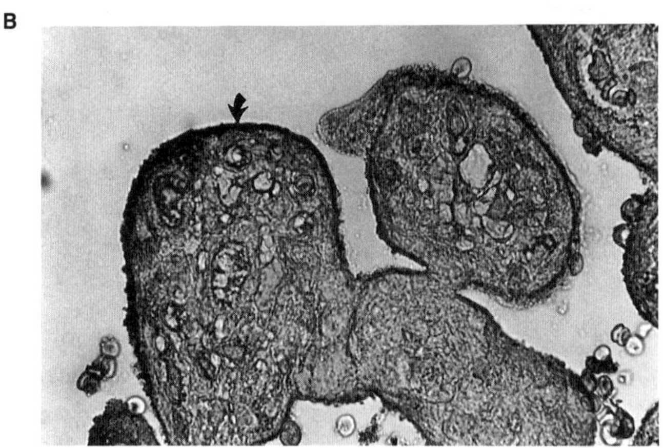

FIGURE 18.2. Indirect immunoperoxidase staining of formalin-fixed placental sections in the chorionic villus region reacted with IgM aPL-positive patient's serum from patient 2 (*A*) or patient 1 (*B*). Serum from patient 2 had antibodies that reacted with both the syncytium and the villous cytotrophoblastic cells (*A*). The syncytial activity was localized to several regions (arrowhead). Cytotrophoblastic staining appeared uniform (arrow). Serum from patient 1 had IgM antibodies that stained only the syncytiotrophoblast (arrow), with very little cytotrophoblastic reactivity (*B*) (×33).

TABLE 18.2. In vivo activity of monoclonal aPLs.

| Injected monoclonal | Fetal weight (mg) | Placental weight (mg) |
|---|---|---|
| IgM anti-dsDNA | 151 ± 56[a] | 91 ± 13 |
| D11A4 ($CL^+/PS^-$) | 119 ± 47 | 89 ± 25 |
| BA3B5C4 ($CL^+/PS^+$) | 149 ± 48 | 89 ± 20 |
| 3SB9b ($CL^-/PS^+$) | 81 ± 27[b] | 63 ± 15[b] |

[a] Mean ± SD.
[b] Decreased fetal and placental weights for 3SB9b were significant at $P < 0.001$.

cytotrophoblastic cells (Fig. 18.2A). The syncytial activity generally did not span the entire syncytium, although several regions were observed in which the staining was very intense. Cytotrophoblastic staining appeared uniform, with most, if not all, cytotrophoblast staining in an apparent cytoplasmic pattern. Serum from patient 1 had IgM antibodies that stained only the syncytiotrophoblast, with very little cytotrophoblastic reactivity (Fig. 18.2B). Serum from patient 3 possessed IgM that stained only the villous cytotrophoblast, with little apparent syncytiotrophoblastic staining.

## *In Vivo Activity of Monoclonal aPLs*

Although still preliminary, very intriguing observations were made when monoclonal aPLs were tested in pregnant mice (Table 18.2). Animals injected with a negative-control IgM mAb against dsDNA or with D11A4 ($CL^+/PS^-$) had only 8% and 10% absorptions, respectively, and normal weights in comparison with saline-injected controls. Animals injected with 3SB9b ($CL^-/PS^+$) and BA3B5C4 ($CL^+/PS^+$) each showed markedly increased absorption (33% and 30%), although only 3SB9b induced decreased fetoplacental size.

## Discussion

Phospholipid distribution in normal membranes is asymmetric; PS is expressed primarily on the cytoplasmic surface of the plasma membrane, and CL is confined to mitochondria (16–22). Neither phospholipid, therefore, is normally accessible to aPLs. The key to understanding the pathophysiology of the aPL syndrome is in comprehending the normal physiological conditions during which antigenic phospholipid conformations are made accessible to circulating aPLs. Although CL would only be accessible after cell death and fragmentation, there are several physiological circumstances under which plasma membrane PS and PE are externalized.

Erythrocytes and lymphocytes undergoing senescence express surface PS that may, in part, be the recognition by which aged cells are removed from the circulation by macrophages (35, 36). Phosphatidylserine is also externalized on platelets during activation and fusion between platelet granules and the plasma membrane (30–32). During intercellular fusion in which myoblasts form myotubules, both PS and PE are externalized, being present at low concentrations (approximately 35% and 25%, respectively, of the total phospholipid) in the outer leaflet in fusion-incompetent cells and increasing to 65% and 45% within the outer leaflet of fusion-competent myoblasts (33, 34). Phosphatidylserine and PE are normally exposed, therefore, under conditions of senescence and membrane fusion, either during degranulation or syncytium formation.

In order to investigate antigenic phospholipid expression, we have used a set of IgM monoclonal aPLs and aPL-positive sera from patients with the aPL syndrome. Each murine mAb reacted differently with CL and PS; 3SB9b reacted with PS ($CL^{-}/PS^{+}$), D11A4 with CL ($CL^{+}/PS^{-}$), and BA3B5C4 with both CL and PS ($CL^{+}/PS^{+}$).

These reagents were tested against human platelets by flow cytometry (10). Normal resting platelets did not bind aPLs. After activation of platelets with thrombin, both PS-reactive monoclonal aPLs (3SB9b and BA3B5C4) bound to the platelet surface, whereas the CL-reactive monoclonal aPL (D11A4) did not react. An asymmetry of phospholipids in platelet membranes has been described in which no PS and little PE are present on the exterior surface of the resting platelet membrane, although both are present within the plasma membrane (20). Thrombin will cause externalization of approximately 15% of membrane PS, although other platelet activators are much more efficient (30, 31, 37, 41).

Freeze-thawed or sonicated platelets can bind aCL from patients' sera, and phospholipids extracted from platelets inhibit aCL binding (42). The precise platelet phospholipids that were responsible for absorption of aCL included PS, PE, and PI, whereas PC and sphingomyelin were not active. The aCL reactivity described in this study was probably cross-reactive with PS, as is the mAb BA3B5C4 in our study.

It has also been demonstrated that antibodies eluted from platelets obtained from patients with circulating aPLs have aPL activity in ELISAs (43). The aPS reacted well with interior structures of platelets, but virtually no binding was observed on the surface of unactivated platelets (44). The results of these studies strongly suggest that platelet activation is necessary to induce externalization of negatively charged phospholipids and provide the targets for aPLs.

Monoclonal 3SB9b and aPL-containing patients' sera also do not react with the surface of live cultured endothelial cells, but are strongly reactive with a lipid-dependent antigen on vimentin-containing cytoskeletal intermediate filaments (Lin et al., submitted). Additionally, aPL-containing sera can specifically block the binding of 3SB9b to intermediate filaments. Negatively charged phospholipids, especially PS, interact directly or indirectly with a variety of $Ca^{++}$-dependent and -independent cytoskeletal proteins (32, 45–58). This interaction may play a role in maintaining phospholipid asymmetry in plasma membranes of inactive cells and in mediating their externalization during cell activation (50, 55, 56).

Talin, for instance, binds negatively charged phospholipids, is found in the cytoplasm of some resting cells, is associated with the plasma membrane of activated cells, and is hypothesized to form the link between transmembrane proteins and the cytoskeleton and to mediate the exposure of PS during platelet activation (58). Additionally, PS binds to spectin in erythrocyte membranes, where oxidation of spectin sulfhydryl groups results in increased exposure of PS and PE on the erythrocyte surface (48). Thus, the activation of several cytoskeletal proteins has been linked to PS expression on cell surfaces. In endothelial cells these interactions are less well characterized, but, as with platelets, the endothelial cells may need activation by specific agents in order to externalize antigenic PS and make the cell susceptible to aPLs.

During normal placental development the trophoblastic syncytium that separates maternal and fetal compartments expands by intercellular fusion from the underlying villous cytotrophoblastic cells (59). Our results suggest that trophoblastic cells express antigenic PS, which undergoes epitope modulation as the cell gains fusion competence. The BA3B5C4 ($CL^+/PS^+$) reacted strongly with the prefusion cytotrophoblast. The 3SB9b ($CL^-/PS^+$) reacted with the syncytiotrophoblastic layer of formalin-fixed and frozen placental tissue. The apical surface of the syncytiotrophoblastic layer reacted in all sections, regardless of the method of preparation, age, or region of the placenta. Other studies have demonstrated that both trophoblast and choriocarcinoma trophoblast models express the 3SB9b-reactive epitope on their surface after undergoing differentiation (Katsuragawa et al., submitted). These observations provide direct evidence that aPS will react with normal syncytiotrophoblast and may potentially directly mediate pathological effects at the critical fetomaternal interface.

The trophoblastic layer directly in contact with the maternal circulation is most reactive with aPS, rather than with aCL. The differential reactivity of 3SB9b and BA3B5C4 suggests that the antigenic conformation involving PS on the cytotrophoblast is altered into a different conformation as a result of fusion into the syncytium. The modulation of PS antigenicity between the cytotrophoblast and syncytiotrophoblast may be due to conformational changes occurring during PS redistribution and reassociation with an evolving array of specific proteins or lipids during differentiation of the cytotrophoblast.

The conformational nature of phospholipid epitopes has been strongly suggested by studies on PE, whose bilaminar forms are not antigenic, yet whose inverted hexagonal presentations are strongly reactive with aPLs

(60). Phosphatidylethanolamine may be unusual, however, because of its ease in assuming different conformations. In plasma membranes, the vast majority of the phospholipid appears to be naturally in bilaminar form.

Accessory molecules may help stabilize hexagonal antigenic forms (61). The apparent conformational nature of phospholipid-dependent antigenic determinants was first experimentally supported by reports that aCL may, in fact, be directed against a complex of CL and an ~50-kd plasma protein cofactor, *β2 glycoprotein I* (β2-GPI or *apolipoprotein H* [apo H]) (62). β2-GPI can induce CL to undergo bilaminar-to-hexagonal transition (63). Most likely, this protein stabilizes CL in an antigenic conformation that is not expressed on pure CL (64, 65). The aPL binding to phospholipids may also be dependent on other proteins, including prothrombin (66). Interactions with β2-GPI have not been thoroughly examined for PS, and there is some evidence that PS only becomes hexagonal at low pH (67). It is highly probable, therefore, that through interactions with other molecules, such as proteins or other lipids, bilaminar phospholipids can assume nonbilayer forms at localized sites in the plasma membrane. All experiments described in our study were performed in the presence of serum; therefore, both CL and PS are likely maintained in protein-dependent conformations.

Pregnancy is characterized by the replacement of the maternal vascular endothelial cells of spiral arterioles by extravillous cytotrophoblast and by progressive movement of the cytotrophoblast up the spiral arteriole to the inner intramyometrial segments. Monoclonal aPS also reacted with extravillous cytotrophoblast; BA3B5C4 with cytoplasmic structures and 3SB9b with the plasma membrane region. The staining of extravillous cytotrophoblast is interesting because of the inflammatory process in the maternal spiral arteries in these patients. This reaction may be due to direct injury by aPLs to the extravillous cytotrophoblast that lines the spiral arteries and may lead to thrombosis.

Not all patients with aPLs may be susceptible to pregnancy loss. The two different forms of PS-reactive (3SB9b and BA3B5C4) and CL-reactive (BA3B5C4 and D11A4) mAbs that we investigated cannot be differentiated on PS or CL ELISAs. They can, however, be differentiated by the pattern of reactivity against human placenta. Our in vivo experiment suggests that the syncytiotrophoblast-reactive specificity may be the most important; 3SB9b, the monoclonal aPL reactive with the syncytiotrophoblast, causes significant reductions in placental and fetal weights, whereas neither BA3B5C4 nor D11A4 affects placental or fetal size. Although 3SB9b and BA3B5C4 have completely different specificities for placenta, they are indistinguishable in ELISAs using PS as an antigen. Similarly, BA3B5C4 reacts strongly with cytotrophoblast, and D11A4 is totally unreactive, although they cannot be differentiated on ELISAs using CL.

Patients have similar specificities (Vogt et al., submitted). All 3 aPL-positive patients we investigated had similar IgM antibodies against CL and PS. Each of the 3, however, reacted differently against placenta; one reacted primarily against the syncytiotrophoblast, a second reacted primarily against cytotrophoblast, and the third had both specificities. Only the BA3B5C4-like reactivity was absorbed with PC/PS vesicles (unpublished observations). Although these data were produced with a limited number of patients, they suggest that the various obstetric effects of the aPL syndrome may depend on the fine specificity of the aPLs against placenta, rather than reactivity in ELISAs.

How may aPLs interfere with trophoblast function? Our data support the hypothesis that PS accessibility on trophoblast is related to differentiation and intertrophoblastic fusion. In addition, the PS-dependent epitope modulates in relationship to the differentiation and fusion process; the BA3B5C4 epitope is expressed on the prefusion villous cytotrophoblast. The postfusion syncytiotrophoblast expresses the 3SB9b epitope, primarily on the apical surface.

The relationship of PS expression with differentiation and intertrophoblastic fusion has been studied using an in vitro model; forskolin induced fusion of the choriocarcinoma line BeWo (39). BA3B5C4 reacted with prefusion BeWo, but weakly with postfusion cells; 3SB9b reacted strongly only after forskolin treatment. Preliminary data from our laboratory suggest that 3SB9b will prevent intertrophoblastic fusion in vitro, whereas fusion proceeds normally in the presence of BA3B5C4 and D11A4 (unpublished observations). Antiphospholipid antibody-induced obstetrical complications, therefore, may be mediated by aPL-mediated inhibition of the cytotrophoblast fusion process, either at the prefusion cytotrophoblastic step or on the postfusion syncytium, resulting in placental damage. Inefficient fusion of the cytotrophoblast to the growing syncytium may result in defective placentation and ineffective trophoblastic transport mechanisms.

The possibility of direct damage to the trophoblastic layer by aPLs is supported by recent histologic observations on placentas from aPL-positive patients who experienced intrauterine fetal deaths (4). The most significant difference between placentas from aPL-negative and aPL-positive patients was decreased vasculosyncytial membranes. In addition, every aPL-positive patient's placenta that had evidence of thrombosis also had an increase in syncytial knots. Although the effect on syncytial membranes was attributed to hypoxia secondary to placental thrombosis and infarction, a counterhypothesis is that thrombosis occurred secondary to trophoblastic damage, allowing free transplacental passage of maternal aPLs.

To our knowledge, our studies provide the first demonstration of potential sites of aPL reactivity within the normal human placenta. Furthermore, the data presented in this study support the proposal that different relative aPL specificities within a patient will dictate the range of pathological conditions observed. Thus, aPLs reactive with PS may have direct cytopathic effects on the placenta, and aPLs reactive with CL may be unrelated to placental pathophysiology.

*Acknowledgments.* This work was supported by National Institutes of Health Grant HD-23697. The authors would like to thank Dorene Johnson at the Foundation for Blood Research for providing the patients' sera.

## *References*

1. Rote NS, Walter A, Lyden TW. Antiphospholipid antibodies: lobsters or red herrings? Am J Reprod Immunol 1992;28:31–7.
2. Rote NS. Alloantibodies, autoantibodies, and disorders of pregnancy. In: Naz R, ed. CRC monographs: immunology of reproduction. Boca Raton, FL: CRC Press, 1993:145–68.
3. DeWolf F, Carreras LO, Moerman P, Vermylen J, Van Assche A, Renaer M. Decidual vasculopathy and extensive placental infarction in a patient with repeated thromboembolic accidents, recurrent fetal loss, and a lupus anticoagulant. Am J Obstet Gynecol 1982;142:829–34.
4. Out HJ, Kooijman CD, Bruinse HW, Derksen RHWM. Histopathological findings in placentae from patients with intra-uterine fetal death and antiphospholipid antibodies. Eur J Obstet Gynecol Reprod Biol 1991;41:179–86.
5. Lochshin MD, Druzin ML, Goei S, et al. Antibody to cardiolipin as a predictor of fetal distress or death in pregnant patients with systemic lupus erythematosus. N Engl J Med 1985;313:152–6.
6. Althabe O, Labarrere C, Telenta M. Maternal vascular lesions in placentae of small-for-gestational-age infants. Placenta 1985;6:265.
7. Sheppard BL, Bonnar J. An ultrastructural study of utero-placental spiral arteries in hypertensive and normotensive pregnancy and fetal growth retardation. Br J Obstet Gynaecol 1981;88:695.
8. Abramowsky CR, Vegas ME, Swinehart G, Gyves MT. Decidual vasculopathy of the placenta in lupus erythematosus. N Engl J Med 1980;303:668–72.
9. Branch DW, Rote NS, Dostal DA, Scott JR. Association of lupus anticoagulant with antibody against phosphatidylserine. Clin Immunol Immunopathol 1987;42:63–75.
10. Rote NS, Ng AK, Dostal-Johnson DA, Nicholson S, Siekman R. Immunologic detection of phosphatidylserine externalization during thrombin-induced platelet activation. Clin Immunol Immunopathol 1993;66:193–200.
11. Triplett DA, Brandt JT, Musgrave KA, Orr CA. The relationship between lupus anticoagulants and antibodies to phospholipid. JAMA 1988;259:550–4.
12. Lubbe WF, Butler WS, Palmer SJ, Liggins GC. Lupus anticoagulant in pregnancy. Br J Obstet Gynaecol 1984;91:357–63.
13. Branch DW, Scott JR, Kochenour NK, Hershgold E. Obstetric complications associated with the lupus anticoagulant. N Engl J Med 1985;313:1322–6.

14. Derue GJ, Englert JH, Harris EN, Gharavi AE, Morgan SH, Hull RG. Fetal loss in systemic lupus: association with anticardiolipin antibodies. J Obstet Gynaecol 1985;5:207–9.
15. Branch DW, Scott JR. Clinical implications of anti-phospholipid antibodies: the Utah experience. In: Harris EN, Exner T, Hughes GRV, Asherson RA, eds. phospholipid-binding antibodies. Boca Raton, FL: CRC Press, 1991: 335–46.
16. Ioannou PV, Golding BT. Cardiolipins: their chemistry and biochemistry. Prog Lipid Res 1979;17:279–318.
17. Daum G. Lipids of mitochondria. Biochim Biophys Acta 1985;822:1–42.
18. Dale MP, Robinson NC. Synthesis of cardiolipin derivatives with protection of the free hydroxyl: its application to the study of cardiolipin stimulation of cytochrome c oxidase. Biochem J 1988;27:8270–5.
19. Op den Kamp JAF. Lipid asymmetry in membranes. Annu Rev Biochem 1979;48:47–71.
20. Schick PK, Kurica KB, Chacko GK. Localization of phosphatidyl ethanolamine and phosphatidyl serine in the human platelet plasma membrane. J Clin Invest 1976;57:1221–6.
21. Sune A, Bette-Bobillo P, Bienvenue A, Fellmann P, Devaux PF. Selective outside-inside translocation of amino-phospholipids in human platelets. Biochem J 1987;26:2972–8.
22. Beleznay Z, Zachowski A, Devaux PF, Navazo MP, Ott P. ATP-dependent aminophospholipid translocation in erythrocyte vesicles: stoichiometry of transport. Biochem J 1993;32:3146–52.
23. Gharavi AE, Harris EN, Asherson RA, Hughes GRV. Anticardiolipin antibodies: isotype distribution and phospholipid specificity. Ann Rheum Dis 1987;46:1–6.
24. Branch DW, Rote NS, Dostal DA, Scott JR. Association of lupus anticoagulant with antibody against phosphatidylserine. Clin Immunol Immunopathol 1987;42:63–75.
25. Branch DW, Rote NS, Scott JR. Demonstration of lupus anticoagulant antigens using an enzyme-linked immunoabsorbent assay (ELISA). Ann NY Acad Sci 1986;475:370–2.
26. Branch DW, Rote NS, Scott JR. The demonstration of the lupus anticoagulant by an enzyme-linked immunoabsorbent assay (ELISA). Clin Immunol Immunopathol 1986;39:296–307.
27. Scott JR, Rote NS, Branch DW. Immunologic aspects of recurrent abortion and fetal death. Obstet Gynecol 1987;70:645–56.
28. Triplett DA, Brandt JT. Lupus anticoagulants: misnomer, paradox, riddle, epiphenomenon. Hematol Pathol 1988;2:121–3.
29. Lechner K. Lupus anticoagulants and thrombosis. In: Verstraete M, Vermylen J, Lijnen R, Arnout J, eds. Thrombosis and haemostasis. Leuven, Belgium: Leuven University Press, 1987:525–47.
30. Bevers EM, Comfurius P, Zwaal RFA. Changes in membrane phospholipid distribution during platelet activation. Biochim Biophys Acta 1983;736:57–66.
31. Thiagarajan P, Tait JF. Binding of annexin V/placental anticoagulant protein I to platelets: evidence for phosphatidylserine exposure in the procoagulant response of activated platelets. J Biol Chem 1990;265:17420–3.

32. Zwaal RFA, Bevers EM, Comfurius P, Rosing J, Tilly RH, Verhallen PF. Loss of membrane phospholipid asymmetry during activation of blood platelets and sickled red cells; mechanisms and physiological significance. Mol Cell Biochem 1989;91:23–31.
33. Sessions A, Horowitz AF. Myoblast aminophospholipid asymmetry differs from that of fibroblasts. FEBS Lett 1981;134:75–8.
34. Sessions A, Horowitz AF. Differentiation related differences in the plasma membrane phospholipid asymmetry of myogenic and fibrogenic cells. Biochim Biophys Acta 1983;728:103–11.
35. Fadok VA, Voelker DR, Campbell PA, Cohen JJ, Bratton DL, Henson PM. Exposure of phosphatidylserine on the surface of apoptotic lymphocytes triggers specific recognition and removal by macrophages. J Immunol 1992;148:2207–16.
36. McEvoy L, Williamson P, Schlegel RA. Membrane phospholipid asymmetry as a determinant of erythrocyte recognition by macrophages. PNAS, USA 1986;83:3311.
37. Comfurius P, Bevers EM, Zwaal RFA. The involvement of cytoskeleton in the regulation of transbilayer movement of phospholipids in human blood platelets. Biochim Biophys Acta 1985;815:143–8.
38. Lyden TW, Vogt E, Ng AK, Johnson PM, Rote NS. Monoclonal antiphospholipid antibody reactivity against human placental trophoblast. J Reprod Immunol 1992;22:1–14.
39. Lyden TW, Ng AK, Rote NS. Modulation of phosphatidylserine epitope expression on BeWo cells during forskolin treatment. Placenta 1993;14:177–86.
40. Cowchock FS, Reece EA, Balaban D, Branch DW, Plouffe L. Repeated fetal losses associated with antiphospholipid antibodies: a collaborative randomized trial comparing prednisone with low-dose heparin treatment. Am J Obstet Gynecol 1992;166:1318–23.
41. Schorer AE, Wickham NW, Watson KV. Lupus anticoagulant induces a selective defect in thrombin-mediated endothelial prostacyclin release and platelet aggregation. Br J Haematol 1989;71:399–407.
42. Khamashta MA, Harris EN, Gharavi AE, et al. Immune mediated mechanism for thrombosis: antiphospholipid antibody binding to platelet membranes. Ann Rheum Dis 1988;47:849–54.
43. Out HJ, de Groot PG, van Vliet M, de Gast GC, Nieuwenhuis HK. Antibodies to platelets in patients with anti-phospholipid antibodies. Blood 1991;77:2655–9.
44. Maneta-Peyret L, Freyburger G, Bessoule J-J, Cassagen C. Specific immunocytochemical visualization of phosphatidylserine. J Immunol Methods 1989;122:155–9.
45. Glenney JR. Calpactins: calcium-regulated membrane-skeletal proteins. Bioessays 1987;7:173–5.
46. Isenberg G. Actin binding proteins-lipid interactions. J Muscle Res Cell Motil 1991;12:136–44.
47. Fukami K, Furuhashi K, Inagaki M, Endo T, Hatano S, Takenawa T. Requirement of phosphatidylinositol 4,5-bisphosphate for α-actinin function. Nature 1992;359:150–2.

48. Haest CWM, Plasa G, Kamp D, Deuticke B. Spectrin as a stabilizer of the phospholipid asymmetry in the human erythrocyte membrane. Biochim Biophys Acta 1978;509:21–32.
49. Wagner MC, Barylko B, Albanesi JP. Tissue distribution and subcellular localization of mammalian myosin I. J Cell Biol 1992;119:163–70.
50. Zwaal RFA. Scrambling membrane phospholipids and local control of blood clotting. NIPS 1988;3:57–61.
51. Comfurius P, Bevers EM, Zwaal RFA. Interaction between phosphatidylserine and the isolated cytoskeleton of human blood platelets. Biochim Biophys Acta 1989;983:212–6.
52. Zhuang Q, Stracher A. Purification and characterization of a calcium binding protein with "synexin-like" activity from human blood platelets. Biochem Biophys Res Commun 1989;159:236–41.
53. Burgener R, Wolf M, Ganz T, Baggiolini M. Purification and characterization of a major phosphatidylserine-binding phosphoprotein from human platelets. Biochem J 1990;269:729–34.
54. Wolf M, Baggiolini M. Identification of phosphatidylserine-binding proteins in human white blood cells. Biochem J 1990;269:723–8.
55. Calleau M, Herve P, Fellman P, Devaux PF. Transmembrane diffusion of fluorescent phospholipids in human erythrocytes. Chem Phys Lipids 1991;57:29–37.
56. O'Halloran T, Beckerle MC, Burridge K. Identification of talin as a major cytoplasmic protein implicated in platelet activation. Nature 1985;317:449–51.
57. Beckerle MC, Miller DE, Bertagnolli ME, Locke SJ. Activation-dependent redistribution of the adhesion plaque protein, talin, in intact human platelets. J Cell Biol 1989;109:3333–46.
58. Heise H, Bayerl Th, Isenberg G, Sackmann E. Human platelet P-235, a talin-like actin binding protein, binds selectively to mixed lipid bilayers. Biochim Biophys Acta 1991;1061:121–31.
59. Kliman HJ, Feinman MA, Strauss JF III. Differentiation of human cytotrophoblasts into syncytiotrophoblasts in culture. Troph Res 1987;2:407–21.
60. Rauch J, Janoff AS. Phospholipid in the hexagonal (II) phase is immunogenic: evidence for immunorecognition of nonbilayer lipid phases in vivo. PNAS, USA 1990;87:4112–4.
61. Galli M, Barbui T, Zwaal RFA, Comfurius P, Bevers EM. Editorial: antiphospholipid antibodies: involvement of protein cofactors. Haematologica 1993;78:1–4.
62. Galli M, Comfurius P, Maassen C, et al. Anticardiolipin antibodies (ACA) directed not to cardiolipin but to a plasma protein cofactor. Lancet 1990; 335:1544–7.
63. Rauch J, Janoff AS. Role of monoclonal antibodies in understanding the interactions between anti-phospholipid antibodies and phospholipids. In: Harris N, Exner T, Hughes GRV, Asherson RA, eds. Phospholipid-binding antibodies. Boca Raton, FL: CRC Press, 1991:108–22.
64. Chamley LW, McKay EJ, Pattison NS. Cofactor dependent and cofactor independent anticardiolipin antibodies. Thromb Res 1991;61:291–9.
65. McNeil HP, Simpson RJ, Chesterman CN, Krilis SA. Anti-phospholipid antibodies are directed against a complex antigen that includes a lipid-binding

inhibitor of coagulation: $\beta_2$-glycoprotein I (apolipoprotein H). PNAS, USA 1990;87:4120–4.
66. Janoff AS. Relationship of lipid and antibody: phospholipid three dimensional structure determines antigenicity. In: Lochshin M, ed. Proc Antiphospholipid Antibody/Lupus Anticoagulant Workshop. NIH, Sept 25, 1991.
67. Bevers EM, Galli M, Barbui T, Comfurius P, Zwaal RFA. Lupus anticoagulant IgG's (LA) are not directed to phospholipids only, but to a complex of lipid-bound human prothrombin. Thromb Haemost 1991;66:629–32.

# 19

# Immunotherapy for Recurrent Spontaneous Abortion

CAROLYN B. COULAM

*Recurrent spontaneous abortion* (RSA) is a significant health problem affecting 2%–5% of reproducing couples (1). Although genetic, anatomic, and hormonal factors have been implicated in the etiology of RSA, a sizable proportion remains unexplained (2). An immunologic cause has been suggested for more than 80% of otherwise unexplained RSAs (3). Immunotherapy in various forms has been proposed as a treatment for these couples (4–12).

Skepticism about an immunologic cause of RSA had been expressed because of lack of direct incidence of an immunologic mechanism involved in the process of pregnancy loss and conflicting results reported from clinical trials using immunotherapy in the treatment of RSA. Data are accumulating to provide evidence of a direct immunologic role in the mechanism of recurrent pregnancy loss and to answer the controversy about the efficacy of immunotherapy for treatment of RSA (13–15). This chapter presents a rationale for immunotherapy by reviewing reported data that support an immunologic etiology in some women experiencing RSA and discusses options for treatment in these individuals.

## Rationale for Immunotherapy

The original rationale for immunotherapy in the treatment of RSA was to present an allogenic stimulus to the maternal immune system that would evoke an appropriate response to antigens on the fetal trophoblastic cells needed to protect the pregnancy from failure. It was hypothesized that this protective response involved the production of blocking antibodies (16). Subsequently, it was shown that the fetal trophoblast that envelops the fetus and forms the fetomaternal interface is not susceptible to transplantation immunity (17).

Trophoblastic cells are resistant to lysis by cytotoxic T lymphocytes, *natural killer* (NK) cells, and antibody-dependent cytotoxicity (18, 19). They are, however, susceptible to activated NK cells that have the asialo $GM1^+$ marker (*lymphokine-activated killer* [LAK]) cells (20). Suppression of the activation of NK to LAK cells is necessary for successful pregnancy to occur. A number of cytokines have been shown to prevent LAK cell activation and abortion in mice. These cytokines include *interleukin-3* (IL-3), *granulocyte-macrophage colony stimulating factor* (GM-CSF) (21, 22), *transforming growth factor β2* (TGFβ2) (20), and a 34-kd protein produced by $CD8^+$ cells with progesterone receptors (23). Production of these cytokines has been studied more locally at the fetodecidual interface, but there is also evidence of systemic production. The dominant T cell in human decidua is $CD8^+$ (20). The $CD8^+$ cells in pregnancy may acquire progesterone receptors and in response to progesterone secrete a 34-kd factor that suppresses NK-LAK cell activity (23). The T cells produce IL-3 and GM-CSF. GM-CSF is also produced by non-T cells (21, 22). Interleukin-3, the only T cell-specific cytokine, has not been detected in the decidua later than 3–4 days after mating (17). The $CD8^+$ cell activity seems to be present during the pre-preimplantation and preimplantation phase of pregnancy (20). After implantation a non-T, non-B cell population appears. These cells have a $CD56^+$ $CD16^-$ phenotype and produce TGFβ2 (25–28). TGFβ2 can inhibit activation of NK to LAK cells (17).

By in situ hybridization using the *Ped* G1G2 probe that selectively detects TGFβ2-producing cells, a subpopulation of patients with recurrent pregnancy loss has been identified that lacks these cells (26–28). Further, an infiltrate of $CD56^+$ $CD16^+$ LAK cells has been noted in placental bed biopsies of incipiently aborting patients with a history of RSA (25, 29).

Results from these studies suggest a local immune response in the mechanism of RSA. Other studies have provided support for the concept that a systemic immune response is involved in RSA. Women experiencing repeat RSA have circulating embryotoxins resembling *interferon γ* (IFNγ) (30). Normally pregnant women have an increase in circulating T cells that express TJ6 protein and $CD8^+$ cells that express progesterone receptor. Antibodies to TJ6 or $CD8^+$ cells administered in the murine model at or about the time of implantation ablate the pregnancies (17, 31). Taken together, these observations suggest that both a local and systemic immune response are evoked. The local response is directed toward preventing the activation of NK to LAK cells. The systemic response may help in controlling further migration of effector cells.

Both the local and systemic immune responses are mediated through the production of cytokines. The stimulation of the production of cytokines necessary to protect pregnancy from immunologic attack is not antigen specific. Data indicate that this protective effect can be induced by immunization with white blood cells (4, 5), stimulation of the reticuloendothelial system by intralipid (11), and psychotherapy (32, 33). Evidence also exists that stress-induced abortions can be corrected by immune stimulation (34), suggesting a final common pathway of protection from pregnancy loss. To further test the effectiveness of non-antigen-specific immunologic stimuli in the treatment of RSA, prospective, randomized, double-blinded-controlled clinical trials using pooled seminal plasma and *intravenous immunoglobulin* (IVIG) were initiated. An update of the progress of these trials is presented in the following paragraphs, and the live birth rates obtained in these trials are compared with other forms of therapy.

## Seminal Plasma Treatment for RSA

The rationale for using seminal plasma for treatment of RSA is provided from the concept that the mammal mother responds to antigens present not only in the trophoblast, but also in seminal plasma (14, 35). In this way, immunization could occur at the time of mating, making the environment at the time of implantation of the blastocyst favorable for successful pregnancy (15, 36). To test this hypothesis, a randomized placebo-controlled trial using vaginal suppositories containing either seminal plasma or lubrication jelly was conducted. We now report the preliminary results of that clinical trial obtained when the code was broken after half of the sample was achieved.

### *Patients*

From July 1, 1987, to July 31, 1991, 87 women were randomized into a placebo-controlled blind clinical trial using vaginal capsules containing seminal plasma or lubrication jelly. One patient was randomized to seminal plasma, but never started using the vaginal capsules and was not included in the analysis. Of the remaining 86 women, 43 received seminal plasma vaginal capsules, and 43 received a placebo. Eighty-three women had a history of RSA characterized by 2 or more losses with the same partner. Seventy-six couples had an obstetrical history of no pregnancies progressing farther than 20 weeks of gestation (primary abortion), and 7 couples experienced 3 (secondary abortion). An unexplained infertility of greater than 3 years' duration was diagnosed in 3 couples.

The obstetrical history of each of the women was received, and the number of total pregnancies, live births, stillbirths, abortions, ectopic pregnancies, hydatidiform moles, and partners for each pregnancy was recorded. All couples were investigated with chromosome analysis, *hysterosalpingography* (HSG) and hysteroscopy, luteal phase endometrial biopsy, and serum progesterone timed with ovulation documented by ultrasonic monitoring of folliculogenesis, *anticardiolipin antibody* (ACA), activated partial thromboplastin time, HLA typing, assays of the maternal sera for the presence of both complement-dependent and -independent antipaternal antibodies utilizing lymphocytotoxicity assays, and mixed lymphocyte culture reactions (17). All couples with a diagnosis of chromosomal, anatomic, endocrinologic, and autoimmunologic etiology of recurrent pregnancy loss were excluded from the study. There was no significant difference between age (32.4 vs. 31.8 years), gravidity (4.4 vs. 4.2 pregnancies), and HLA sharing between the seminal plasma and placebo groups.

## *Vaginal Capsules*

Gelatin capsules were filled with 1 mL of pooled plasma or lubrication jelly (placebo). Semen was collected from healthy donors previously screened for HIV antibody, *hepatitis B surface antigen* (HBsAG), chlamydia, and gonorrhea. After liquefaction the semen specimen was centrifuged at 12,000 × g for 5 min. The cell-free seminal plasma supernatants were individually frozen at −70°C and quarantined for 6 months. After 6 months semen donors were tested for HIV antibody, and HBsAG was confirmed. They were thawed and pooled. Gelatin capsules were filled with 1 mL of pooled seminal plasma in a cold room at 4°C and then immediately frozen and maintained at −70°C until used. Placebo capsules were filled with 1 mL of lubrication jelly.

## *Protocol Procedure*

Women were instructed to insert 1 capsule intravaginally on days 7, 14, and 21 of their menstrual cycle. If by day 28 there was no menses, the patient increased the treatment with the application of 2 capsules per week. On day 35 of her cycle, a serum determination of *human chorionic gonadotropin* (hCG) was performed. If a positive result was obtained, the use of 2 vaginal capsules per week was continued until week 36 of pregnancy. If a negative hCG resulted, therapy was discontinued until the start of the next cycle.

## *Implantation Rates*

Women who used vaginal capsules for at least 3 months were evaluated for the occurrence of pregnancy within one year of using the capsules. Pregnancy rates were compared between women receiving seminal plasma and placebo. Pregnancy was documented by the appearance of an intrauterine gestational sac visible on transvaginal ultrasonic examination by week 6 of gestation.

Figure 19.1 shows the pregnancy rates among 86 women who participated in the study. Nineteen (22%) women did not achieve pregnancy

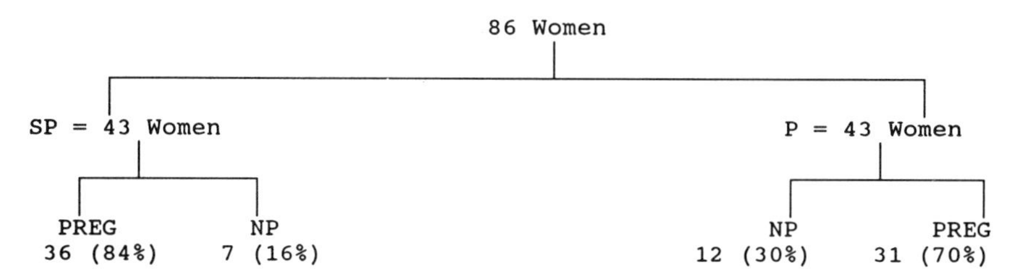


FIGURE 19.1. Pregnancy rates among 86 women participating in a randomized, placebo-controlled clinical trail using vaginal capsules containing seminal plasma (SP) or lubrication jelly (P). The number and percentage (%) of women achieving pregnancy (PREG) are compared to those not becoming pregnant (NP) within one year of participation in the study.

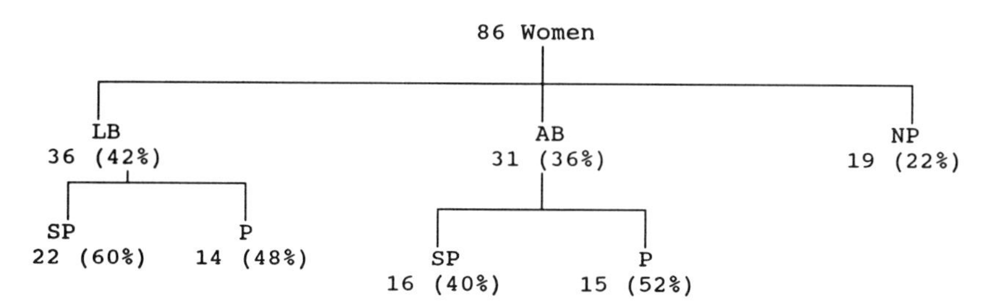


FIGURE 19.2. Pregnancy outcome among 86 women participating in a randomized, placebo-controlled clinical trial using seminal plasma (SP) or lubrication jelly (P) in vaginal capsules. The number and percentage (%) of women not achieving pregnancy (NP), delivering a live birth (LB), and recurrently aborting (AB) are noted.

within one year. Those women receiving seminal plasma became pregnant more frequently than those receiving placebo (84% vs. 70%; $P = 0.08$).

## Pregnancy Outcome

The number of live births, spontaneous abortions, and ectopic pregnancies was recorded. The rate of live births and recurrent pregnancy loss was compared between women receiving seminal plasma and placebo. Data were analyzed using repeated measure design ANOVA, followed by Student's t-test (1-tail). Significance was defined as $P = 0.05$.

The pregnancy outcomes of women participating in the study are shown in Figures 19.2 and 19.3. The 86 women participating in the study produced 67 pregnancies, of which 36 (54%) ended in live birth and 31 (46%) in recurrent abortion (Fig. 19.2). Those women receiving seminal plasma had a 60% live birth rate compared to 48% for those receiving placebo. Secondary aborting women and women with maternal anti-

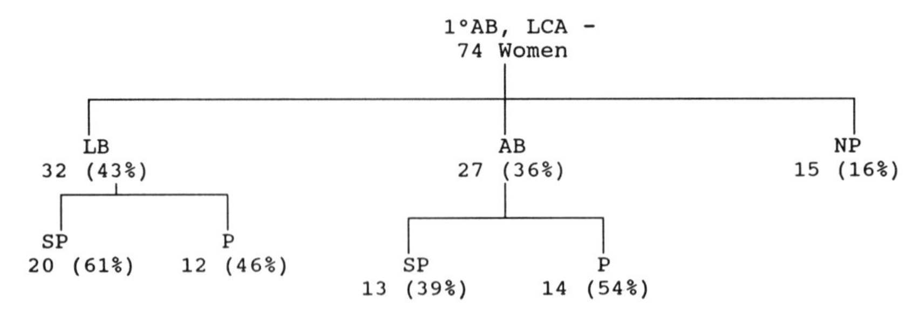


FIGURE 19.3. Pregnancy outcome among 74 primary aborting women who do not demonstrate the presence of maternal antipaternal antibodies ($LCA^-$) and who participated in a randomized, placebo-controlled trial.

TABLE 19.1. Difference in rates of live births between women treated with seminal plasma (SP) or lubrication jelly (P) compared with obstetrical history and laboratory test.

| Clinical findings | *n* | Live birth (%) SP | P | Diff | *P* |
|---|---|---|---|---|---|
| Total | 86 | 60 | 48 | 12 | 0.296 |
| 2° Ab[a] | 7 | 33 | 100 | −67 | |
| 1° Ab[b] | 79 | 60 | 46 | 14 | 0.236 |
| $LCA^{+}$[c] | 5 | 50 | 50 | 0 | |
| $LCA^{-}$ | 74 | 51 | 46 | 15 | 0.199 |
| <2 HLA[d] | 19 | 50 | 57 | −7 | |
| >2 HLA | 55 | 63 | 42 | 21 | 0.136 |

[a] 2° Ab = secondary abortion.
[b] 1° Ab = primary abortion.
[c] LCA = maternal antipaternal lymphocytotoxic antibody.
[d] HLA = HLA sharing between mates.

paternal lymphocytoxic antibodies did not respond to seminal plasma treatment (Table 19.1). When 74 women with primary abortion and absence of maternal antipaternal lymphocytotoxic antibodies were evaluated, 32 of 59 pregnancies (54%) ended in live births and 27 (46%) in recurrent abortion (Fig. 19.3). Live births occurred in 61% of women receiving seminal plasma compared with 46% of women receiving placebo.

## Conclusion

Treatment with seminal plasma enhances both implantation and live birth rates by 14%–15%. That the increase in the live birth rate is the same as the implantation rate suggests that the increase in the live birth rate seen after treatment with seminal plasma could be explained by the increase in the implantation rate. The nature of the substance in seminal plasma that enhances implantation rates is not known. Animal studies suggest that its source is the seminal vesicle (37–40). Compounds present in semen that have been shown to enhance implantation include polyamines (41) and *transforming growth factor α* (TGFα) (42). However, the source of TGFα in semen has not been reported, and polyamines appear to be produced by spermatozoa (41).

# IVIG Treatment for RSA

A prospective, randomized, placebo-controlled clinical trial using IVIG for treatment of RSA is currently ongoing, and the results of the clinical trial are not available. However, while conducting the trial, a striking observation was made in the ultrasonographic findings of those pregnancies that were lost. The following section is a description of those findings.

## *Patients*

Women experiencing 2 or more consecutive spontaneous abortions with the same partner were offered the opportunity to participate in an *Institutional Review Board* (IRB)-approved, randomized, placebo-controlled trial using IVIG or albumin (placebo). The obstetrical history of each of the women was obtained, and the number of total pregnancies, live births, stillbirths, abortions, ectopic pregnancies, hydatidiform moles, and partners for each pregnancy was recorded. All couples were investigated with chromosome analysis, hysterosalpingography and hysteroscopy, luteal phase endometrial biopsy, and serum progesterone timed with ovulation documented by ultrasonic monitoring of folliculogenesis, ACA, activated partial thromboplastin time, HLA typing, assays of the maternal sera for the presence of both complement-dependent and -independent antipaternal antibodies utilizing lymphocytotoxicity assays, and mixed lymphocyte culture reactions.

All couples with a diagnosis of chromosomal, anatomic, endocrinologic, and autoimmunologic etiology of recurrent pregnancy loss were excluded from the study. Also excluded were women less than 18 years and greater than 45 years of age, women with a history of IgA deficiency, or women with a hypersensitivity to immunoglobulin. Each woman had blood screened for the presence of HIV antibody and HBsAG.

## *Protocol*

Patients entering the study are randomized; one-half receives IVIG, and the remaining half receives albumin infusions. Each patient receives an intravenous infusion in the follicular phase of the cycle when pregnancy is desired. Patients are randomized to receive either Sandoglobulin (550 mg/kg per month) or albumin in an intravenous infusion. The patient receives the infusion every 27 days until pregnancy or for 4 months. If the patient is not pregnant in 4 months, she is dropped from the study and replaced with another patient. Once conception occurs, the patient receives an infusion every 28 days until delivery or until weeks 28–32 of gestation.

## *Statistical Analysis*

Statistical analysis was performed with Fisher's exact test. A *P*-value of significance of <0.05 was assigned.

## *Results*

Figure 19.4 summarizes the ongoing results of the prospective, randomized, placebo-controlled trial. To date, 90 women have been enrolled in the study. Study medication has been discontinued in 29 (32%) women who failed to conceive within 4 cycles after IVIG infusion. Nine women are trying to conceive, and 52 women have achieved pregnancies. The

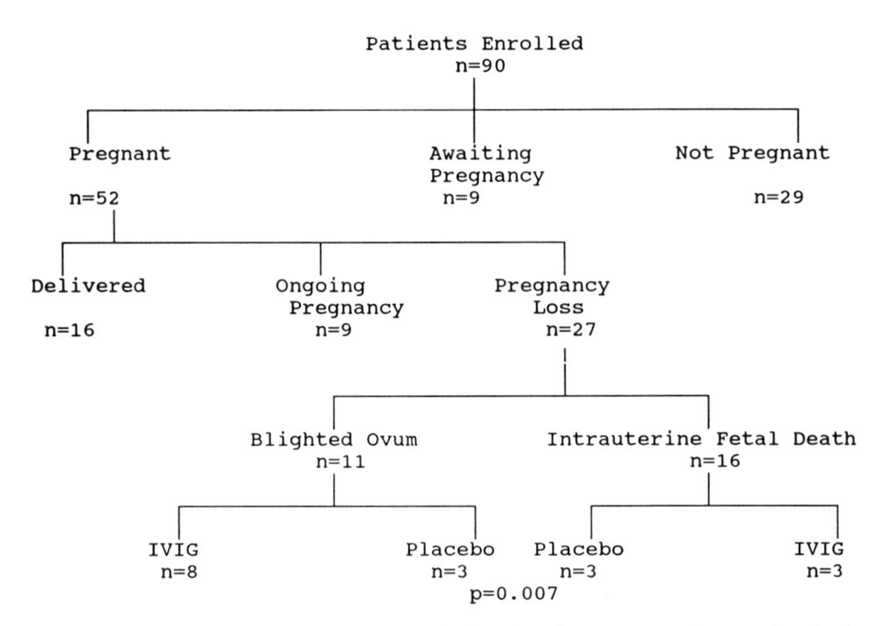


FIGURE 19.4. Summary of patients enrolled to date in a prospective, randomized, placebo-controlled clinical trial using IVIG for treatment of RSA. The differences in ultrasonographic findings between pregnancies lost during IVIG and placebo are significant ($P = 0.007$).

outcomes of the 52 pregnancies include 16 deliveries, 9 ongoing pregnancies, and 27 pregnancy losses.

Of the 27 pregnancy losses, 11 (41%) occurred after infusion with IVIG, and 16 (59%) occurred after placebo ($P$ = NS). Ultrasonographic examination of the 27 pregnancy losses revealed 11 empty gestational sacs (blighted ova) and 16 intrauterine fetal deaths (fetal pole present with no fetal cardiac activity). Of 11 blighted ova pregnancy losses, 8 (73%) were in women receiving IVIG and 3 (27%) were in women receiving placebo. Sixteen pregnancies ended with intrauterine fetal deaths: 13 (81%) in women receiving placebo and 3 (19%) in women receiving IVIG. Of 11 pregnancy losses occurring in women receiving IVIG, 8 (73%) were the result of blighted ova, and 3 (27%) were intrauterine fetal deaths. Sixteen pregnancy losses occurred in women receiving placebo: 3 (19%) were the result of blighted ova, and 13 (81%) were intrauterine fetal deaths. The difference in the frequency of blighted ova between IVIG- and placebo-treated women was significant ($P = 0.007$).

### *Comment*

Analysis of the ultrasonic findings of pregnancies lost during the ongoing, randomized, placebo-controlled clinical trial suggests that IVIG is efficacious in preventing intrauterine fetal death, but not blighted ovum. Forty-one percent of the pregnancies lost in this study manifested as blighted ova pregnancy losses. This frequency compares with 43% observed in women with no history of previous pregnancy loss (43). Among women with a history of RSA, 55% abort their next pregnancy after documentation of fetal cardiac activity without treatment (43). With IVIG treatment only 27% of the losses occurred after the establishment of fetal cardiac activity. Analysis of these data suggests that IVIG prevents intrauterine fetal death after the establishment of fetal cardiac activity, but does not prevent the occurrence of blighted ovum.

## Comparison of Various Treatments for RSA

In addition to the above-described treatments, a number of other forms of immunotherapy have been studied. The treatment most widely used has been immunization using allogenic leukocytes from the husband. The efficacy of paternal leukocyte immunization has been questioned because of conflicting results reported from clinical trials designed to treat RSA (2–5).

To answer the question of the effectiveness of husband white blood cell immunization for treatment of RSA, the Ethics Committee of the American Society for Reproductive Immunology has undertaken a worldwide prospective collaborative observation study (13–15). The results of

TABLE 19.2. Summary of results of clinical trials using various types of treatment for RSA.

| Treatment | $n$ | Live births (%) |
|---|---|---|
| Husband leukocytes | 822 | 73 |
| Autologous leukocytes | 99 | 58 |
| Trophoblast membrane infusion | 118 | 58 |
| Intralipid | 20 | 70 |
| IVIG | 52 | 67 |
| Psychotherapy | 135 | 85 |
| Saline | 32 | 67 |
| None | 304 | 39 |

this study will be available this year. It is anticipated that the analyses and conclusions of this study will impact clinical medicine. Table 19.2 summarizes the published literature of live birth rates reported after various types of therapy (44). Intralipid and trophoblast membrane infusions (11) and psychotherapy (32, 33) have been reported to result in live birth rates similar to those reported for other forms of immunotherapy (44). When the published results are pooled and compared by chi-square analysis, all treatments listed in Table 19.2, including saline, are significantly better than no treatment ($P < 0.005$) (44).

## Summary

Recurrent spontaneous abortion is a significant health concern. Efficacious and safe therapies are needed. A number of forms of immunotherapy have been studied. Data from these studies taken collectively suggest that a nonspecific alteration in regulatory cytokines might be involved in the mechanism of all forms of immunotherapy. It is not clear which forms of treatment act to enhance implantation or prevent intrauterine fetal death after implantation. The overall live birth rates appear similar for all treatments studied. Thus, the following questions arise. First, what is the best method of achieving a protective response with the fewest side effects and least cost? Second, who should be treated with what kind of therapy? Sensitive and specific markers are needed to identify women who will be most likely to respond to putatively different treatments.

## *References*


1. Coulam CB. Unification of immunotherapy protocols. Am J Reprod Immunol 1991;25:1–6.
2. Coulam CB. Unexplained recurrent pregnancy loss: epilogue. Clin Obstet Gynecol 1986;29:999–1004.

3. McIntyre JA, Coulam CB, Faulk WP. Recurrent spontaneous abortion. Am J Reprod Immunol 1989;21:100–4.
4. Mowbray JF, Lidlee H, Underwood JL, Gibbings C, Reginald PW, Beard RW. Controlled trial of treatment of recurrent spontaneous abortion by immunization with paternal cells. Lancet 1985;1:941–9.
5. Gatenby PA, Cameron K, Simes RJ, et al. Treatment of recurrent spontaneous abortion by immunization with paternal lymphocytes: results of a controlled trial. Am J Reprod Immunol 1993;29:88–94.
6. Ho H, Gill TJ III, Hsuish HJ, Jiang JJ, Lee TY, Hsish CY. Immunotherapy for recurrent spontaneous abortion in a Chinese population. Am J Reprod Immunol 1991;25:10–5.
7. Cauchi MN, Lemi D, Young DE, Klosa M, Pepperell RJ. Treatment of recurrent spontaneous aborters by immunization with paternal cells—controlled trial. Am J Reprod Immunol 1991;25:16–7.
8. Coulam CB, Peters AJ, McIntyre JA, Faulk WP. The use of intravenous immunoglobulin for the treatment of recurrent spontaneous abortion. Am J Reprod Immunol 1990;22:78.
9. Mueller-Eckhardt G, Heine O, Neppert J, Kunzel W, Mueller-Eckhardt C. Prevention of recurrent spontaneous abortion by intravenous immunoglobulin. Vox Sang 1989;56:151–4.
10. Mueller-Eckhardt G, Huni O, Poltrin B. IVIG to prevent recurrent spontaneous abortion. Lancet 1991;1:424.
11. Johnson PM, Ramsden GH, Chia KV. Trophoblast membrane infusion (TMI) in the treatment of recurrent spontaneous abortion. In: Beard RW, Sharp F, eds. Early pregnancy loss, mechanisms and treatment. Ashton-under-Tyne, UK: Peacock Press, 1988:389–96.
12. Coulam CB. Treatment of recurrent spontaneous abortion. Am J Reprod Immunol Microbiol 1988;14:149.
13. Coulam CB, Clark DA. Report from the Ethics Committee for Immunology: American Society for Immunology of Reproduction. Am J Reprod Immunol 1991;26:93–5.
14. Coulam CB, Clark DA, Beer AE. Report from the Ethics Committee for Immunology. Am J Reprod Immunol 1992;28:3–5.
15. Coulam CB. Report from the Ethics Committee for Immunology. Am J Reprod Immunol 1993;30:45–7.
16. Rocklin RE, Kitzmiller JL, Carpenter CB, Garovoy MR, David JR. Maternal-fetal relation: absence of an immunologic blocking factor from the serum of women with chronic abortions. N Engl J Med 1976;295:1209–13.
17. Clark DA. Controversies in reproductive immunology. Crit Rev Immunol 1991;11:215–47.
18. Drake BL, Head JR. Murine trophoblast can be killed by lymphokine-activated killer cells. J Immunol 1989;143:9–14.
19. King A, Loke YW. Human trophoblast and JEG choriocarcinoma cells are sensitive to lysis by IL-2–stimulated decidual NK cells. Cell Immunol 1990; 129:435–48.
20. Clark DA, Flanders KC, Banwatt D, et al. Murine pregnancy decidual produces a unique immunosuppressive molecule related to transforming growth factor β-2. J Immunol 1990;144:3008–14.

21. Chaouat G, Menu E, Athanassakis I, Wegmann TG. Maternal T cells regulate placental size and fetal survival. Reg Immunol 1988;1:143–8.
22. Chaouat G, Menu E, Clark DA, Minowsky M, Dy M, Wegmann TG. Control of fetal survival in CBA × DBA/2 mice by lymphokine therapy. J Reprod Fertil 1990;89:447–58.
23. Szekeres-Bartho J, Autran B, Debre P, Andreu G, Denver L, Chaouat G. Immunoregulatory effects of a suppressor factor from healthy pregnancy women's lymphocytes after progesterone induction. 1989;122:281–94.
24. Starkey PM. In: the natural immune system: the natural killer cell. IRL Press, 1992:206.
25. Michel M, Underwood J, Clark DA, Mowbray J, Beard RW. Histologic and immunologic study of uterine biopsy tissue of incipiently aborting women. Am J Obstet Gynecol 1989;161:409–14.
26. Clark DA, Lea RG, Podor T, Daya S, Banwatt D, Harley CB. Cytokines determining the success or failure of pregnancy. Ann NY Acad Sci 1991; 626:524–44.
27. Clark DA, Banwatt D, Fulop G, Quarrington C, Croy BA. Genetic aspects of spontaneous abortion (resorption) in normal and immunodeficient (SCID) mice. Am J Reprod Immunol 1992;27:48–9.
28. Clark DA, Lea RG, Underwood J, et al. A subset of recurrent first trimester aborting women show subnormal TGFβ2 suppressor activity at the implantation site associated with miscarriage. J Immunol Immunopharmacol 1992; 12:83.
29. Clark DA, Lea RG, Denburg J, et al. Transforming growth factor beta related suppressor factor in mammalian pregnancy decidual: homologies between the mouse and human in successful pregnancy and in recurrent unexplained abortion. In: Chaouat G, Mowbray J, eds. Cellular and molecular biology of the materno-fetal relationship. 1991;212:171–9.
30. Ecker JL, Laufer MR, Hill JA. Measurement of embryotoxic factors is predictive of pregnancy outcome in women with a history of recurrent abortion. Obstet Gynecol 1993;81:84–7.
31. Beaman KD, Chang JA, Nichols TC, Angkachatchi V. Differential expression of the immunoregulatory pregnancy associated factor TJ6. J Immunol Immunopharmacol 1992;12:74.
32. Stray-Pedersen B, Stray-Pedersen S. Recurrent abortion: the role of psychotherapy. In: Beard RW, Sharp F, eds. Early pregnancy loss, mechanisms and treatment. Ashton-under-Tyne, UK: Peacock Press, 1988:433–40.
33. Tupper C, Weil RJ. The problem of spontaneous abortion. Am J Obstet Gynecol 1962;83:421–9.
34. Clark DA, Banwatt D, Manuel J. Immunotherapy of stress-triggered abortion in mice. Am J Reprod Immunol 1992;27:35.
35. Kajino T, Torry DS, McIntyre JA, Faulk WP. Trophoblast antigens in human seminal plasma. Am J Reprod Immunol Microbiol 1988;17:91–4.
36. Thaler CJ. Immunologic role for seminal plasma in insemination and pregnancy. Am J Reprod Immunol 1989;21:147–51.
37. Carp HJA, Serr DM, Mashiach S, Nebel L. Influence of insemination on the implantation of transferred rat blastocysts. Gynecol Obstet Invest 1984;18: 194–8.

38. O WS, Chen HQ, Chow PH. Effects of male accessory sex gland secretions on early embryonic development in the golden hamster. J Reprod Fertil 1988;84:341–4.
39. Pang SF, Chow PH, Wong TM. The role of the seminal vesicles, coagulating glands and prostate glands on the fertility and fecundity of mice. J Reprod Fertil 1979;56:129–32.
40. Queen K, Dhabuwala CB, Pierrepoint CG. The effect of the removal of the various accessory sex glands on the fertility of male rats. J Reprod Fertil 1981;62:423–6.
41. Porat O, Clark DA. Analysis of immunosuppressive molecules associated with murine in vitro fertilized embryos. Fertil Steril 1990;54:1154–61.
42. Yu SM, Lobb KD, Clark DA, Younglai EV. Identification of a TGFα-like growth factor in human seminal plasma. Fertil Steril 1993.
43. Stern JJ, Coulam CB. Mechanism of recurrent spontaneous abortion, I. Ultrasonographic findings. Am J Obstet Gynecol 1992;166:1844–52.
44. Coulam CB, Coulam CH. Update on immunotherapy for recurrent pregnancy loss. Am J Reprod Immunol 1992;27:124–7.

# 20

# New Horizons in the Evaluation and Treatment of Recurrent Pregnancy Loss

ALAN E. BEER, JOANNE Y.H. KWAK, ALICE GILMAN-SACHS, AND KENNETH D. BEAMAN

During the past 15 years we have evaluated couples with unexplained *recurrent spontaneous abortion* (RSA), couples with normal pregnancies, and women who electively terminate their pregnancies early in gestation and have found major differences in the alloimmune and autoimmune status in women with unexplained RSA (1–3). In studies of over 1500 couples in this latter category we have shown (i) a higher incidence of HLA-DR and HLA-DQ antigen sharing compared to fertile controls (4); (ii) a higher incidence of HLA-DR and -DQ homozygosity in the male partners (5); (iii) a markedly decreased level of maternal alloantibody to paternal T and B lymphocytes compared to fertile control couples (2, 6); (iv) a strikingly higher incidence of antiphospholipid antibodies to *phosphatidylserine* (PS), *phosphatidylinositol* (PI), *phosphatidylglycerol* (PG), and *cardiolipin* (CL) that increases in incidence and titer with each subsequent pregnancy loss in three distinct populations of women studied (United States, Kuwait, and Colombia) (3, 7); and (v) an 89% incidence of subsequent pregnancy loss following lymphocyte immune therapy in autoimmune women untreated for the autoimmune abnormalities (2).

This chapter presents data of the molecular genetic analysis of couples in circumstances where optimal alloimmune and autoimmune therapies were conducted. The results of this study indicate an increase in the HLA-DQα4.1 (0501) antigen in mothers who have repeated pregnancy losses despite optimal alloimmune or autoimmune therapies and a very strong association (84%) with maternal HLA-DQα4.1 (0501) gene product and the presence of antiphospholipid antibodies in the mother. A survey of live-born infants from other couples treated in the same manner did not show this association, except for a higher incidence of HLA-DQα4.1 (0501) homozygotes in first-born children of couples who subsequently become secondary aborters. These data may indicate that couples sharing HLA-DQα4.1 (0501) are at risk to produce fetuses that may be autoimmune unacceptable to the mother and, as a result, fail to stimulate the appropriate alloimmune responses. The most consistent assayable autoimmune response in the mother is to phospholipid antigens that are shown to function as adhesion molecules for myoblast to myoepithelial formation, for fibroblast to fibroblastic syncytium formation (8, 9), and possibly for cytotrophoblast to syncytiotrophoblast differentiation.

## Materials and Methods

Two hundred and seventy women with 3 or more RSAs received alloimmune and autoimmune testing, including antiphospholipid and antinuclear antibody monitoring. Women who lacked alloimmune recognition were immunized with lymphocytes, and all developed adequate levels of antipaternal antibody prior to attempting another pregnancy. Patients with autoimmune abnormalities were treated with preconception aspirin, heparin, and prednisone when indicated, as previously reported (2).

An HLA molecular genetic evaluation of class II HLA-DQ antigens of the couple was done utilizing DNA isolation, amplification, and appropriate oligonucleotide probes. HLA-DQ$\alpha$ typing was performed using 10 *sequence-specific oligonucleotide probes* (SSOs). Nucleotide sequences for PCR primers and SSOs were provided by Cetus Corporation (Emeryville, CA). All sequences were synthesized by Synthecell Corporation (Rockville, MD). DNA (0.5–1 μg) was subjected to amplification by PCR using primers GH26 and GH27. Amplification was performed for 28 cycles of denaturation at 95°C for 2 min, annealing at 37°C for 2 min, and extension at 72°C for 2 min in 100-μL PCR buffer (Cetus Corp., Emeryville, CA) containing 1 unit Taq polymerase (Cetus Corp.) and 4 pmol of each primer. The first cycle was preceded by 5 min at 94°C, and the last cycle was followed by 7 min at 72°C. Five microliters of the PCR reaction was denatured in 0.4 M NaCL, added to 0.2X SSO and dot-blotted with g-$^{32}$P-ATP, and hybridized to filter-bound DNA for 12 h at 42°C in 0.5 M NaCL, 50 mM NaPO4, 1% SDS, 5X Denhardts, and 200-μg/mL salmon sperm DNA. Membranes were washed twice at room temperature for 30 min in 5X SSO and 1% SDS and twice at 42°C for 30 min each in 2X SSO and 1% SDS. X-ray film was exposed to each blot. Genotype assignments were made on the basis of combinations of positive signals with the 10 probes (10). Autoantibody titers to nuclear and phospholipid antigens were assayed every 2 weeks during pregnancy, at the time of delivery, and at the time of a repeated pregnancy loss, as previously reported (3).

Reproductive immunophenotypes of lymphocyte subsets by flow cytometry were conducted in women during pregnancy and at the time of pregnancy termination with a live-born delivery or a repeat abortion.

Two-color immunofluorescence analysis of lymphocyte markers was performed. Peripheral blood was stained directly with FITC or PE-labeled monoclonal antibodies (Ortho Diagnostics, Raritan, NJ). Antibodies to CD4, CD8, CD3, CD19, CD56, CD14, CD16, and HLA-DR were used in this study. Immunofluorescence analysis of the lymphocyte population was performed on a Profile II Flow Cytometer (Coulter Diagnostics, Miami Lakes, FL). The instrument was aligned and calibrated daily as described by the manufacturer to standardize the fluorescence intensity measurements. Data were displayed on a 4-decade log scale. Cells within the lymphocyte cell gate, as determined by forward-angle light scatter and side-angle light scatter, were evaluated for fluorescence after reaction with the antibodies. Mouse immunoglobulin isotype controls were run concurrently with lymphocytes from all individuals.

## Results

### *Pregnancy Outcome Post-Immunological Evaluation and Therapy*

Women entering the program for recurrent pregnancy losses were evaluated and treated for their alloimmune and autoimmune disorders according to the flow diagram shown in Figure 20.1. The reproductive

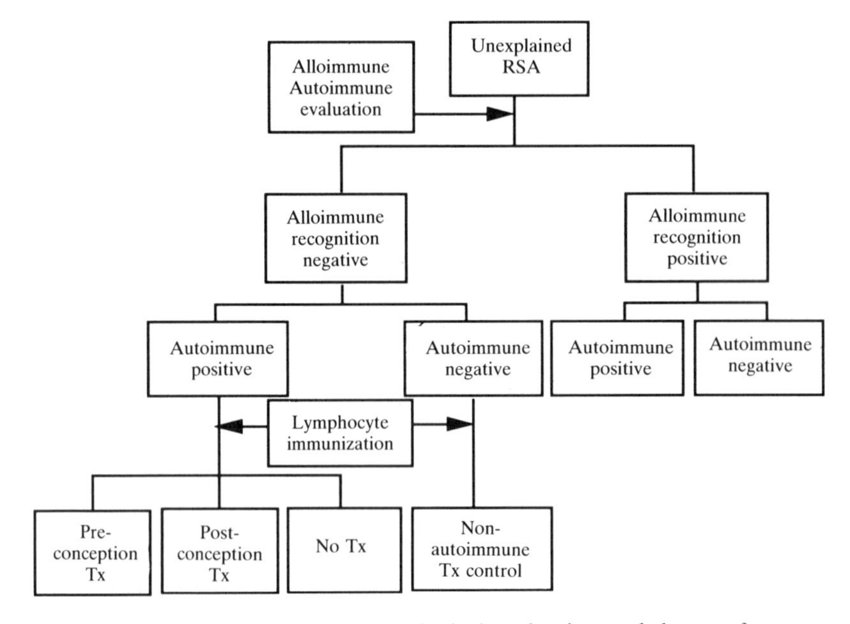


FIGURE 20.1. Flow diagram of immunological evaluation and therapy for women with recurrent pregnancy losses.

outcome in the nonautoimmune women treated only with lymphocyte immunization was 84%. The infants were delivered at term and were of normal body weight. Seventy-five percent of autoimmune women treated with lymphocyte immunization and preconception aspirin, heparin, and prednisone, if indicated, delivered normally grown infants at 37 weeks of gestation. In contrast, if the autoimmune therapy was initiated at the time of a positive pregnancy test, the live birth rate postconception was reduced to 44.1%, and deliveries occurred at 35 weeks of gestation. If autoimmune women were not given any autoimmune medications during pregnancy, 89% miscarried their pregnancies an additional time (2). These data suggest a high incidence of autoimmunity to phospholipid antigens and DNA antigens in women with an alloimmune etiology for their RSA, as previously reported by us (3, 11). The presence of untreated autoimmunity in lymphocyte-treated women is associated with repeated pregnancy losses in most. In order to understand the mechanisms involved, couples losing infants an additional time while receiving optimal treatments underwent molecular genetic HLA-DQ evaluation.

## *Antiphospholipid Antibody Titers*

Figures 20.2 and 20.3 plot the dynamics of IgM and IgG *anticardiolipin antibody* (aCL) in women who were aCL positive entering their pregnancy

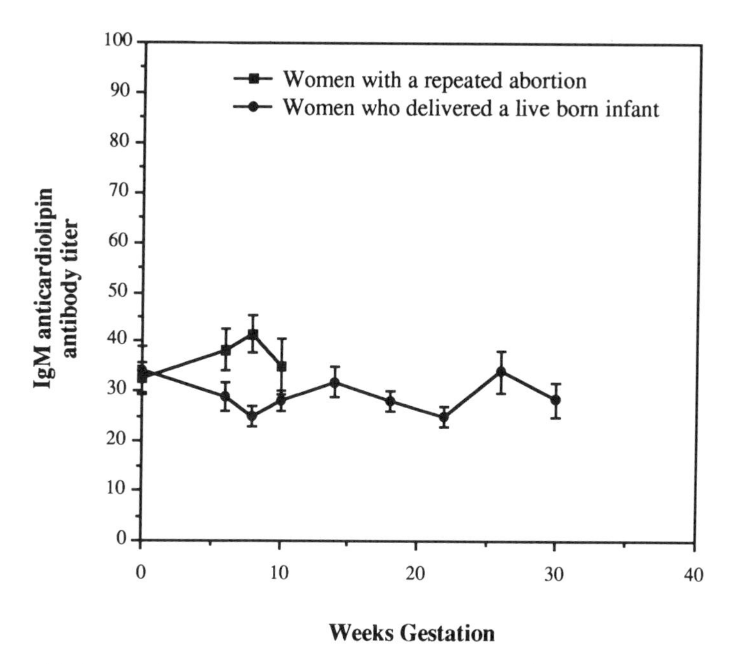


FIGURE 20.2. Tracing of IgM aCL titers during index pregnancy in women with recurrent pregnancy losses.

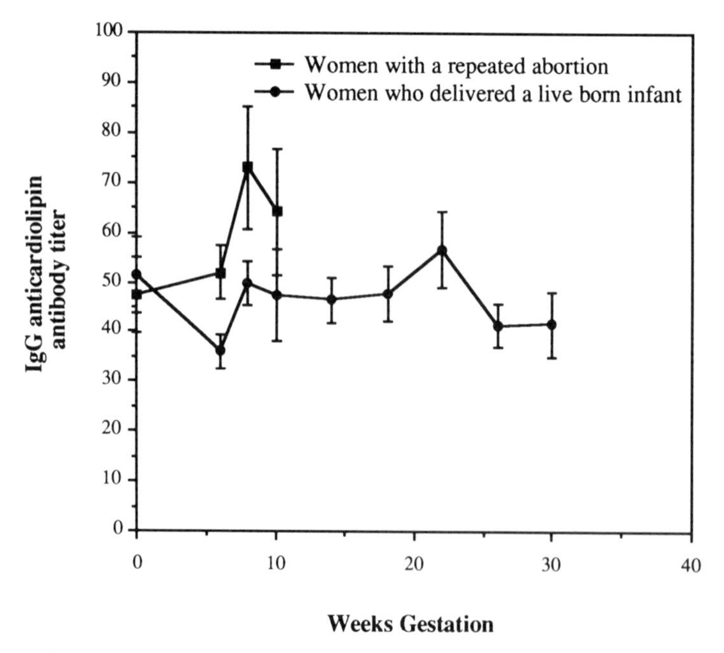


FIGURE 20.3. Tracing of IgG aCL titers during index pregnancy in women with recurrent pregnancy losses.

under study. Down-regulation of the antibody titers occurred in women who delivered a live-born child, and significant elevations were seen in those who were to abort their infants an additional time. Regardless of the *antiphospholipid antibody* (aPL) studied (PS, PI, PG, and CL), elevations in the titer predicted a subsequent pregnancy loss that occurred by 10 weeks in all studied (12). This increase in titer and poor pregnancy outcome occurred despite optimal alloimmune and autoimmune treatment. The dynamics of this process were reminiscent of Rh antibody responses in women with Rh disease and suggested that fetuses evoking the response in their mothers carried certain antigens lacking in those not evoking an elevation in the aPLs.

## *Immunophenotype of Circulating Lymphocyte Subsets*

Women were studied every 2 weeks for lymphocyte subsets, including $CD56^+$ NK lymphocytes, as described above. Figure 20.4 lists preliminary data on women whose pregnancies were completed. Prior to a repeat pregnancy loss with optimal treatment and prior to an infant experiencing intrauterine growth retardation, elevations in $CD56^+$ lymphocytes were seen consistently in peripheral blood. None of the autoimmune therapies except *intravenous immunoglobulin G* (IVIG) lowered

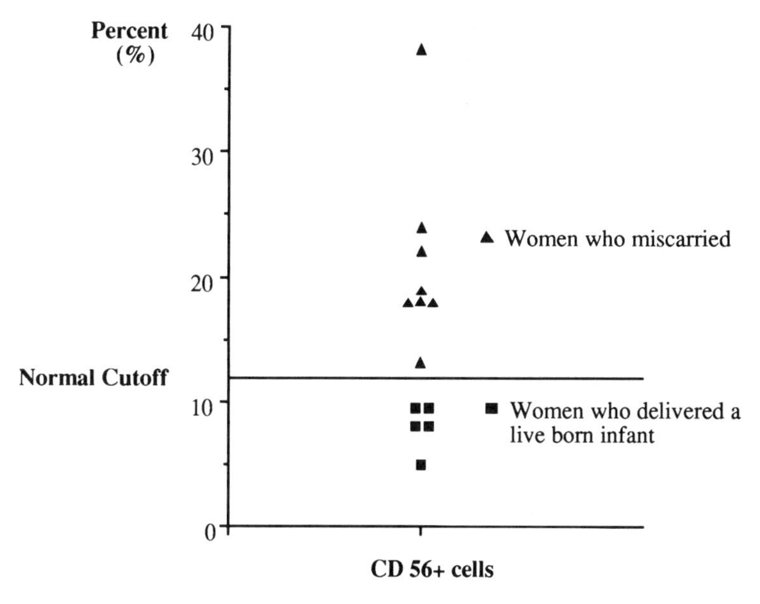


FIGURE 20.4. $CD56^+$ cell population in peripheral blood in women who miscarried the index pregnancy and women who delivered a baby.

the concentration of these cells in the maternal circulation. Following spontaneous termination of pregnancy in most women, the circulating levels of these cells returned to normal.

Of interest was the finding that all women with elevations in $CD56^+$ lymphocytes and a repeat pregnancy loss while receiving optimal alloimmune and autoimmune therapies were aPL positive. Whether these $CD56^+$ lymphocytes are also $CD16^+$ and *interleukin-2* (IL-2) receptor positive is a question under study. Cells of this phenotype can damage or retard trophoblast growth and development (13). Whether these circulating NK cells originate from the maternal decidua is also under investigation.

## *HLA-DQα Analysis*

All couples experiencing a repeat miscarriage following lymphocyte therapy and optimal autoimmune treatment were analyzed for HLA-DQ alleles. Couples experiencing a live birth following the same therapy were also analyzed. Figure 20.5 documents a strong association with maternal HLA-DQα4.1 (0501) in women who experienced a repeat miscarriage. Mothers who were HLA-DQα4.1 (0501) positive developed high positive aPL titers. Mothers who were lacking HLA-DQα4.1 (0501) did not develop aPLs following their pregnancy loss. The role of the fetal genotype in HLA-DQα4.1 (0501) mothers is the subject of another report. In

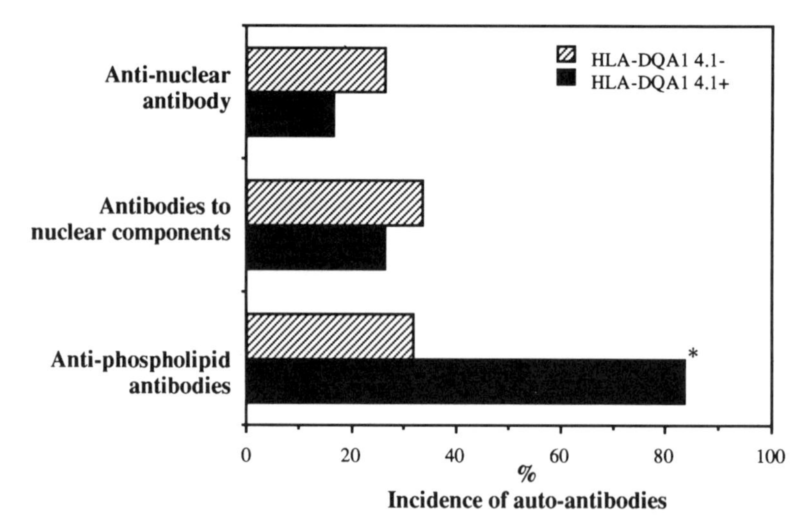


FIGURE 20.5. Incidence of autoantibodies in recurrent spontaneous aborters according to HLA-DQα4.1 (0501). The incidence of aPLs in women with HLA-DQα4.1 (0501) was significantly higher than in women without HLA-DQα4.1 (0501) ($P < 0.05$). There was no relationship between the presence of ANA and antibodies to nuclear components (DNA, histone, and polynucleotides).

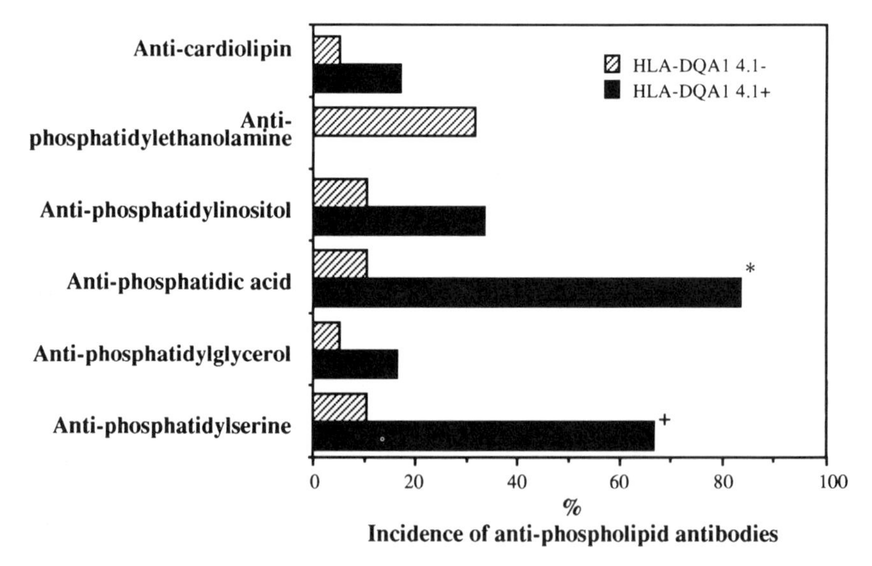


FIGURE 20.6. Incidence of aPLs in recurrent spontaneous aborters according to HLA-DQα4.1 (0501). The incidence of aPA in women with HLA-DQα4.1 (0501) was significantly higher than in women without HLA-DQα4.1 (0501) ($P < 0.005$). The incidence of aPS in women with HLA-DQα4.1 (0501) was also significantly higher than in women without HLA-DQα4.1 (0501) ($P < 0.05$).

live-born infants no homozygous HLA-DQα4.1 (0501) live-born infants were identified; however, many first-born infants of couples who became secondary aborters were HLA-DQα4.1 (0501) homozygous. Further studies are ongoing to determine if certain HLA alleles of first-born children of secondary aborting couples initiate the aPL syndrome that prejudices subsequent pregnancies in a manner similar to the Rh isoimmunization syndrome so well described. Figure 20.6 plots the nature of aPLs and the percentage positive in HLA-DQα4.1 (0501)-positive and HLA-DQα4.1 (0501)-negative women. Anticardiolipin antibody was not the most common aPL detected in RSA women.

## *IVIG Therapy*

Fifteen women with a history of RSA entered a pregnancy under study and experienced severe pregnancy complications, primarily *intrauterine*

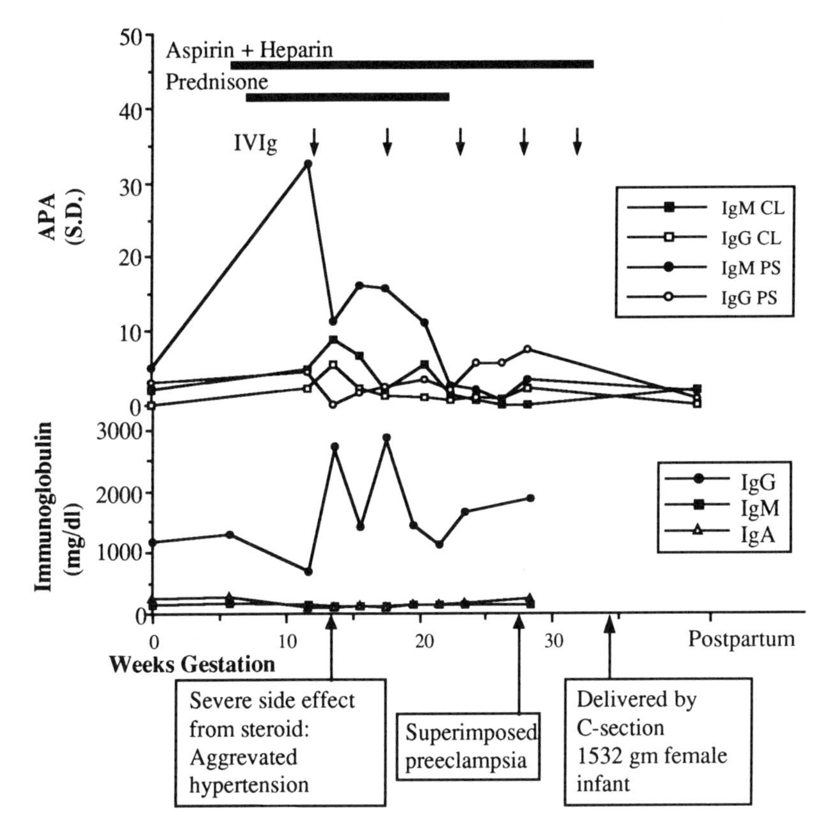


FIGURE 20.7. Tracing of aPLs (APA) and IgG, IgM, and IgA immunoglobulin levels during pregnancy of 34-year-old woman (gravida 4, spontaneous abortion 4) who received immunoglobulin G infusion therapy.

TABLE 20.1. Reproductive outcome after IVIG therapy.

| Patient | Weeks gestation | Delivery method | Indication for delivery | Sex | Weight (gm) | APGAR | Comments |
|---|---|---|---|---|---|---|---|
| 1 | 32 | C/S[a] | IUGR[b]<br>PROM[c] | M | 1007.2 | 6/7 | IUGR |
| 2 | 35 | C/S | IUGR<br>Fetal distress | M | 1675.1 | 7/8 | IUGR |
| 3 | 34 | C/S | IUGR<br>Fetal distress<br>Severe preeclampsia | F | 1532.2 | 6/8 | IUGR |
| 4 | 38 | C/S | POP[d] | F | 3203.8 | 9/9 | AGA[e] |
| 5 | 36 | SVD[f] | — | M | 2748.8 | 8/9 | AGA |
| 6 | 39 | C/S | Overt DM<br>Preeclampsia | F | 3289.5 | 9/10 | AGA |
| 7 | 36 | C/S | Fetal distress | F | 2778.7 | 7/9 | AGA |
| 8 | 36 | C/S | Twin | M | 2748.8 | 9/9 | AGA |
| | | | Breech | M | 2807.7 | 9/9 | AGA |
| 9 | 37 | C/S | Twin | M | 2664.9 | 9/9 | AGA |
| | | | | M | 2225.3 | 8/9 | AGA |
| 10 | 35 | SVD | — | F | 2268.0 | 9/9 | AGA |
| 11 | 36 | C/S | Placenta previa | F | 1814.3 | 8/9 | IUGR |
| 12 | 34 | C/S | Preterm labor<br>Fetal distress | M | 2379.4 | 9/9 | AGA |

[a] C/S = cesarean section.
[b] IUGR = intrauterine growth retardation.
[c] PROM = premature rupture of membrane.
[d] POP = persistent OP position.
[e] AGA = appropriate for gestational age.
[f] SVD = spontaneous vaginal delivery.

*growth retardation* (IUGR) or a subsequent pregnancy loss with optimal alloimmune and autoimmune therapies. When the fetuses began to show slowing of growth and the mother developed both elevations in autoantibodies as well as elevations in $CD56^+$ NK circulating cells, IVIG therapy was initiated at the dosage of 400 mg/Kg/day for 3 days each month through 28 weeks of gestation. Alloimmune, autoimmune, and immunophenotype analysis were done every 2 weeks. Figure 20.7 documents the immunoprofile of a patient who was treated, and Table 20.1 lists the pregnancy outcome of women receiving IVIG. Women not receiving IVIG experienced *intrauterine fetal death* (IUFD) of their infants. Studies are under way to identify women at risk for IUGR and repeat abortions with optimal alloimmune and autoimmune therapies and candidates for IVIG therapy. Women who (i) are aPL high positive with titers of greater than 1:640, (ii) are HLA-DQα4.1 (0501) positive, and (iii) have elevations of circulating $CD56^+$ NK cells prior to pregnancy appear to be at highest risk for subsequent IUGR.

TABLE 20.2. HLA-DQα alleles in couples with unexplained infertility.

| | | | Failed | HLA-DQα alleles | | | | | | | | |
|---|---|---|---|---|---|---|---|---|---|---|---|---|
| Age | G | SAB[a] | IVF[b] | DQα ♀ | | DQα ♂ | | LIT[c] | aPL[d] | ANA[e] | ANUA[f] | Comment |
| 34 | 2 | 2 | 4 | 1.1 | 3.0 | 1.3 | 3.0 | DONOR[g] | + | − | − | Ongoing pregnancy |
| 41 | 0 | 0 | 1 | 1.1 | 1.3 | 1.2 | 2.0 | + | + | − | − | No pregnancy |
| 31 | 0 | 0 | 6 | 1.2 | 2.0 | 1.2 | 3.0 | DONOR | + | − | − | Full term |
| 31 | 0 | 0 | 3 | 1.2 | 1.2 | 1.2 | 1.2 | + | + | + | + | No pregnancy |
| 41 | 0 | 0 | 6 | 1.2 | 4.0 | 1.1 | 4.0 | + | − | + | − | No pregnancy |
| 34 | 0 | 0 | 2 | 1.3 | 4.0 | 1.1 | 1.2 | DONOR | + | + | − | Full term |
| 32 | 0 | 0 | 6 | 1.3 | 1.3 | 1.2 | 1.3 | + | + | − | − | Full term |
| 36 | 3 | 2 | 2 | 3.0 | 4.0 | 2.0 | 2.0 | + | + | − | + | Ongoing pregnancy |
| 30 | 1 | 1 | 12 | 3.0 | 4.0 | 1.3 | 2.0 | + | + | + | − | Ongoing pregnancy |
| 33 | 0 | 0 | 5 | 3.0 | 3.0 | 1.1 | 2.0 | + | − | − | − | No pregnancy |
| 39 | 0 | 0 | 2 | 3.0 | 4.0 | 1.1 | 3.0 | + | + | + | − | Chemical pregnancy SAB |
| 36 | 0 | 0 | 2 | 4.1 | 4.1 | 1.1 | 4.0 | + | + | − | − | No pregnancy |

[a] SAB = spontaneous abortion.
[b] IVF = in vitro fertilization.
[c] LIT = lymphocyte immunotherapy with paternal lymphocytes.
[d] aPL = antiphospholipid antibody.
[e] ANA = antinuclear antibody.
[f] ANUA = antibodies to nuclear antigens (DNA, histone, and polynucleotide).
[g] DONOR = lymphocyte immunotherapy with donor lymphocytes.

### *Primary Idiopathic Infertility Followed by Repeated Pregnancy Failure with In Vitro Fertilization*

We have studied couples with primary infertility, including those who entered advanced reproductive technology programs. All couples were successful in creating healthy fertilized oocytes in culture, and all lost their pregnancies, some having been through 11 failed *in vitro fertilization* (IVF) cycles. Although data are still very preliminary, Table 20.2 lists available data on 12 women who finished immunological evaluation and alloimmune and autoimmune therapies. Since not all have been treated or have entered a cycle of conception posttreatment, adequate controls for this study are not available. Based on available HLA-DQα analysis, couples who continue to be unsuccessful share HLA-DQα4.1 (0501) alleles. We are currently conducting immunogenetic screening of idiopathic infertile couples who repeatedly fail at advanced reproductive technologies to determine if certain immunogenetic, alloimmune, and autoimmune parameters create an unlucky fetomaternal match and a hostile maternal environment for successful pregnancy.

## Discussion

Based on all data available, the following comprehensive hypothesis is offered for recurrent pregnancy loss in some couples. This may serve as a stimulus for new and focused research in the field to identify couples with a true immune etiology for their infertility and repeated pregnancy losses and may allow the identification of those that will profit from alloimmune or autoimmune therapy.

Paternal HLA-G on the trophoblast may present peptides that differ from the maternal HLA-G expression (Fig. 20.8). Possibly, these recruit $CD8^+$ suppressor cells from the decidua for initial immunoregulation (14), while large granulated lymphocytes of the NK lineage are instructed through trophoblast hormones to produce cytokines essential for the up-regulation of HLA-G, as well as for the growth and development of trophoblast cells (15, 16). This response is attended by an alloimmune IgG antibody response that is identified first by maternal antibody binding to paternal B cells and then to paternal T cells. The dynamics of the alloimmune response have been reported by us (6, 17). The possible benefits of this alloantibody response to the fetus and placenta have been the subject of many reports and much debate (17–21).

### *Antiphospholipid Antibodies*

There are potential consequences to the lack of alloimmune recognition by the mother during pregnancy that may be hazardous to her health (21,

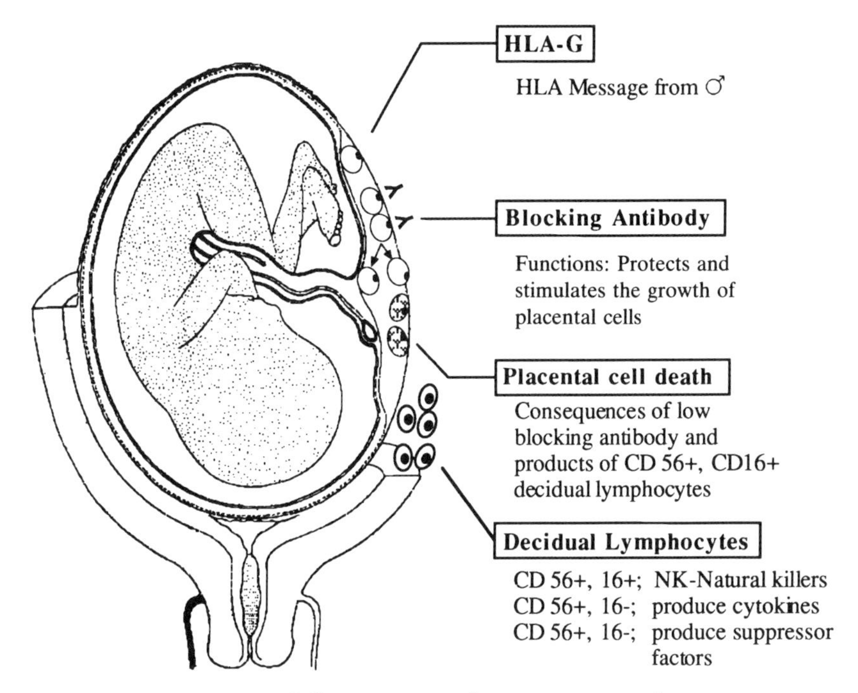


FIGURE 20.8. HLA compatibility as a cause for recurrent spontaneous pregnancy loss. The consequences are (i) inadequate blocking antibody formation, (ii) ineffective camouflage of placenta, (iii) failure of placental cells to grow and divide, and (iv) possible death of placental cells.

22). Women who experience repeated pregnancy losses are at increased risk to develop autoantibodies to phospholipid antigens of the type shown to function as adhesion molecules in cell fusion to syncytial structures (myoblasts, fibroblasts, and cytotrophoblast). The best-studied phospholipids in this regard are PS and PE. The hierachy of the maternal immune response to phospholipids during a failed pregnancy is PS, PI, CL, PG, PE, and *phosphatidic acid* (PA). Figure 20.9 outlines this hypothesis of autoimmunity developing as a consequence of recurrent pregnancy loss. Studies of normal women do not show these autoantibodies developing during a normal pregnancy (Fig. 20.10). Figure 20.10 lists the data of women experiencing a normal pregnancy whose aPL antibodies were titered each trimester and at delivery. Elevations are not seen in titers above 25 in the women surveyed. There is no increase in the antibody titers during normal pregnancy. There was no increase in titers with increasing gravidity in normal women experiencing normal pregnancy. For this reason, a titer of 1:50 is considered a positive.

It is plausible that when a pregnancy fails in a woman because of the lack of alloimmune recognition, an autoantibody response is initiated that

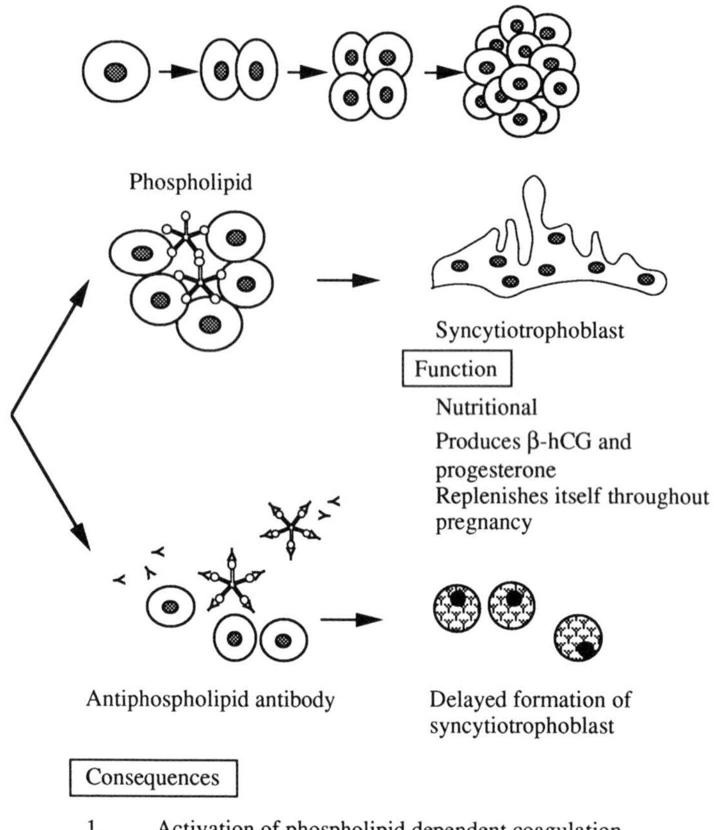


FIGURE 20.9. Phospholipids as a fusion molecule during placental development and neutralization of phospholipids by aPLs.

targets phospholipid adhesion molecules essential for placental development and integrity. Women who are HLA-DQ and -DR compatible with their fetuses and mothers who are HLA-DQα4.1 (0501) and who gestate fetuses with this allele appear to be most prone to develop autoantibodies

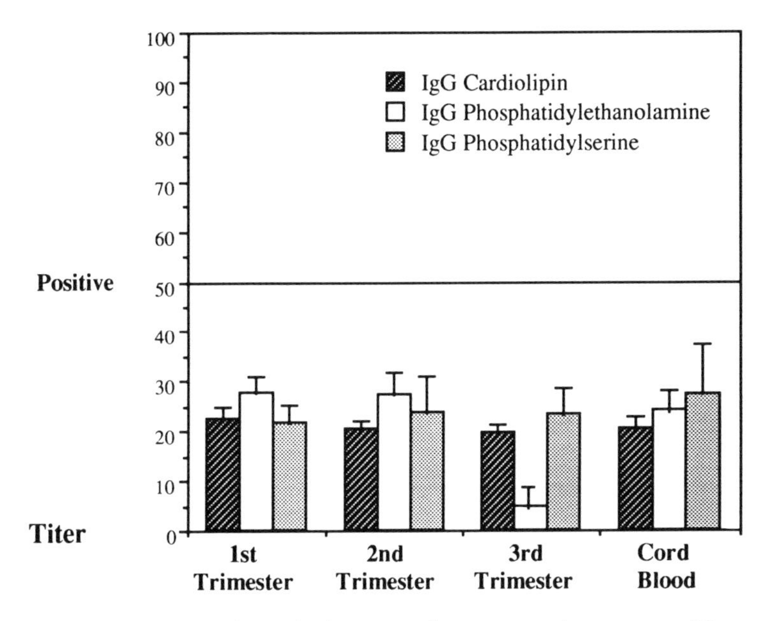


FIGURE 20.10. The aPL titers during normal pregnancy in women with normal pregnancy history ($n = 25$).

to phospholipid antigens when the pregnancy fails. The fact that HLA-DQα4.1 (0501) homozygotes are often first born and then followed by siblings who miscarry supports this line of reasoning. Some women with repeated exposure develop aPLs that interfere with coagulation homeostasis through alterations in natural fibrinolysins and alterations in phospholipid-dependent coagulation. Only 7% of women ever develop aCL, and less than 0.5% develop an autoimmune disorder later in their lives. The role of low-dosage aspirin and preconception low-dosage heparin for this disorder has been discussed and published (2).

## *Antinuclear Antibodies*

Repeated exposure of the mother to decaying pregnancies is associated with an autoimmune response to DNA and histones, leading often to a low-titered *antinuclear antibody* (ANA)-positive test with a speckled pattern. Although not as pronounced as the mother's response to phospholipid antigens, there is an increasing incidence of anti-DNA antibodies observed with each spontaneous pregnancy loss. The role of prednisone therapy in the treatment of these women during pregnancy has been widely discussed and published (23, 24). Figure 20.11 outlines this possible consequence of recurrent pregnancy loss.

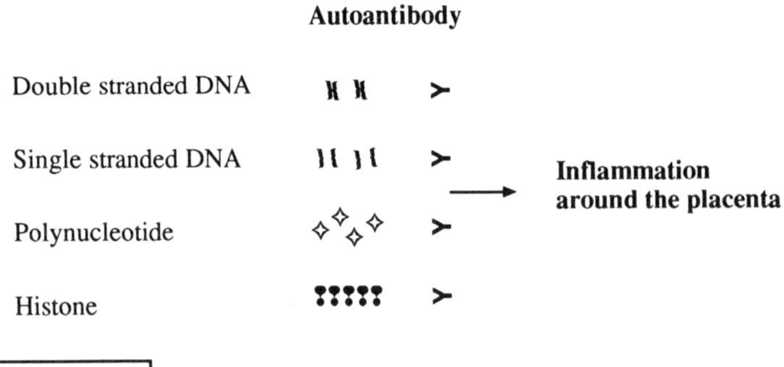


Consequences

ANA + , Speckled pattern

Autoantibody to DNA leads to inflammation in the placenta

Autoimmune disease screening in the woman is negative

- no evidence of lupus or rheumatoid arthritis

Treatment

Prednisone (steroid therapy) preconception started and continued until 24 weeks gestation.

FIGURE 20.11. Autoimmune response to nuclear components of fetal origin.

## *Antisperm Antibodies*

Four percent of couples with recurrent pregnancy losses develop antisperm antibodies and subsequently fail to achieve a pregnancy. We have found that these couples can be identified by an antisperm antibody flow cytometric assay that detects maternal IgG or IgA antibody binding to fresh gradient-separated spermatozoa. The higher incidence of antisperm antibodies in RSA women has been reported, but is poorly understood (25). Figure 20.12 outlines this possible consequence of recurrent pregnancy loss.

This hypothesis of reproductive failure progressing to infertility is a plausible one. When studies are conducted in women with an autoimmune disorder, such as rheumatoid arthritis, it has been found that exacerbation of the autoimmune disorder in the mother during pregnancy involves a fetus that is HLA-DR and -DQ compatible with her. Amelioration of the disease is associated with a fetus that is HLA-DR and -DQ incompatible with the mother (26). Studies are under way to determine if maternal-fetal compatibility with regard to the HLA-DQα4.1 (0501) gene is of higher frequency in mothers experiencing a worsening of their rheumatoid arthritis during pregnancy. Findings such as this may

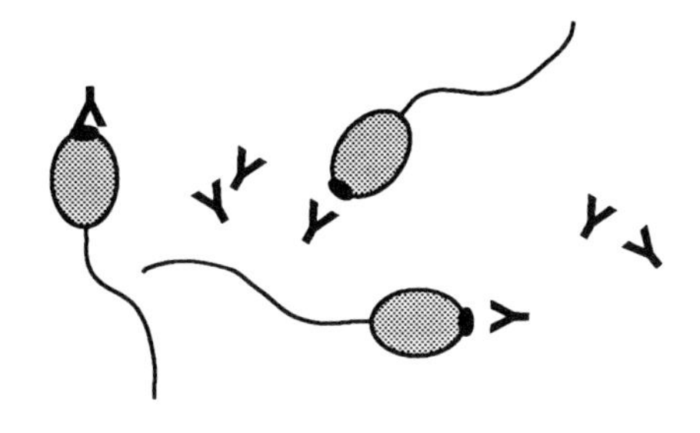

Consequences

Sperm antibody test positive

Sperm antibody positive by flow cytometer

Couples unable to conceive normally

Multiple failed pregnancy through IVF, GIFT or ZIFT

Treatment

1. Immunization with paternal white blood cells to stimulate blocking antibody.
2. Preconception treatment with low dosage aspirin, heparin, and prednisone.
3. Some couples conceive normally following this treatment.
4. Others return for IVF, GIFT or ZIFT and become parents.

FIGURE 20.12. Autoimmune response to sperm antigen.

permit pregnancy counseling for autoimmune women and their spouses prior to a pregnancy.

## Summary

Reproductive immunology has focused on alloimmune networks since the birth of the speciality (27). It is very possible that a fetus and its placenta must be autoimmune acceptable to the mother before alloimmune networks are initiated. The higher incidence of autoimmune disorders in women compared to men may in part be explained by the line of reasoning outlined in this chapter. The higher death rates from cardiovascular complications seen in multigravid women compared to nonparous women may also relate to a higher incidence of autoantibodies in the former (28).

A major concern relating to offspring born of couples treated with immunotherapy and autoimmune therapy may not surface for 20 or 30 years. If homozygosity with regard to certain HLA alleles is associated with hypofertility in some couples and if certain HLA antigens predispose to autoimmune responses and even autoimmune diseases, these possible complications in the offspring will not be assayable at the present time. A plea is made for long-term follow-up for autoimmune and reproductive complications of all live-born children in immunotherapy programs into adulthood and well into their reproductive careers. An understanding of the issues and possible mechanisms discussed in this chapter may lead to advances in testing and therapy for autoimmune diseases and to a deeper knowledge of the role that autoimmune responses may play in chronic organ or tissue rejection.

"Innovators are rarely received with joy and established authorities launch into condemnation of newer truths, for at every crossroads to the future are a thousand self-appointed guardians of the past" (Betty MacQuilty, *Victory Over Pain: Morton's Discovery of Anaesthesia*).

## *References*


1. Beer AE, Quebbeman JF, Ayers JWT, Haines RF. Major histocompatibility complex antigens, maternal and paternal immune responses, and chronic habitual abortions in human. Am J Obstet Gynecol 1981;141:987–99.
2. Kwak JYH, Gilman-Sachs A, Beaman KD, Beer AE. Reproductive outcome in women with recurrent spontaneous abortions of alloimmune and autoimmune etiologies; pre vs post conception treatment. Am J Obstet Gynecol 1992;166:1975–87.
3. Kwak JYH, Gilman-Sachs A, Beaman KD, Beer AE. Autoantibodies in women with primary recurrent spontaneous abortion of unknown etiology. J Reprod Immunol 1992;22:15–31.
4. Beer AE, Zhu X, Semprini AE, Quebbeman JF. Pregnancy outcome in human couples with idiopathic recurrent abortions: the role(s) of female serum, mixed lymphocyte culture blocking factors, potentiating factors, and local uterine immunity before and after paternal leukocyte immunization. In: Talwar GP, ed. Immunological approaches to contraception and promotion of fertility. Plenum, 1986:393–406.
5. Kwak-Kim JYH, Beaman KD, Gilman-Sachs A, Kerns K, Ober C, Beer AE. HLA immunogenetic evaluation of couples with primary and secondary recurrent spontaneous abortions. Soc Gynecologic Investigation, 37th annu meet. St. Louis, Missouri, 1990:340.
6. Lubinski J, Vrodoljak VJ, Beaman KD, Kwak JYH, Beer AE, Gilman-Sachs A. Characteristics of antibodies induced by paternal lymphocyte immunization in couples with recurrent spontaneous abortion. J Reprod Immunol 1993.
7. Bahar AM, Kwak JYH, Beer AE, et al. Antibodies to phospholipids and nuclear antigens in non-pregnant women with unexplained spontaneous recurrent abortions. J Reprod Immunol 1993.
8. Session A, Horwitz A. Myoblast aminophospholipid asymmetry differs from that of fibroblasts. FEBS Lett 1981;134:75–8.

9. Session A, Horowitz A. Differentiation related differences in the plasma membrane phospholipid asymmetry of myogenic and fibrogenic cells. Biochim Biophys Acta 1983;728:103–11.
10. Ehrlich HA, Bugawan TL. Principles and applications for DNA amplification. In: Erlich HA, ed. PCR technology. New York: Stockton Press, 1989: 193–208.
11. Beer AE, Kwak JYH, Beaman KD, Gilman-Sachs A. Antiphospholipid antibodies in women with recurrent pregnancy losses: elicitation, expression and therapy. In: Dondero F, Johnson PM, eds. Reproductive immunology; vol 79. Rome: Raven Press, 1993:265–72.
12. Kwak JYH, Barini R, Gilman-Sachs A, Beaman KD, Beer AE. Down regulation of maternal antiphospholipid antibodies during early pregnancy and pregnancy outcome. Soc Gynecologic Investigation, 40th annu meet. Toronto, Canada, 1993:107.
13. Loke YW, Kina A, Chumbley G, Holmes N. Does HLA-G influence trophoblast susceptibility to lysis by decidual large granular lymphocytes? In: Dondero F, Johnson PM, eds. Reproductive immunology; vol 79. Rome: Raven Press, 1993:145–8.
14. Sanders SK, Giblin PA, Kavathas P. Cell-cell adhesion mediated by CD8 and human histocompatibility leukocyte antigen G, a nonclassical major histocompatibility complex class 1 molecule on cytotrophoblasts. J Exp Med 1991; 174:737–40.
15. Bulmer JN. Decidual leucocytes in normal and pathological human pregnancy. In: Dondero F, Johnson PM, eds. Reproductive immunology; vol 79. Rome: Raven Press, 1993:129–32.
16. Johnson PM, Deniz G, McLaughlin PJ, Hampson J, Christman SE. Functional properties of cloned CD3-decidual leucocytes and loss-affinity cytokine receptor expression on human trophoblast. In: Dondero F, Johnson PM, eds. Reproductive immunology; vol 79. Rome: Raven Press, 1993:141–4.
17. Reed E, Beer AE, King DW, Suciu-Foca N. The alloantibody response of pregnant women and its suppression by soluble HLA antigens and anti-idiotypic antibodies. J Reprod Immunol 1991;28:103–13.
18. Wegman TG. Fetal protection against abortion: is it immunosuppression or immunostimulation? Annu Rev Immunol 1984;135D:309–12.
19. Clark DA, Chaput A, Tutton B. Active suppression of host versus graft reaction in pregnant mice, VII. Spontaneous abortion of CBA × DBA/2 fetuses in the uterus of CBA/J mice correlates with deficient non-T suppressor cell activity. J Immunol 1986;136:1668–74.
20. Chaouat G, Menu E, Clark DA, Dy M, Minkowski M, Wegmann TG. Control of fetal survival in CBA × DBA/2 mice by lymphokine therapy. Reprod Fertil 1990;89:447–58.
21. Beer AE, Kwak JYH. Recurrent spontaneous abortions immunological evaluation. In: Current therapy in obstetrics and gynecology. Philadelphia: Saunders, 1993.
22. Galstian A, Beer AE, Roberts JM, Coulam CB, Faulk WP. Immunology of preeclampsia. In: Coulam CB, Faulk WP, McIntyre JA, eds. Immunological obstetrics; vol 1. NY and London: W.W. Norton, 1992:502–16.
23. Branch W, Scott JR, Kochenour NK, Hershgold E. Obstetric complications associated with the lupus anticoagulant. N Engl J Med 1985;313:1322–6.

24. Lockshin MD, Qamar T, Druzin ML. Antibody to cardiolipin, lupus anticoagulant and fetal death. J Rheumatol 1987;14:259–62.
25. Haas GGJ, Kubota K, Quebbeman JF, Jijon A, Menge AC, Beer AE. Circulating antisperm antibodies in recurrently aborting women. Fertil Steril 1986;43:209–15.
26. Nelson JL, Hughes KA, Smith AG, Nisperos BB, Branchaud AM, Hansen JA. Maternal-fetal disparity in HLA class II alloantigens and the pregnancy-induced amelioration of rheumatoid arthritis. N Engl J Med 1993;329:466–71.
27. Billingham RE, Beer AE. Reproductive immunology: past, present, and future. Perspect Biol Med 1984;27:259–75.
28. Ness RB, Harris T, Cobb J, et al. Number of pregnancies and the subsequent risk of cardiovascular disease. N Engl J Med 1993;328:1528–33.

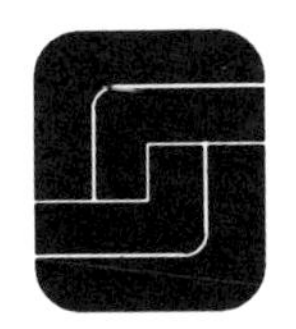

# Author Index


**A**

Allen, W.R., 153
Anderson, D.J., 14
Antczak, D.F., 153

**B**

Bazer, F.W., 37
Beaman, K.D., 316
Becker, C., 214
Beer, A.E., 316
Bikoff, E.K., 201

**C**

Chen, H.-L., 136
Clark, D.A., 125
Cohen, P.E., 104
Coulam, C.B., 303

**D**

Demeur, C., 184
Donaldson, W.L., 153

**F**

Flanders, K.C., 125
Flickinger, C.J., 254

**G**

Garcia, F.U., 136
Garcia-Lloret, M., 99
Gill, T.J., III, 170
Gilman-Sachs, A., 316
Girr, M., 184
Grünig, G., 153
Guilbert, L., 99
Guillaudeux, T., 184

**H**

Herr, J.C., 254
Hill, J.A., 23
Hirte, H., 125
Ho, H.-N., 170
Hooshmand, F., 214
Howards, S.S., 254
Hunt, J.S., 136

**J**

Johnson, H.M., 37
Johnson, P.M., 3

**K**

Kanbour-Shakir, A., 170
Kunz, H.W., 170
Kwak, J.Y.H., 316
Kydd, J., 153

**L**

Lala, P.K., 57
Le Bouteiller, P., 184
Lin, H., 99
Lou, Y.-H., 239

Lyden, T.W., 285
Lysiak, J.J., 57

M

Maher, J.K., 153
Manuel, J., 125
Mosmann, T., 99
Mowbray, J., 125

N

Ng, A.K., 285

O

Ott, T.L., 37

P

Pollard, J.W., 104
Pudney, J., 14

R

Robertson, E.J., 201
Robertson, S.A., 82
Roby, K.F., 136
Rodriguez, A.-M., 184
Rote, N.S., 285

S

Seamark, A.C., 82
Seamark, R.F., 82
Smith, H., 239
Spencer, T.E., 37
Starkey, P., 125

T

Tung, K.S.K., 239

U

Underwood, J., 125

V

Vince, G., 125
Vogt, E., 285
Voland, J.R., 214

W

Wegmann, T.G., 99

Y

Yang, Y., 136
Yui, J., 99

# Subject Index


ABC superfamily of transporters, 202, 208
Abortion
  prevention of, 101. *See also* Immunotherapy
  secondary, 305, 307-308, 316-317, 319-323
  spontaneous, 6, 102, 127-128, 163-164, 174, 177-180, 214, 217-218, 230-232, 303-305, 307-309, 316-327, 329-330
Acetylcholine receptor (AChR) peptides, 246-247, 249-250
Acrosome, 7
β-actin, 85, 205-206
Activins, 140
Adenocarcinoma, 45-46, 142
Adhesion formation, 25, 27-28
Adhesion molecules, 317, 327-328
Adrenalitis, 239
Aging process, 177, 181
Albumin, 309-311
Alkaline phosphatase, 67, 189
Allantochorion, 158, 161, 163-164
Allantois, 203
Alloantibodies, 4, 156-157, 159, 316
Allograft, 3, 160
Alloimmune disease, 318-319
Alloimmune response, 101, 317, 326, 331
Alloimmune therapy, 101, 316, 318, 320-321, 324-326
American Society for Reproductive Immunology, study, 311-312
Amerindians, 178
Amino acid sequences, 41-42, 47, 61-62
Aminoacetylfluorene (AAF), 177
Amniochorion, 60, 187
Amnion, 58, 66, 69-70, 203
Amnion invasion assay, 70
Amniotic cavity, 58
Amniotic fluid, 141
Amphiregulin (AR), 61-62, 64-66, 68
Androgen, 107, 110
Androgenetic zygote, 173-174
Angiogenesis, 59, 93, 105, 141
Antibodies
  to amphiregulin, 64-65
  to cardiolipin (aCL), 285-291, 293-298, 305, 309, 316, 319-320, 322-323, 327, 329
  to cytokeratin, 291
  to DNA, 289-290, 292, 322, 325, 329-330
  to endometrium, 24-26
  to epidermal growth factor receptor, 68
  to F4/80 antigen, 86, 88-89
  to granulocyte macrophage colony stimulating factor, 90
  to heavy chain, 157, 215
  to histone, 322, 325, 329-330
  to HIV, 5, 15, 17, 19, 306, 309
  to Ia antigen, 86, 89
  to interferons, 48, 245
  to interleukin-10, 100
  maternal, 154, 156, 159-163, 165
  to β2 microglobulin, 188, 191
  monoclonal, 8-9, 64, 82, 88-89, 100, 153-154, 156-159, 163, 172,

Antibodies (*cont.*)
188–191, 215, 219–220, 287–288, 290–296, 318
to nuclear antigens, 289, 317, 319, 322, 325, 329
to oxytoxin, 40
to p50, 194–196
paternal, 306–309, 317
to phosphatidic acid, 322, 327
to phosphatidylethanolamine, 322, 327, 329
to phosphatidylglycerol, 316, 320, 322, 327
to phosphatidylinositol, 316, 320, 322, 327
to phosphatidylserine (aPS), 287–291, 293–298, 316, 320, 323, 327, 329
to phospholipids, 25, 285–297, 316–317, 319–325, 328–329
polyclonal, 64–65, 194–195, 290
to polynucleotides, 322, 325, 330
Rh, 320
secondary responses, 160, 162, 164
to sperm, 254–269, 274–275, 277–281, 330–331
to TIMP-1, 72
to TJ6 protein, 304
to transforming growth factors, 64–65, 68–69, 72, 126–130
to trophoblast, 154, 188
to tumor necrosis factor, 138, 245
to urokinase-type plasminogen activator, 71
to ZP3, 245–247, 250
Antigens. *See also* MHC class I; MHC class II; MHC class III
ATPase, 243
endometrium, 24–26
hepatitis B, 306, 309
HLA-DQα4.1, 316–317, 321–326, 328–330
HML-1, 9
maternal, 188, 316, 326
NDOG-5, 67
oocyte, 243
ovary, 25, 239, 250
paternal, 154–157, 159–162, 164, 171–174, 201, 204, 214, 221, 311, 326
proliferating cell nuclear (PCNA), 68
63D3, 67
transgenic, 240
zona pellucida, 239
ZP3, 239, 245–247, 250
Antiluteolytic activity, 37–38, 40–42, 44, 46–50
Antinuclear antibody (ANA) test, 322, 325, 329–330
Antiphospholipid antibody (aPL) syndrome, 285, 288, 293–294, 297–298
Antisense probe, 142–143, 208
Antiserum. *See* Antibodies
hyperimmune, 277, 279–281
Antiviral activity, 42, 44, 47, 49, 140
AP-1, 44
Apolipoprotein H (apo H), 296
Apoptosis, 102, 240
Aprotinin, 71
Arachidonic acid, 39–40, 93
Arterioles, spiral, 285, 296
Arthritis, 99, 127, 330
Ascites, 130
Asialo-GM1 marker, 127, 304
Aspirin, 289, 317, 319, 323, 328, 331
Astrocytes, 138
Atherosclerosis, 254
Atherosis, 285–286
ATP-dependent translocation, 202, 208
ATPase, 243
Atresia, follicles, 106
Autoantibodies, 25, 27, 239–240, 243, 249–251, 255–259, 285, 289, 317, 322, 324, 327–328, 330–331
Autoantigens, 277–280, 282
Autocrine activity, 63, 68–69, 73, 90, 92, 105, 136–138, 143–144
Autoimmune disease, 99, 229–230, 239–240, 243–246, 316, 318, 329–330, 332
Autoimmune response, 20, 25–26, 306, 309, 316, 327–330, 332
Autoimmune therapy, 316–321, 324–326, 331
Autoimmune women, 316, 318–319, 331
Autoradiography, 91
Azoospermia, 15, 273–274, 276

Baboon, 26, 38
Bacteriuria, 18
*BamHI*, 218–219, 229
Band-shift assay, 194
Basement membrane, 70–71, 109, 111–112
B cells, 9, 20, 25, 245, 250–251, 316, 326
Betaglycan receptors, 62
BeWo cell line, 6, 43, 90–91, 287, 297
Binding proteins, 62–63, 66–67, 73
Binucleate cells, 155, 159, 161, 163
Biopsy
  breast cancer, 131
  endometrial, 14, 24, 305
Biotinylation, 220, 290
Blastocyst, 4, 8, 58–59, 72, 85, 90, 92–93, 115, 117, 141, 156–157
Blood
  fetal, 57, 102, 153, 155, 210, 230
  maternal, 3–4, 57–59, 102, 153, 155, 159–160, 210, 215–216, 232, 290, 320–321
  sinusoids, 222, 224–225, 229
Blood-epididymis barrier, 259
Blood-testis barrier, 16, 117, 259
Bone marrow, 60, 91, 126–127, 136
Brain, 210
Breast cancer, 46, 61, 131, 136, 144
Bulbourethral glands, 15, 19
Buserelin, 27

Cachexia, 137
cAMP, 45, 111
Cancer
  AE-7 cells, 46
  breast, 46, 61, 131, 136, 144
  chorionic, 5–6, 70, 74, 90, 143, 174
  decidual, 60
  embryonal, 193, 202
  endometrial, 46, 142
  fostering of, 136
  hepatic, 140
  metastatic, 71, 141–142
  ovarian, 62, 130, 141–142
  prostate, 130
  resistance to, 170–171, 174, 177, 179–181, 184
  tissue wasting, 137
  transforming growth factor $\beta$ in, 126, 130–131
Carcinogens
  chemical, 177–179
  resistance to, 177–179
Carcinoma. *See* Cancer; Choriocarcinoma
Cardiolipin, 286–287, 296
Caruncle, 155–156
Cat, MHC antigen, 157
Cathepsin D, 62
Cattle. *See* Cow
Caucasians, 178
CBA × DBA system, 218
CCE cell line, 208–209
CD lymphocyte proteins
  CD3, 3, 6–7, 9–10, 318
  CD4, 9, 15–16, 18–19, 153, 156, 161, 163, 241–245, 318
  CD5, 241–244
  CD8, 4, 7, 9, 15–18, 153, 156, 161, 163, 218, 226–227, 231, 241–244, 304, 318, 326
  CD16, 3, 6, 9–10, 318, 321, 327
  CD45, 86, 89, 156, 188
  CD56, 3, 6, 9–10, 128, 318, 320–321, 324, 327
  others, 7–8, 102, 318
cDNA, 5, 8–9, 41, 46, 84–85, 90, 211
Cell culture, 66–73, 102, 107, 128
Cell death, 102, 293
Cell membrane, 5–8, 46
Cervix, 14, 113
Cesarean section, 188, 289, 323–324
Chemical toxicity, 239
Chemotaxis, 25, 108, 115–116
Chick, in vitro assays, 69
Chimera, 60
Chinese couples, 175, 178–179
Chinese hamster ovarian cells, 141–142
Chlamydia, 306
Chorioallantoic membrane, 69
Choriocarcinoma, 5–6, 70, 74, 90, 143, 174, 295
Choriocarcinoma cell line, 185–189, 194, 196, 287, 297
Chorion, 38, 58, 66, 158–163, 187, 203

Chorionic gonadotropin (CG), 37–38, 41, 50, 67, 90, 92, 110, 128, 158–160, 162–163, 188, 306, 328
Chorionic villi, 4, 6–7, 58–59, 65–67, 164, 215, 290–292
Chromatin, structure of, 185, 188, 194
Chromosome analysis, 305–306, 309
Chromosome 6, 5–6, 137, 170, 184
Chromosome walking, 174
Chromotubation, 27
*Cis*-acting regulation, 185-193, 202
Clonal deletion, 240, 243
Clonal expansion, 249
Cloning, 8–9, 46, 164
Coagulability, 328–329
Coccidioidomycosis, 99
Collagen, 70, 127, 269–270
Collagenase, 70–73, 92
Colombian women, 316
Colony stimulating factor I (CSF-I), 83, 104–110, 112–117, 136–137, 140
Colony stimulating factor I receptor (CSF-I-R), 108, 110, 112
Complement system, 7–8, 24–25
Complotypes, 170
Concanavalin A, 100, 128
Conceptus, 37–38, 40–41, 44, 49, 173
  pregnancy signaling, 37–38, 40, 49–50
Conceptus secretory proteins (CSP), 49–50
Congenital anomalies, 179–180
Connective tissue, 266–267, 270, 272
Consensus motif, 43, 46, 61
Contraceptive vaccine, 239, 245
Corpus luteum (CL), 41–42, 44, 48–50, 105–106, 138–139
  maintenance of, 37–40, 50
*Corynebacterium*, 99
Cotyledon, 155–156
Cow
  corpus luteum, 48
  cytokines, 47–48
  estrous cycle, 38-39, 42, 44, 47–48
  lens capsule, 69
  MHC, 156
  placenta, 155
  pregnant, 47–48, 86
Cowpers gland, 15, 19
CpG islands, 185–188, 190
Cross-reaction, 41, 47, 279, 287, 294
Cumulus cells, 108–109
Cutter enzymes, 185–188, 192–193
Cyclooxygenase, 39
Cytochrome C, 126
Cytokeratin, 65, 67, 90
Cytokine cascade, 24
Cytokines, 3–4, 6, 9–10, 19–20, 24–27, 38, 44, 50, 99, 231. *See also* specific cytokines
Cytoplasmic domain, 83, 91
Cytoskeletal intermediate filaments, 294–295
Cytotoxic T cells. *See* T cells, cytotoxic
Cytotoxic T lymphocyte (CTL) response assay, 126, 129
Cytotoxicity, 3–4, 8, 15, 25–26, 140
Cytotrophoblast, 4–7, 9, 58–59, 64–66, 90, 154, 181, 184, 187–189, 191–196, 215, 287, 290–293, 295–297, 317, 327–328

Danazol, 27
Decidua, 3–10, 57–61, 64–66, 72–74, 100, 114, 116, 126–130, 139, 187, 194–195, 216, 222–223, 304, 321, 326–327
  thrombosis, 285–286
Decidualization, 38, 59–60, 72, 115
Deciduoma, 60
Decorin, 62, 66
Dendritic cells, 18, 89
$^{125}$I-deoxyuridine, 70
Diestrous, 39, 48, 139
Diethylnitrosamine (DEN), 177–179
Dimethylbenzathracine (DMBA), 178
Dioxin, 26
Disease resistance, 170, 181
Disulfide bonds, 43, 62
DNA
  amplification, 317
  *cis*-acting, 45
  promoter, 46
  proviral, 5
  recombinant, 41–42
DNAse, 84–85
Dog, MHC antigen, 157
Dolphin, 42

Donkey-in-horse pregnancy, 162–164
*Drosophila*, 193

Eclampsia, 176, 178. *See also* Pre-eclampsia
*EcoR*I, 208, 219
Ectoplacental cone, 70, 90, 127, 203
Edema, 92, 222, 224
ELISA, 100, 254-256, 258-266, 269, 274-275, 278, 287-288, 294, 296-297
Embryo. *See also* Preimplantation
  death of, 171
  developmental stages, 26, 37, 90, 172, 203
  malformation, 93
Embryo transfer, 163, 172, 174–175, 179, 217-218
Embryonic stem cell lines, 208, 231
Endocrine activity, 27, 48, 57
Endoderm, 203
Endoglycosidase F, 62
Endometrial cup, 158–164
Endometrioma, 27
Endometriosis, 23–28
  NK cells and, 25–26
  reproductive failure and, 23–24
  stages of, 24, 27
Endometrium, 3, 5-6, 8-9, 42, 44-46, 138, 163
  cancer of, 27, 45–46, 142
  ectopic, 23–24, 26
  epithelium, 8, 37–40, 155, 163
  estrogen receptors, 38, 45–48
  glands in, 24, 37, 161, 163
  GM-CSF production, 82–83, 89, 92–93, 115
  interferon receptors, 42
  leukocytes in, 85–89, 92–94, 114–116
  progesterone receptors, 40, 45–46
  prostaglandin F (PGF) production, 37–40, 44–45, 47–49
Endonuclease, 185
Endoplasmic reticulum (ER), 201–202
Endothelium, 86, 91, 140, 286-287, 294-296
Endotoxins, 114, 137–138
Enhancer A, 192–194
Enzyme-linked immunoassay. *See* ELISA
Eosinophils, 84, 86–89, 92–93, 162
Epidermal growth factor (EGF), 24, 61–64, 66, 68, 72–73, 105–106, 126, 137
Epidermal growth factor receptor (EGF-R), 24, 61-64, 68, 72-73, 106
Epididymis, 16–17, 112–113, 116
  epithelium, 270, 274, 276–277
  inflammation of, 17, 269, 271-278, 281
Epithelioid cells, granuloma, 266-267
Epithelium, 14, 16–20, 259
  endometrial, 8, 37–40, 155, 163
  glandular, 37, 45, 65, 83, 115
  GM-CSF production, 82–83, 89, 92–94
  IL-6 production, 91-92
  uterine, 58–59, 82, 89, 91–94, 114–115
*Equus* hybrids, 162
Erythrocytes, 141, 294–295
*Escherichia coli*, 17
Estradiol, 40, 47–48, 108, 115
Estrogen, 38–39, 45, 49–50, 83–84, 91–94, 104, 139–142
Estrogen receptors (ER), 38, 40, 45–48
Estrous cycle, 37–39, 42, 44–49, 83–85, 91, 94, 108, 114
Ewe. *See* Sheep
Exocrine effect, 48
Exocytosis, 40
Extinguisher regions, 192
Extracellular matrix (ECM), 65-66, 70–71

F9 cell line, 202, 208–209
Factor VIII, 67
Fallopian tubes, 26–27
Ferguson reflex, 48
Fertility
  male, 19–20
  reduced, 14, 23, 27, 102, 105, 111, 113, 116–117, 138, 157, 175–178, 181, 332
Fertilization, 8, 25–26, 58, 90, 108, 114, 245

Fertilization (*cont.*)
 in vitro (IVF), 173, 175, 179, 325–326, 331
Fetomaternal compatibility, 128, 184, 204, 321, 326, 328, 330–331
Fetus. *See also* Intrauterine growth retardation
 blood, 57–59, 102, 153, 155, 210
 heart rate, 286, 311
 intrauterine death (IUFD), 285–286, 297, 310–312, 324
 resorption, 94, 101, 217–218, 222, 290
 vestigial, 174
 waste products, 57, 155
 weight reduction, 293, 296
Fibrinolysins, 328–329
Fibroblast growth factor (FGF), 105–106
Fibroblast growth factor receptor (FGF-R), 106
Fibroblasts, 25, 86, 93, 102, 194–195, 269–270, 317, 327
Fibronectin, 70–71, 90
Fibrosis, 127
Flow cytometry, 188–191, 317–318, 330–331
Follicle, development of, 48, 104–109, 117, 138, 140, 305, 309
Follicle stimulating hormone (FSH), 107–108, 110–112
Follicle stimulating hormone receptor (FSH-R), 106, 110
Forskolin, 287, 297
Freund's adjuvant, 218, 239, 245–247

G-protein, 83
Gastritis, 241–243
Gastrulation, 203
Gazelle, 42
Gel electrophoresis, 174, 185, 195
Gene
 c-*fos*, 44, 214, 217–220, 228, 230
 CSF-I, 116
 $D^d$, 217–229, 231
 deletions, 175, 177
 *dorsal*, 193
 estrogen receptor, 46
 β-globin, 60
 GM-CSF, 117
 *grc*, 170, 174–177, 181
 HAM, 208
 interferons, 41–46
 lethal, 171
 MHC. *See* MHC
 MHC-linked, 174–177, 181, 201–202
 β2 microglobulin, 202–203, 207–208, 218, 220
 multidrug resistance (*Mdr*), 202
 *nu*, 240–243
 *Ped*, 177, 181
 progesterone receptor, 46
 *rcc*, 177–178
 transforming growth factors, 108, 114, 116, 136–144
 TP-p7, 44
 tumor suppressor, 181
Genome
 homology, 42–43, 45
 imprinting, 154, 171, 173–174, 188
 library, 174
 rearrangements of, 185
 sequencing, 190
Germ cell depletion, 259, 273
Gestational sac, 306, 310–311
Giraffe, 42
Glands, endometrial, 24, 37, 161, 163
Glioblastoma, 130
Glomerulonephritis, 254
Glycosylation, 41, 45, 71, 201, 229
Goat, 40–41, 46, 155
 estrous cycle, 38–39, 44, 47–48
Golgi apparatus, 268, 272
Gonadotropin releasing hormone (GnRH) agonists, 27
Gonadotropins, 105, 110, 113, 116
Gonorrhea, 306
Graft rejection, 162, 201, 218
Granulocyte macrophage colony stimulating factor (GM-CSF), 9, 43–44, 82–83, 85–94, 102, 104–106, 108–109, 115–116, 125, 137, 304
Granulocyte macrophage colony stimulating factor receptor (GM-CSF-R), 83–86, 90–91
Granulocytes, 18, 59, 83, 85–86, 89, 92–94
Granuloma, spermatic, 259, 266–268, 281

Granulosa cells, 105–109, 116, 138–139
*grc*, 170, 174–177, 181
Growth hormone, 43
Growth, reduced, 174–175, 177, 181, 293, 296
Gynecogenetic zygote, 173–174

Hamster egg penetration, 8, 20
Heart, 210–211, 221
Heavy chain, 5, 201–202, 208, 216
HeLa cells, 140
Helper T cells. *See* T cells, helper
Hematopoiesis, 60, 91, 136, 203
Hemorrhage, 4, 222
Heparan sulfate proteoglycan, 66, 70
Heparin, 289, 317, 319, 323, 328, 331
Hepatoma, 140
HHK cell line, 185–186, 188, 191
High-performance liquid chromatography (HPLC), 126–128, 130
*Hind*III, 5, 185–187, 192–193, 219
Histamine, 92
HIV, 5, 15, 17, 19, 99, 102, 306, 309
HLA antigens. *See also* MHC
  HLA-G, 5–7, 9, 326–327
  homozygosity, 316, 323, 325–329, 332
  sharing, couples, 306, 308, 316, 325–326
  sharing, maternofetal, 128, 326, 328, 330–331
HLA-DQα4.1 antigen, 316–317, 321–326
Homology
  amino acid, 47, 62
  cross species, 42–43, 46
  interferons, 42–43, 45
Horse
  conceptus, 158–159, 161
  estrous cycle, 48
  genes, 42
  hybrids, 162–164
  luteolysis, 48
  pregnant, 158–160, 163
  trophoblasts, 158–160, 163–165
Human immuodeficiency virus. *See* HIV
Hutterites, 178
Hybridization, 185–188, 192–193
  in situ, 65, 67, 100–101, 108, 138–143, 164, 202–210, 215–216
Hydatidiform mole, 174, 305, 309
Hypersensitivity, 99, 309
Hypertension in pregnancy, 285, 287, 323
Hysterectomy, 37, 45
Hysterosalpingography (HSG), 27, 305, 309
Hysteroscopy, 305, 309

IgA, 16–18, 289, 309, 323, 330
IgG, 17–18, 25, 50, 100, 246–248, 250, 286, 288–289, 319–320, 323, 326, 329–330
IgM, 18, 286, 288–294, 296, 319, 323
Immune response, 24, 28, 101, 153, 159–161
  anamnestic, 156, 160, 162, 164
  male, 16, 17–18, 254
Immune system, maternal, 153–155, 164–165, 204, 214, 218, 228–229, 303–304
Immunity
  cell-mediated, 25, 99
  humoral, 25, 99
Immunodeficiency syndrome, 178. *See also* HIV
Immunofluorescence, 24, 194–195, 246–247, 280, 318
Immunohistochemistry, 7, 45, 64, 72, 86, 101, 105, 127, 139–141, 159, 163, 216, 219–225, 229, 280, 282
Immunoperoxidase, 247, 290–292
Immunoreactivity, 64–66, 68, 143, 277, 280
Immunoregulation, 5, 9, 26, 44, 60, 104, 125–127, 130, 153, 156, 326
Immunotherapy, 303–309, 311–312, 332
  intravenous immunoglobulin G (IVIG), 305, 309–312, 320, 323–325
  paternal lymphocytes, 304, 311–312, 316–325, 331
  seminal plasma, 305–309
Implantation, 26, 37–38, 49, 57–59, 86, 90, 92, 94, 115–116, 126, 144, 164, 204, 304
  rate, 306–309, 312

Imprinting, genomic, 154, 171, 173–174, 188
Infertility, 19–20, 330
  endometriosis and, 23, 25–28
  idiopathic, 305, 325–326
  male, 175, 178, 254, 264
Inflammatory response, 24–25, 28, 86, 89, 92–94, 101–102, 127, 217, 222, 247, 273–275, 285, 296, 330
Inheritance
  maternal, 188
  paternal, 156–157, 163, 188, 221
Inhibins, 140
Inositol phospholipid (IP), 39, 45, 48
Insulin, 62
Insulin-like growth factor I (IGF-I), 62, 105, 137
Insulin-like growth factor II (IGF-II), 61, 62–63, 66–68
Insulin-like growth factor binding proteins (IGFBPs), 62–63, 66–67, 73
Insulin-like growth factor receptor (IGF-R), 62–63, 73
Insulitis, 239
Integrins, 9, 70, 73
Interferon (IFN)
  consensus sequence, 5
  receptors, 42, 44–47, 50
  recombinant, 42–45, 47
  -stimulated gene factors (ISGF), 45–46
  -stimulated response elements (ISREs), 45–46
Interferon $\alpha$ (IFN$\alpha$), 41–43, 45–47, 49–50, 216
Interferon $\beta$ (IFN$\beta$), 42–43, 45, 216
Interferon $\gamma$ (IFN$\gamma$), 4, 9, 19–20, 26, 49–50, 94, 100–102, 104, 106, 114, 125–126, 185, 231, 245, 304
Interferon $\tau$ (INF$\tau$), 38–48, 50
Interleukin-1 (IL-1), 19, 26, 43, 86, 104–109, 111–112, 115, 125, 137
Interleukin-1 inhibitor, 93
Interleukin-1 receptor, 106
Interleukin-2 (IL-2), 26, 60, 101, 125–126, 243, 245
Interleukin-2 receptor, 9, 60, 321
Interleukin-3, 304
Interleukin-4, 43, 100–101, 243, 245
Interleukin-5, 100–101
Interleukin-6, 91–92, 101, 104–105, 108, 137
Interleukin-6 receptor, 92
Interleukin-8, 137
Interleukin-10, 100, 125, 127
Interstitial cells, 15, 104, 106, 110, 112, 116–117, 215, 217, 247, 270–274, 276, 278
Intralipid infusion, 305, 312
Intrauterine fetal death (IUFD), 324
Intrauterine growth retardation (IUGR), 285–287, 320, 323–324
Intravenous immunoglobulin G (IVIG) therapy, 305, 309–312, 320, 323–325
Isoantibodies, 279–282
Isoschizomer restriction enzymes, 188, 190

JAR cells, 43, 90–91, 143, 185–189, 194–196
JEG cells, 6, 90–91, 143, 188–190
Juxtacrine activity, 83

Kes Kummer Tauregs, 178
Kidney, 211, 221–222
Kuwait women, 316

Lactogen, 38, 90
Laemmli extract, 277–279
Laminin, 70
Laparoscopy, 23
Leprosy, 99
Leukemia inhibitory factor (LIF), 83, 137, 144
Leukocytes, 3, 6–10, 24–25, 27, 64–65, 82, 84, 86–88, 92, 94, 105, 113, 130, 137, 273–274
  decidual granulated, 3, 6–10
  male reproduction, 15–16, 19
  maternal, 153, 155–156, 161–162, 164
  paternal, 304, 311–312, 316–325, 331
Leukocytospermia, 15
Leydig cells, 15, 19–20, 105, 110–113, 116, 221

*Listeria*, 99, 114
Littré glands, 15
Liver
adult, 210, 221
fetal, 6, 203, 209-211
Llama, 42
Lubrication jelly, 305-308
Lung, 221
Lupus anticoagulant, 298
Lupus erythematosis, 99, 286, 288-289, 330
Luteinization, 107
Luteinizing hormone (LH), 38, 110-112
Luteinizing hormone receptor, 37, 106-107
Luteolysis, 37-41, 48-49, 106
Lymphoblastoid cell line, 185-186, 188, 191
Lymphocyte analysis, 317-318
Lymphocytes, 15-20, 26-27, 59-60, 82, 92-93, 114, 127-128, 157, 161, 266-268, 270, 272-273, 294. *See also* CD proteins; Leukocytes
Lymphocytotoxicity assay, 306-309
Lymphohemopoiesis, 82-83, 90, 94, 144
Lymphoid tissue, 100, 130, 161, 208, 250
Lymphokine-activated killer (LAK) cells, 60, 125-126, 304
Lymphokines, 50, 231
Lymphoma, 178, 205-208
Lymphotoxin, 137
Lysozyme, 24

*Macaca radiata*, 17
Macroglobulin, 62
Macrophage colony stimulating factor (M-CSF), 9-10
Macrophages, 4, 9, 15-20, 24, 26-28, 59, 64, 67, 272-273, 294
and cytokines, 82-86, 106, 117, 137
in ovary, 104-109, 116-117
in oviduct, 114, 116-117
proliferation of, 104, 106, 109-110, 115-116
recruitment, 93-94, 108-109, 116, 125
in semen, 113, 117
in testis, 104-105, 110-113, 116
tumor necrosis factor $\alpha$ production, 137-138, 140
in uterus, 82-86, 89, 92-93, 104, 114-117
Major histocompatibility complex. *See* MHC
Malaria, 99
Mast cells, 270
Maternal-fetal compatibility, 128, 326, 328, 330-331
Maternal-paternal incompatibility, 326-327
Mating, 83-86, 92-94, 111, 114-115, 126, 137
Matrigel invasion, 70, 73
Meiosis, block, 175
Melanoma, 91
Membrane attack complex (MAC) inhibitory factor, 7-8
Membrane cofactor protein (MCP), 7-8
Menopause, 139
Menstrual cycle, 24, 37-38, 59, 114, 136, 306
Menstruation, 64, 139
retrograde, 23, 28
Mesenchyme, 4, 65, 102, 138-139, 141
Mesoderm, 59, 203
Mesometrial triangle, 114
Metalloproteases, 70-71
Metastasis, 17, 141-142
Methylation, 163, 185-188, 190, 192
Methylene blue, 27
Metrial gland cells, 153
MHC. *See also* Heavy chain
antigen sharing in humans, 176-179
class I, 3-8, 154-157, 159-160, 162-164, 170-173, 175, 181, 184-188, 192-194, 196, 201-211, 214-217, 221, 227-230, 232
class II, 3-4, 20, 24, 104, 154-157, 159, 163, 170-171, 180-181, 208, 214, 317
class III, 170, 180-181
homozygote deficiency, 176-178
interspecies differences, 170-171, 180-181
linked genes, 174-177, 181
structure of, 170, 180-181, 249

$\beta$-microglobulin, 5, 157, 164, 185, 191, 193, 201–203, 207–208, 210, 215, 227
Microglobulin knockout mice, 7, 218, 220, 226–228, 230–231
Mink lung assay, 129–130
Miscarriage, 285, 319, 321. *See also* Pregnancy, loss
Mitochondria, 286, 293
Mitogenic activity, 61, 64
Mitomycin, 246
Monkey, 17, 254
Monoclonal antibodies. *See* Antibodies, monoclonal
Monocytes, 59, 89, 127
Morphogen, 193
Morula, 90
Mouse
  abortion in, 217–218, 304
  athymic, 240–243
  autoimmunity in, 241–242
  cytokines and, 42–43, 82–86, 90–91
  decidua, 60, 127
  embryo, 203–207, 210–211
  genetic crosses, 127, 204–206, 127–128, 177, 181, 218–223, 226–228, 231, 244
  IL-10 deficient, 125
  MHC in, 154, 174–175, 177, 180–181, 193, 201, 203–211, 214, 216
  microglobulin in, 203–204, 207–208, 210
  microglobulin knockout, 7, 218, 220, 226–228, 230–231
  nuclear transplantation, 173–174
  osteopetrotic (*op/op*), 104, 107–114, 116–117, 140
  ovariectomized, 83, 86–89, 91–92, 250
  ovary, 106–109, 138–140, 240
  phospholipid antibodies and, 290–291, 293
  placenta, 90, 116, 127, 139, 215–216, 220, 222–225
  pregnant, 60, 82, 84, 93, 99–100, 102, 114, 126–127, 136, 220, 290, 293
  pseudopregnant, 219
  testis, 15, 141
  thymectomized (D3TX), 240, 243–244
  tooth-lid factor, 61
  transgenic, 144, 214, 217–228, 231
  uterus, 127, 136–138, 140
  vasectomized, 86, 254
  ZP3 immunized, 239, 245–246, 248–250
mRNA for
  estrogen receptors, 39, 45–47
  growth factors, 64–67, 73, 82–83, 106, 108, 115–116, 127
  growth factor receptors, 84–86, 90–91, 108
  HAM, 208–211
  interferons, 41, 43, 47, 106
  interferon receptors, 45
  interleukins, 91–92, 100–101, 106, 111
  JAR, 186
  LMP2, 211
  MHC class I, 164, 202–208, 211, 215–216, 229
  microglobulin, 202–204, 208, 210–211, 216
  oxytocin, 48
  oxytocin receptors, 39
  P450, 111
  progesterone receptors, 39, 46–47
  TIMP, 71–73
  tumor necrosis factor $\alpha$, 106, 115–116, 136, 138–140, 142–143
Mucosal system, 153, 155
Mule, 162
Mullerian tumor, 142
Muscle, 210
Musk ox, 42–43
*Mus musculus*, *Mus caroli* system, 217–218
Myasthenia gravis, 239, 249
Myeloid cells, 84–85
Myoblasts, 294, 317, 327
Myometrium, 45, 58, 86, 114–115, 296
Myotubules, 294

Natural killer (NK) cells, 6, 9, 25–26, 28, 60, 93, 99, 125–127, 153, 157, 215, 218, 304, 320–321, 324, 326–327
Neoplasms, 126
Neutrophils, 84, 89, 92–93, 162, 266

Northern analysis, 67, 73, 138-139, 186, 221
NRK fibroblasts, 126, 129-130
Nuclear transplantation, 173-174
null cells, 127

Oligonucleotide, 194, 317
  antisense, 71, 143
Oligosaccharides, 71
Oligospermia, 15
Oocyte, 9, 26, 114, 117, 138-139, 245, 326
  complement regulation, 7-8
  development, 105, 108-109
Oophoritis, autoimmune, 239-250
Orchitis, 15
Osteoclasts, 105
Osteopetrosis, 105, 140
Ovariectomy, 46, 83, 86-89, 91-92, 140-141, 250
Ovary, 37, 250
  autoimmune disease, 239-250
  cancer, 62, 130, 141-142
  cytokines and, 105-106, 108, 136, 138-141, 144
  failure of, 239-240
  macrophages, 104-109, 116-117
Oviduct, 114, 116-117, 136, 138, 141, 173
Ovulation, 26, 38, 106-109, 114, 158, 177, 305, 309
Ovum, blighted, 310-311
Oxytocin, 37-40, 42, 44-49
  -antagonist, 40
  -neurophysin, 40
Oxytocin receptors (OTR), 39-40, 44-45, 47-48

PAGE, 126, 277, 279
Palindrome, 192
Pancreas, 221-222
Paracrine activity, 9, 24, 38-39, 44, 63, 69, 73, 83, 105, 136, 143-144
Parity, 227-229, 231-232
Parturition, 4, 101-102, 139, 156
PCR, 63, 84-85, 101, 138, 317
Pelvic inflammatory disease, 24
Peptide mimicry, 245-247, 249-250
Peptide transporters, 201-202, 208, 211
Peroxidase, 65, 86-88, 220, 289-290
pGEM, 127, 208-210, 218
Phagocytosis, 17, 27, 104, 113-114, 117, 266
Phospholipids, 39-40, 67, 287, 294, 328
  externalization of, 288, 294-295
  phosphatidic acid (PA), 287-288
  phosphatidylcholine (PC), 287-288, 294
  phosphatidylethanolamine (PE), 287-288, 294-296
  phosphatidylglycerol (PG), 287-288
  phosphatidylinositol (PI), 8, 287-288, 294, 327
  phosphatidylserine (PS), 287-288, 293-297, 327
Phosphorylation, 46, 83
Pig, 19, 38, 42, 48-49, 157
Pituitary, 39-40, 140
Placebo-controlled trial, IVIG, 309-311
Placebo-controlled trial, seminal plasma, 305-309
Placenta, 4-5, 7, 102, 117, 127-128, 136-137, 139-141, 188. *See also* Cytotrophoblast; Syncytiotrophoblast; Trophoblast
  damaged, 285-286, 297-298, 330
  developing, 203-204, 210-211
  epitheliochorial, 155, 157
  first trimester, 184, 190, 193-195
  hemochorial, 155
  interferon $\gamma$, 100-101
  invasion by, 57-61, 63, 66-67, 69-74
  MHC and, 153-157, 159-165, 170, 172, 181, 201, 205-206
  microglobulin in, 227-228
  normal human, 289-290, 292-293, 296-297
  previa, 324
  protection of, 214, 217
  rejection, 214, 228, 232
  retained, 156-157
  term, 184, 187-189, 192-193
  thrombosis, 285-286, 296-297
  tiny, 174, 293, 296
  transgenic, 222-225
  tumors, 59, 69-70
Placental bed giant cells, 58-59, 69
Placentation, 37-38, 83, 93, 125, 153, 158, 297

Placentome, 155–156
Plasma cells, 25, 266–268, 272–273
Plasma membrane, 287, 293–296
Plasmid pGEM, 127, 208–210, 218
Plasmin, 62, 72
Plasminogen, 71
  activator, 92
  activator inhibitor (PAI), 72
Platelet activating factor (PAF), 39
Platelet-derived growth factor (PDGF), 24, 105
Platelets, 286–287, 294–295
Polyacrylamide-gel electrophoresis. *See* PAGE
Polymerase chain reaction. *See* PCR
Posttranscriptional regulation, 164
Posttranslational control, 211
Postvasectomy, 254–265, 269, 271, 275–282
Postvasovasostomy, 254, 259–269, 271–275, 278, 280
Prednisone, 317, 319, 323, 329–331
Preeclampsia, 74, 176, 178, 285–286, 323–324
Preembryos, 90
Pregnancy. *See also* Trimester 1
  breech, 324
  cytokines in, 99–102, 125–130, 136, 139
  ectopic, 305, 307, 309
  immunization against, 214, 231
  loss. *See* Abortion, spontaneous
  maintenance of, 27, 57, 60, 125–130, 184
  MHC compatibility, 156–157, 162
  normal, 128, 144, 327, 329
  rate, 49, 111, 306–307, 311
  recognition of, 37–38, 40, 42–44, 47–50
  twin, 324
Preimplantation, 4, 7, 82, 85, 92–93, 116, 138, 141, 157, 304
Premature ovarian failure (POF), 239–240, 249
Premature rupture of membrane, 324
Primates, 17, 25–26, 37–38, 254
Primers, 84–85, 90
Probe
  actin, 205–207
  antisense, 142–143, 208
  *grc*$^+$, 174–175
  HLA, 185–188, 192–193
  microglobulin, 207
  *Ped* G1G2, 304
  pGEM-HIPP, 208–210
  RNA, 202–208
  sequence-specific oligonucleotide, 317
Progesterone, 37–40, 45–47, 49, 83, 91–92, 104, 106–108, 115, 139–142, 304–305, 309, 328
  block, 39–40
  receptors (PR), 38–39, 44–47, 50, 304
  replacement therapy, 46
Prolactin, 41, 60, 91, 107
Proliferating cell nuclear antigen (PCNA), 68
Promoter
  c-*fos*, 214, 217–220, 228, 231
  interferon genes, 43–44, 46
  region, 192–194
  viral, 185
Prostaglandin E (PGE), 37, 40, 49, 60, 86, 93, 104, 114
Prostaglandin F (PGF), 37–42, 44–45, 47–49, 86
Prostaglandins, 25, 37, 92, 137
  synthesis of, 104
Prostate, 17–18, 113
Proteasome, 202, 208, 211
Protein
  secondary structure, 43
  transmembrane, 7, 45, 61–62, 202, 295
Proteoglycans, 62, 66, 70
Proteolysis, 61–62
Prothrombin, 296
Prothrombosis, 285
Pseudopregnancy, 172, 174, 219
Psychotherapy, 305, 312
Puberty, 110, 239

Rabbit, 42, 46, 154
Radiolabeling, 67, 70–71
Rat, 19, 45, 60, 64, 110, 127, 188
  chemical carcinogenesis in, 177–180
  genetic crosses, 171–172, 174–175, 177–179
  *grc* gene, 170, 174–177, 181
  Lewis, vasectomized, 254–264, 267, 269, 271–281

Lewis, vasovasostomized, 254, 259-264, 269, 271-275, 278, 280-281
MHC and, 154, 170-171, 177, 180-181, 201, 215
oophoritis, 239
ovary, 106, 138-140
placenta, 139, 171, 216-217
Sprague-Dawley, vasectomized, 254, 264-268, 281
Sprague-Dawley, vasovasostomized, 254, 264-268, 281
Recurrent spontaneous abortion (RSA), 6, 174, 177-180, 303-305, 307-312, 316-327, 329-330
Relaxin, 38
Repression of transcription, 184-188, 192
Resistance
to cancer, 170-171, 174, 177, 179-181, 184
to chemical carcinogens, 177-179
to disease, 170
Resorption of embryos, 125, 127, 138
Restriction enzymes, 185, 188, 190, 208, 218
Restriction fragment-length polymorphism (RFLP), 5-6
Retained placenta, 156-157
Retinoic acid, 202, 209
Retrovirus, 3-5, 243
Reverse transcriptase-polymerase chain reaction (RT-PCR), 63, 84-85, 138
Reverse transcription, 5, 84-85
RGD binding site, 62-63, 73
Rh disease, 320, 323
Rhesus monkeys, endometriosis, 25-26
Rheumatoid arthritis, 99, 127, 330
RNA. *See also* mRNA
double-stranded, 218
polymerase II, 46
retroviral, 4
total, 65, 205-207, 209-210
RNAse protection assay, 204-207, 209-211

Sandoglobulin, 305, 309-311
Second messenger, 39
Segregation distortion, 170, 175
Semen, 8, 14-15, 83, 94. *See also* Sperm
GM-CSF stimulating factor, 83, 91, 93-94
toxic effect, 15
Seminal fluid, 83, 94, 137, 173, 305-309
Seminal vesicles, 17-18, 83, 113, 309
Seminiferous tubules, 15-16, 111-112, 117, 222, 259-260, 262, 280, 282
Senescence, 294
Sensitization, maternal, 156-157, 162
Septic shock, 137
Sequence-specific oligonucleotide probes, 317
Serine protease, 7, 67, 71
Sertoli cells, 16, 110-112, 117, 259
Serum response element, 218-219
Sexually transmitted pathogens, 14, 17-19
Sheep, 40-41, 44-47, 100
estrogen receptors, 46-48
estrous cycle, 38-39, 44-45
MHC antigens, 156
progesterone receptors, 46-47
Sialic acid, 62
Sialidase, 62
Signal transduction, 8, 45-46, 62, 83, 91, 240
Signaling, 37-38, 41, 46-47, 49, 82, 117, 163
fetomaternal, 3, 6, 9-10, 38, 49
Southern analysis, 42, 185-187, 192-193, 219-220
Spectin, 295
Sperm, 7-8, 15-17, 20, 26, 83, 110, 112-113, 185, 215, 254-259, 266-267, 269-271, 275-277, 279, 309, 317
abnormal, 113, 117
absence of, 15, 273-274, 276
antibodies to, 16, 330-331
antigens on, 8, 16, 280
oocyte interaction, 8-9
tail, 280, 282
Spermatids, 259-260, 280
Spermatocytes, 26, 259-260
Spermatogenesis, 19-20, 111-112, 175, 222, 282
Spermatogonia, 259
Sphingomyelin, 294
Spleen, 205-208, 221, 243

Splenocytes, 100, 107, 241–242
Spongiotrophoblast, 154, 210, 215–216, 229
Stem cells, 10, 60, 208, 231
Steroid hormones, 57, 82, 104–105, 114–115, 136, 139–140. *See also* Estrogen; Progesterone
Steroidogenesis, 19–20, 26, 107–111, 116
Stillbirth, 305, 309
Stomach, autoimmunity, 241–242
Stroma, 9, 24, 38, 59–60, 83, 85–87, 89, 92–93, 114–115, 138, 140, 142, 290
Suppression
  constitutive, 171–172
  inducible, 171–174
Suppressor cells, 16, 126–128, 130
Suppressor factor, 125–130
Susceptibility, to cancer, 170–171, 174, 177–181
SV40, 68
Swainsonine, 71
Syncytiotrophoblast, 3–5, 58–59, 64–65, 69, 102, 184, 188–189, 193–196, 215, 287, 290–293, 295–297, 317, 328
Syncytium, 90, 292–295, 297, 317, 327

Taiwanese couples, 175, 178–179
Talin, 295
T cell receptor (TCR), 240, 249
T cells, 9, 28, 60, 93. *See also* CD lymphocyte proteins
  autoreactive, 240, 243–244
  cytotoxic, 3–4, 6–7, 16, 25, 114, 126, 155, 160, 162, 164, 201, 216, 218, 229–230, 240, 244–245, 249–251, 304, 326
  helper, 3–4, 16, 20, 99–100, 164, 240, 243–245, 250
  paternal, 316, 326
Teratocarcinoma cells, 202, 208–209
Testicular alteration, 259–262, 264, 271, 273–275, 277, 280–281
Testicular biopsy score count (TBSC), 259–260
Testis
  anatomy of, 15–17
  cytokines in, 141
  interstitial tissue, 15, 104, 110, 112, 116
  macrophages in, 20, 104–105, 110–113, 116
  postvasectomy, 255–262
  small, 259–260, 280
  transgene expression in, 220–222
Testosterone, 15, 27, 110–113, 117
Theca cells, 105–109, 116, 138
Thrombin, 294
Thromboplastic time, 306, 309
Thrombosis, 285–288, 296–297
Thymectomy, 240, 243–244
Thymocytes, 241–243
Thymoma cells, 205
Thymus, 6, 208, 210–211, 240, 243
Thyroiditis, 239
Tight junctions, 16
Tissue inhibitor of metalloprotease (TIMP), 71–73
TJ6 protein, 304
Tolerance, 184, 201, 230, 240, 243–244, 250
Toluidine blue, 270, 272, 274, 282
Tooth-lid factor, 61
Toxoplasmosis, 99
*Trans*-acting regulation, 188, 192–196, 202
Transcription, 4–6, 46, 106, 108, 117, 137–138, 140
  factors, 164, 185, 192–193
  rate, 45–46
  regulation of, 162, 184–188, 192, 202, 217
  start site, 43, 46, 185, 218–220
Transfection, 6, 9, 68, 143, 185, 219, 229
Transformation, tumorigenic, 143
Transforming growth factor $\alpha$ (TGF$\alpha$), 61–63, 66, 68–69, 72–73, 106, 137, 309
Transforming growth factor $\alpha$ receptor (TGF$\alpha$-R), 106
Transforming growth factor $\beta$ (TGF$\beta$), 9, 60–63, 65–66, 68–69, 72–73, 92, 104–107, 115, 125–131, 137, 304
Transforming growth factor $\beta$ receptor (TGF$\beta$-R), 66, 106, 127

Transforming growth factor $\beta$2-like molecule, 125-127, 130-131
Transgene *cfos*/H2-D$^d$, 218-232
Transgenic mice, 144, 214, 217-228, 231
Translation, 137-138, 217
Translocation, MHC region, 170, 180-181
Transmembrane proteins, 7, 45, 61-62, 202, 295
Transplantation
  organ, 154, 162, 165, 171, 201, 214
  protection, 3-4, 303
Transporter peptides, 201-202, 208, 211
Trimester 1, 9, 58, 65-70, 100, 114, 184, 190, 193-195, 215, 286, 288, 329
Trinucleate cells, 155
Trophectoderm, 4, 39, 41, 48, 50, 203
Trophoblast, 3-10, 89, 90-92, 101, 116, 125, 127, 137, 139-140, 171-172, 181, 303, 312, 321
  giant cells, 203-206, 216, 222
  intermediate, 59, 64-67, 70-71, 73
  invasion by, 57-61, 63, 66-74
  labyrinthine, 171-172, 181, 210, 215-216, 222, 224-225, 229, 232
  MHC molecules and, 153-159, 162-165, 171-173, 184-185, 187-188, 190, 193-194, 196, 214-217, 228-232
  multinucleate cell formation, 69
  signaling, 37-38, 42-43, 49-50
  transgenic, 222-225, 228
  tumors, 143, 178-180
Trophoblast-leukocyte common antigens, 7-8
Tuberculosis, 99
Tumor necrosis factor (TNF), 4, 9, 20, 26, 83, 86, 94, 101-102, 104-109, 111-112, 114-117, 125, 136-144, 185, 245
  cancer and, 136-137, 141-144
Tumor necrosis factor receptor (TNF-R), 106-107, 138, 140-142
Tumors, 59, 69-70, 143, 165, 178-180. *See also* Cancer
Tyrosine kinase, 46, 61, 83

Ubiquitin conjugation, 202
Ultrasound, 306, 309-311
Umbilical cord, 58
Urethra, 14, 18-19
Urethritis, 18
Urokinase-type plasminogen activator (uPA), 67, 71-72
Urokinase-type plasminogen activator receptors (uPA-R), 67, 71
Uterus, 57-59, 159, 210. *See also* Endometrium
  cytokine production, 136-140
  epithelium, 82, 89, 91-94, 137
  immune response, 153, 159-162
  leukocytes in, 156-157, 160-161
  lumen, 44, 49, 60, 83, 86-89, 91, 115-116
  macrophages, 104, 114-117

Vaccine, contraceptive, 153, 239, 245
Vas deferens, 16, 113, 254, 264, 266-268
Vasectomy, 15, 86, 110, 254-269, 271, 275-281
Vasopressin, 48
Vasovasostomy, 254, 259-269, 271-275, 278, 280-281
Venereal diseases, 14, 17-19, 306
Vimentin, 67, 294
Virus
  IFN and, 41-44, 47, 49
  immunity, 184
  inducibility, 41-43
  pathology, 164, 239
Viviparity, 3

Western analysis, 126, 130, 194-195, 277, 279, 281

*Xma*III, 219-220

Yolk sac, 158, 203-204, 206, 210

Zona pellucida, 239, 245, 247-248, 250
ZP3 peptide, 239, 245-246, 248-250
Zygote
  androgenetic, 173-174
  gynecogenetic, 173-174